Nanoimprint Lithography: An Enabling Process for Nanofabrication

Weimin Zhou

Nanoimprint Lithography: An Enabling Process for Nanofabrication

Springer

Weimin Zhou
Shanghai Nanotechnology Promotion Center
Shanghai
People's Republic of China

ISBN 978-3-642-34427-5 ISBN 978-3-642-34428-2 (eBook)
DOI 10.1007/978-3-642-34428-2
Springer Heidelberg New York Dordrecht London

Library of Congress Control Number: 2012954429

Printed on acid-free paper

Springer is part of Springer Science+Business Media (www.springer.com)

Foreword

Nanotechnology, biotechnology, and information technology are considered as the three leading technologies in the twenty-first century. Nanotechnology is a core example to lead the next technological revolution, which is of interest in the global scientific and societal field. Its rapid development will cause a worldwide revolution in almost all fields, including life sciences, information technology, environment technology, and energy technology. At present, over 60 countries and regions have established nanotechnology programs at a national level. The Chinese government has attached great importance to the development of nanotechnology and regards the level of its industrialization as the key criteria whether she lies in an advantageous position in the world economy. China released the national nanotechnology program to promote the development of nanotechnology.

All of the subjects of nanotechnology depend on feasible methods to fabricate nanostructures. There are three approaches to nanofabrication: top-down, bottom-up, and their hybrid approaches. Top-down method refers to the fabrication of smaller devices by using larger components, which has been developed based on the conventional IC fabrication process. For example, electron beam lithography or focused ion beam lithography is used to fabricate nanostructures out of larger structures. Bottom-up process is to assemble smaller components into more complex devices. More importantly, the combination of top-down and bottom-up approaches is usually applied in nanofabrication.

Nanoimprint lithography (NIL), invented by Stephen Chou in 1995, has been regarded as a possible nanofabrication technology due to its low cost, high throughput, and high resolution. It has been included on the ITRS Lithography Roadmap at the 32-, 22-, 16-, and 11-nm nodes. NIL has been widely used for the fabrication of biomedicine, high-density memory, photonic crystal, nanodevices, solar cell, sensors, etc., and has aroused immediate attention and widespread research. Some of the nanoimprint lithography products are in the market.

Nanoimprint lithography is one of the advanced fabrication technologies and possibly becomes the fabrication technology for electron and photoelectron industry. Importantly, roll-to-roll, a continuous imprinted process, has been invented to be more suitable for the modern semiconductor manufacturing requirements. In 2005, we established an advanced laboratory of nanoimprint lithography, and some fabrication and characterization facilities were introduced. Based on the laboratory, the first-class R&D and public service platform were also constructed. Some important breakthroughs were made in information and electronics, biotechnology, and energy and environment. Moreover, during the process of these projects, some scientists with comprehensive nanofabrication and management knowledge have been educated.

This book describes the fundamental concepts and processes of nanoimprint lithography. The work described herein includes the author group's research achievements. The characteristics and process of nanoimprint lithography have been remarked in detail. Moreover, some recent applications for nanoimprint lithography have been presented in this book as well. I am sure that scientists and engineers working in different nanotechnology fields will benefit from the book somewhat and obtain inspiration to impel development and applications of nanoimprint lithography.

Professor Xiaoming Niu
President, Shanghai Academy of Science and Technology

Preface

Since Chinese-American scientist Stephen Chou put forward the concept of nanoimprint lithography (NIL) in 1995, it has aroused attentions from both the academia and industrial fields. As one of the most promising nanostructure processing technologies, NIL has the advantages of high resolution, low cost, and mass production. Therefore, NIL has been included in the International Technology Roadmap for Semiconductors (ITRS), for the Lithography Roadmap, at the 32, 22, 16, and 11 nm nodes. Manufacturers of semiconductor equipment and the suppliers of materials and technology have involved in this area of research work. NIL has developed with several types of imprinting methods, such as hot-embossing lithography, UV nanoimprint lithography, and soft nanoimprint lithography. However, NIL is a contact processing technology of nanoreplication, which is different from a noncontact process of conventional optical lithography. There is a body of issues to overcome to fulfill the requirement of the current semiconductor industrial process. But as an emerging nanofabrication method, it is particularly suitable for photonics, organic electronics, and biomedicine fields.

In 2005, we established an advanced laboratory of nanoimprint lithography, and some fabrication facilities (spin-coating, imprint, etching, CVD, etc.) and characterization facilities (SEM, AFM, etc.) were introduced. The content of this book is based on our achievement on NIL. The current status of research for NIL is presented firstly. Next is the process of NIL including stamp fabrication, stamp treatment, the imprinting resist, imprinting process, and etching process. The progress made toward application is demonstrated in the last several chapters. The applications in light-emitting diodes (LED), nonvolatile memory devices, and solar cell are presented in detail.

My thanks go to my beloved supervisor, President Niu Xiaoming, for introducing me to the research work on NIL. I have learnt much on many aspects of NIL in the first-class environment of research which was possible because of his support and initiative. I would like to thank President Min Guoquan and Vice President Fei Licheng for their support in my research work.

I am indebted to Prof. Shi Liyi at Shanghai University; Prof. Song Zhitang at Shanghai Institute of Microsystem and Information Technology, Chinese Academy of Sciences; Profs. Wang Qingkang, Hou Zhongyu, Huang Qiyu, and Yang Bin at Shanghai Jiaotong University; and Prof. Liu Xuan at Shanghai Ocean University for their support and to Prof. Yang Yingge at Shanghai Institute of Scientific and Technical Information for her literature analysis.

I would like to thank many previous and present members of the laboratory for their contributions to the research work of NIL, in particular, Liu Yanbo, Zhang Jing, Li Xiaoli, Zhu Zhaoying, Zhang Jianping, Wang Jinghe, and Zhang Yanping. The author is grateful to all other authors of related papers and some cited figures in the books.

Finally, my thanks go to my wife for her support and understanding. I do not have much free time to accompany her because I often work at night and on weekends.

Shanghai Weimin Zhou

Contents

Chapter 1
Introduction

1.1 Nanofabrication and Nanoimprint Lithography

Nanotechnology refers to the study of manipulating matter on an atomic and molecular scale and deals with materials, devices, or structure fabrication at least one-dimensional patterns from 1 to 100 nm. One nanometer is a billionth of a meter, about 1/80,000 of the diameter of a human hair, or ten times the diameter of a hydrogen atom. The ideas and concepts of nanotechnology started with a talk entitled *There's Plenty of Room at the Bottom* by physicist Richard Feynman at an American Physical Society meeting at the California Institute of Technology on December 29, 1959 [1].

In 1974, Japanese scientist Taniguchi proposed the word "nanotechnology." Prof. H. Gleiter put forward the concept of "nanostructure of solids." The US National Nanotechnology Initiative (NNI) launched in 2000 had invested $0.495 billion to develop nanotechnology. It is making a profound impact on various fields, from materials, electronics, energy, environment, and medicine to our life and has become the popular pursued technology worldwide in recent years. Up to now, more than 60 countries, such as the USA, Japan, Korea, the European Community, Germany, China, Russia, India, Brazil, and several Middle East countries, have paid attention to the technique and promoted research and development by government's funding. The nanotechnology-enabled products are estimated at about $91 billion in the United States and $254 billion worldwide in 2009. It has been estimated that the number of nanotechnology products will double every 3 years, achieving a $3 trillion market by 2020 [2].

Nanotechnology has mainly concerned four components: nanofabrication, nanoelectronic, nano-biology, and nanomaterials. As a core of nanotechnology, nanofabrication merges other subjects together. As we know, various nanostructures must be constructed. Therefore, it is the basis of nanoscience and nanotechnology. Nanofabrication has been defined as a method to design, control, modify, manipulate, and assemble nanoscale components for the purpose of creating a system or product. In 2006, the European Union issued "road nanomanufacturing"

W. Zhou, *Nanoimprint Lithography: An Enabling Process for Nanofabrication*,
DOI 10.1007/978-3-642-34428-2_1, © Springer-Verlag Berlin Heidelberg 2013

and shows the development of the production technology for nanomaterials and nanosurfaces during the next years [3]. US government issued the report entitled *Manufacturing the Future: Federal Priorities for Manufacturing R&D* in 2008 and highlighted the roles of manufacturing (nanofabrication) [4].

Nanofabrication technique consists of top-down approaches and bottom-up approaches and their hybrid. Top-down approaches refer to the conventional or new techniques, such as conventional photolithography, focused ion beam lithography (FIB), e-beam lithography (EBL), atomic force microscopy (AFM) lithography, and nanoimprint lithography, to realize the desired structures. Bottom-up approaches make use of materials, such as nanoparticles, diblock copolymers, and DNA, to directly assemble the patterns. Hybrid approaches are an integration of top-down and bottom-up approaches, which provides an alternative approach to reduce cost and create complex architectures.

Nanofabrication technique is almost as diverse as the materials and applications. Conventional and unconventional nanofabrications are developed. Nanofabrication has been developed to nanomanufacturing, which needs volume and yield of the techniques [5]. Nanoimprint lithography, first proposed by Stephen Chou, is a replicated process for micro/nanostructures, which has the advantages of high resolution, low cost, and high throughput [6]. Although nanoimprint lithography is still in its infancy, it would realize nanomanufacturing technology in the near future.

An intelligentized society needs information system, which depends on the devices of information acquisition, transmission, storage, disposal, and display. The complete emerging devices are an integration with CMOS process along with parallel options and can realize some functionality. The high-performance sensor, memory, and logic devices have been fabricated by nanotechnology process. Conventional lithography fabrication, nonconventional fabrication, and their combination approaches are main fabrication techniques for micro/nanodevices and systems.

1.2 Purpose and Content of the Book

Nanoimprint lithography is an amazing technology and used in modern industry. In fact, the idea is used for the casting of material which has a 6,000-year-old process. The oldest surviving casting is a copper frog from 3200 BC [7]. Today's compact disks with sizes of below 400 nm are fabricated by imprinting technique. Research on nanoimprint lithography becomes an increasing hot topic, and we can expect that many more will focus on the research of nanoimprint lithography in the near future.

The present book aims to present a comprehensive description of nanoimprint lithography. Most of the contents are based on our achievement. The technique consists of imprinting pattern replication and pattern transfer. These include stamp fabrication and stamp treatment, imprint resist, imprint process, and etch transfer. Besides, it has wide application in semiconductor devices, storage memory, solar cell, and biology fields [8].

In the Chap. 2, a general situation of nanoimprint lithography was introduced, including its origin, development, and application. The key issue of nanoimprint lithography is also presented. Based on the introduction, we wish that the readers can quickly access to the field of nanoimprint lithography.

In Chap. 3, stamp material and stamp fabrication approaches have been demonstrated. The material and shape parameters directly affect the template deformation and the quality of transfer pattern. Stamp production can be fabricated by conventional and unconventional methods. These methods have own advantages and disadvantages. The soft stamp technique is becoming a hot issue to meet the large area imprinting patterns.

During demolding process, ease of separation is to achieve a good fidelity of imprint patterns. Demolding force is affected by several factors including adhesion and friction. How to decrease these factors is to ensure easy separation. In Chap. 4, some approaches of decreasing adhesion are presented, which are plasma polymerization, self-assembly monolayer, depositing diamond-like carbon film, and depositing metal nanoparticles. Characterization methods of anti-sticking layer are also presented.

Development of the imprintable resist material is an important component of nanoimprint lithography process. The resist, also called etch barrier, functions as an etch barrier layer. It must satisfy several important characteristics for the different imprint process. In recent years, the requirements of functional devices or systems are increasing. The directed imprintable functional resist material is necessary. The imprinted patterns becoming a permanent part of the device are applicable for the further imprint lithography.

In Chap. 6, the basic principle of nanoimprint was presented. There are three types: thermal, UV curable, and soft nanoimprint lithography. Some other variations of its processes are also elucidated in this chapter. To meet the requirement of large-area imprinting, the roll-to-roll (R2R) processes are intensively considered and developed, which is indeed a continuous imprint technique and used for solar cell, electronics embedded into clothing, large-area flexible displays, and roll-up portable displays in the near future. After imprinting, etching step is necessary for conventional nanoimprint lithography processes. The etching parameter has an impact on the quality of imprinted patterns. Normally, the lift-off and dry etching processes are used in the pattern transfer.

In Chap. 7, modeling and simulation of nanoimprint lithography are emphasized. Understanding of nanoimprint lithography will facilitate high-volume manufacturing of products with high quality and low cost and also enable to provide process guidelines and design rules for industrial application. Furthermore, readers can gain an insight into the principles of nanoimprint lithography.

In the next chapters, the applications of nanoimprint lithography have been demonstrated. Chapter 8 is devoted to detailed analysis of photonic crystals enhancing the extraction efficiency of GaN light-emitting diodes (LEDs) where nanoimprint lithography is used for photonic crystal LEDs. High-density nonvolatile memory device cell arrays can be constructed by nanoimprint lithography, which offers a simple and fast selectivity of novel material, construction design, and process of emerging nonvolatile memory. The application of nanoimprint lithography in the storage memory is given in Chap. 9.

Owning to energy crisis, an alternation to conventional energy source is pressing. Solar cell is one of these energy sources. However, the conversion efficiency is relative low, and enhancing its efficiency is a hot topic. The nanostructures in solar cell can effectively improve the efficiency. Nanoimprint lithography has its own advantage of fabricating these nanostructures. Chapter 10 focuses on nanoimprint lithography in the application of solar cell.

In all, the book will arouse the reader's interest and provide basic principles, approaches, and application for nanoimprint lithography. I wish that scientists and engineers, who are working in fabrication, electronic, material, chemical, and mechanical, would find materials of interest in this multidisciplinary subject.

References

1. Allhoff F, Lin P, Morre D (2010) What is nanotechnology and why does it matter: from science to ethics. Wiley-Blackwell, Chichester
2. Roco MC, Mirkin CA, Hersam MC (2011) Nanotechnology research directions for societal needs in 2020: retrospective and outlook. Springer, Heidelberg
3. Busnaina A (2006) Nanomanufacturing handbook. Taylor & Francis, Boca Raton
4. National Science and Technology Council (U.S.) Committee on Technology (2008) Manufacturing the future: federal priorities for manufacturing R&D. Technical report, National Science and Technology Council
5. Cui Z (2008) Nanofabrication: principles, capabilities and limits. Springer, Heidelberg
6. Chou SY, Krauss PR, Renstrom PJ (1995) Imprint of sub-25 nm vias and trenches in polymers. Appl Phys Lett 67:3114–3116
7. Chastain S (2004) Metal casting: a sand casting manual for the small foundry. PHI, Jacksonville
8. Zhou WM, Min GQ, Zhang J, Liu YB, Wang JH, Zhang YP, Sun F (2011) Nanoimprint lithography: a processing technique for nanofabrication advancement. Nano-Micro Lett 3:135–140

Chapter 2
Principles and Status of Nanoimprint Lithography

2.1 Introduction

Patterning technology is crucial in micro/nanofabrication. Development of photolithography roadmap is agreeable with Moore's law, which claims that the number of transistors per square inch may double approximately every 18 months. However, due to exposure wavelength diffraction limit, the technical complexity and manufacturing costs have been increased dramatically for the nanometer-scale line-width manufacturing. In such a context, next-generation lithography (NGL) has been proposed to replace conventional photolithography.

Nanoimprint lithography (NIL) is a high-throughput, high-resolution patterning method and regarded as an alternative for NGL. It has been a subject of an active research and development since 1995. In 2003, International Technology Roadmap for Semiconductor (ITRS) included nanoimprint lithography as one of the NGL candidates. Up to 2009, it has been included in the ITRS Lithography Roadmap at the 22-, 16-, and 11-nm nodes [1]. Different international conferences on nanoimprint lithography have been held every year; one of them is the International Conference on Nanoimprint and Nanoprint Technology (NNT) [2]. Besides, several commercial companies have devoted to nanoimprint lithography and provide machines, resists, stamp, and the related technology services.

In the last 17 years, there are three different kinds of nanoimprint lithography, including thermal nanoimprint lithography (hot embossing nanoimprint lithography), UV nanoimprint lithography, and soft lithography. Soft stamp imprinting is more popular than the hard one because it has an intimate contact with the surface, which can fulfill large-area imprinting. Recently, roll-to-roll nanoimprint lithography gains popularity for volume fabrication. Although nanoimprint lithography has wide applications in nanodevices, biomedicine, organic devices, etc., there are many issues to overcome, such as mold fabrication and inspection, defects controlling, alignment, and overlay. It can realize nanoscale resolution of patterning, thanks to the facts that it is not limited by the diffraction limit, the scattering effects, and the secondary electrons. However, it belongs to a kind of

W. Zhou, *Nanoimprint Lithography: An Enabling Process for Nanofabrication*,
DOI 10.1007/978-3-642-34428-2_2, © Springer-Verlag Berlin Heidelberg 2013

shaping process of materials, which can be done by deforming imprinting material by mechanical or other factors. Obviously, the nature of nanoimprint lithography renders some challenge. Therefore, from the point of view of manufacturing, these issues, including defect controlling, efficiency, overlay, and process uniformity, are the bottleneck for nanoimprint lithography industry. Although it can be mostly used for low-end products, the high-brightness LEDs and patterned media recording hard disks can be seen as the market for high-end products. In the long term, nanoimprint lithography products can be seen in various fields.

2.2 Photolithography

2.2.1 Photolithography Concepts

Photolithography (or "optical lithography") refers to a process used in microfabrication to selectively remove parts of a thin film or the bulk of a substrate. It uses a light source projected through a patterned mask and onto a substrate. The detailed explanation is as follows: First, a substrate is spin-coated with a thin film of photoresist. Second, a light, typically in the ultraviolet light, is projected onto the photoresist film by shining light through a mask. The exposed area of photoresists is changed. Therefore, the exposed/unexposed area of the photoresist film can be removed leaving a positive/negative-tone resist pattern on the substrate after a development process. The photoresist patterns are regarded as an etch layer in the following fabrication process:

Resolution is an important factor to determine the performance of a lithography process, which is defined as the ability to discern pairs of closely spaced patterns on the wafer. In an optical exposure system, resolution is determined by the diffraction phenomenon of light. The formula for resolution, R, is [3]

$$R = \frac{k\lambda}{NA},$$

where k is the factor that represents specific applications, with a range between 0.6 and 0.8, λ is the wavelength of the radiation source, and NA is the numerical aperture of the lens.

From the above equation, it should be obvious that a decrease in λ will improve the resolution capability of the lithography system and increasing NA of the projection lens also improves resolution. But, increasing NA needs a larger diameter lens, which renders costly.

Therefore, an alternative to improve resolution is decreasing the wavelength λ. However, the requirements of higher resolution and higher throughput and the current lithography tools were no longer able to meet the industry's need.

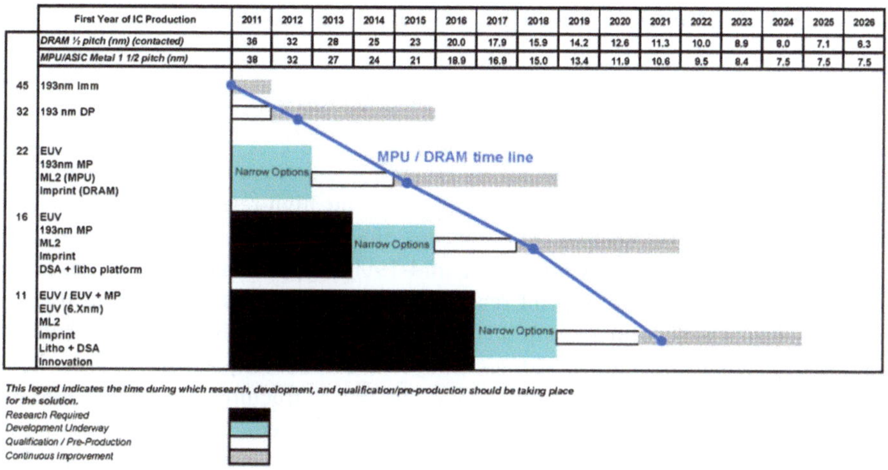

Fig. 2.1 Lithography roadmap in ITRS 2011 edition [4]

2.2.2 *Next-Generation Lithography*

In a conventional optical lithography system, the cost of improving resolution is increasing exponentially. Due to the expensive lithography tools and technique problems, as well as a continuous feature size miniaturization, a number of alternative technologies called next-generation lithography (NGL) are currently being developed. Candidates for next-generation lithography include 193-nm immersion with multiple patterning lithography, extreme ultraviolet lithography (EUV lithography), X-ray lithography, electron beam lithography, focused ion beam lithography, nanoimprint lithography, and maskless lithography. The emerging NGL technologies have been included in International Technology Roadmap for Semiconductors (ITRS) 2011 edition, as shown in Fig. 2.1.

Photolithography is a mainstream technology for IC. The most advanced form of photolithography is 193-nm immersion lithography, in which water replacing air is used as an immersion medium. The technology is used for 45- and 32-nm nodes. Extreme ultraviolet lithography (also known as EUV or EUVL) uses an extreme ultraviolet wavelength (13.5 nm). It is regarded as a promising technology for NGL and faces some challenges. Although EUV was proposed 30 years ago, it still cannot be used in the IC field. One of the challenges of EUV is photoresists problem. Electron beam lithography was the most popular during the 1970s. But the efficiency is low, and it is mainly limited by the throughput, which confines to certain aim. The same is true for focused ion beam lithography.

Nanoimprint lithography is the most popular alternative for NGL, due to its low cost and simplicity, along with its application in some products including LED, hard disk, and microfluidics field.

2.3 Development of Nanoimprint Lithography

2.3.1 Principles of Nanoimprint Lithography

Nanoimprint lithography is a relatively emerging technique and an increasing hot topic as a NGL because of its potential to create high-resolution features with low cost and high throughput. The concept of nanoimprint lithography was first proposed by Prof. Stephen Chou in the mid-1990s [5]. The technique is based on replication idea; first, the imprint resist was coated with a substrate, and then, a high-resolution pattern on a stamp surface is pressed into a resist film at 1:1 scale by means of mechanical contact. The resist layer is cured with pressure with the help of thermal or UV curing. After imprinting, the resist structure may be removed by plasma etching to transfer the resist patterns into the underneath substrate.

It has been shown that imprint lithography resolution can reach below 5 nm [6]. Since the invention of nanoimprint lithography in 1995, it has been increasingly drawing attention from both academia and industry. In 2003, the Technology Review of Massachusetts Institute of Technology (MIT) listed nanoimprint lithography as one of the ten emerging technologies that will change the world [7]. It was also added to ITRS and is lithography candidate for the 22-, 16-, and 10-nm nodes. Nanoimprint lithography conferences have been held continually to promote its further development.

2.3.2 History of Nanoimprint Lithography

In fact, the idea of nanoimprint lithography came from an ancient casting technique. The earliest research of imprinting or molding approaches for microlithography could date back to the 1970s. A patent was filed to use thermal imprint by RCA Corporation at Princeton, NJ in 1977 [8]. The soft nanoimprint technology, called "molded mask method," had been proposed and studied at NTT Laboratories in Japan as early as in the 1970s [9].

However, the research work (thermal imprint lithography, TNIL) of Prof. Stephen Chow has caused great attention. Shortly after a proposal of Stephen Chow, in 1999, Willson's group developed step-and-flash imprint lithography (SFIL), which is another important nanoimprint lithography [10]. The technique has been researched in both academic and industrial studies for its room-temperature operation. In 1998, Stephen Chow proposed roller nanoimprint lithography (RNIL), which can realize continuous imprint lithography, enhance imprinting efficiency, and even imprint on the nonplanar surface [11]. After several years, he proposed direct nanoimprint lithography on the materials, named ultrafast and direct imprint (UDI), with the help of laser [12]. Prof. Guo at University of Michigan has proposed reverse nanoimprint lithography (RNIL) and combined nanoimprint lithography and photolithography (CNP) in 2002 and 2004, respectively [13–15].

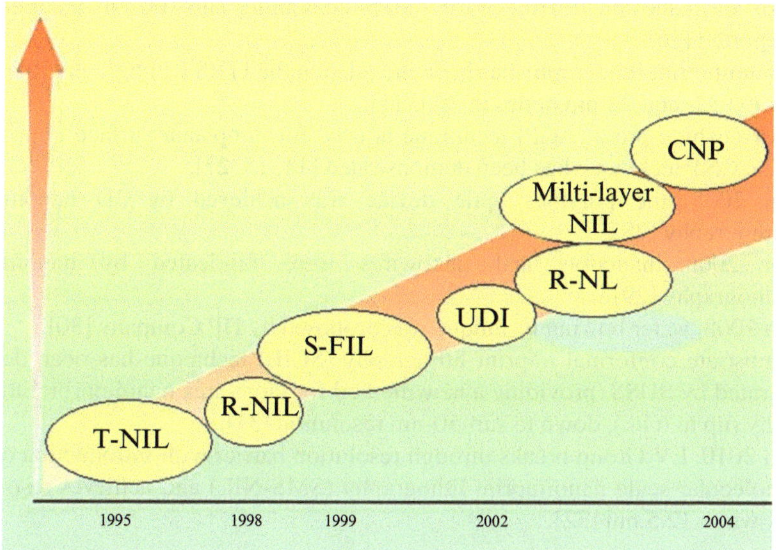

Fig. 2.2 A diagram of nanoimprint lithography development

R-NL refers to the process where the resist is coated on stamp surface and then printed on a substrate with the resist pattern left on. The technique is easy to realize the multilayer imprint lithography layers. CNP is an integration of nanoimprint lithography and photolithography, and this process is easy to fabricate 3D nanostructures. Roll-to roll (R2R) is further done by Guo's group and 1 m/min imprint rate can be achieved on the flexible substrate [16, 17].

Roll-to-roll process has been investigated in many research centers, such as VTT [18], Toshiba Machine [19], and Hitachi [20], for the likelihood of the commercialized production. The diagram of nanoimprint lithography development is shown in Fig. 2.2.

After nanoimprint lithography was proposed, some critical events in nanoimprint lithography development are listed in a chronological sequence.

1. The concept of nanoimprint lithography was proposed by Stephen Chow in 1995 [5].
2. In 1997, 6-nm resolution feature was achieved on the surface of PMMA by nanoimprint lithography [21].
3. The roller nanoimprint lithography was proposed by Stephen Chow in 1998 [22].
4. Nanoimprint lithography can realize alignment accuracy of 1 μm using commercially available tools in 1999 [23].
5. In 2000, 150-mm-diameter wafer-imprinted patterns are achieved [24].
6. The first microfluidic device was realized by nanoimprint lithography in 2001 [25].

7. An anti-adhesion treatment for molds containing sub-100-nm features was reported [26].
8. Nanoimprint lithography has been included in the ITRS Lithography Roadmap at the 32- and 22-nm nodes in 2003 [1].
9. Air cushion press (ACP) technique is used for nonplanar surface imprinting. The CNP technique has been demonstrated [14, 15, 27].
10. In 2005, the polymer optic device was achieved by 3D nanoimprint lithography [28].
11. In 2006, nanodots and nanowires were fabricated by nanoimprint lithography [29].
12. In 2008, water bowing technique was proposed by HP Company [30].
13. Substrate conformal imprint lithography (SCIL) technique has been demonstrated by SUSS, providing a new method for large-area nanoimprint lithography (up to 6 in.), down to sub-50-nm resolution [31].
14. In 2010, EV Group breaks through resolution barrier with introduction of soft molecular scale nanoimprint lithography (SMS-NIL) and achieves resolution down to 12.5 nm [32].

2.3.3 Classification of Nanoimprint Lithography

Nanoimprint lithography has developed with different methods since 1995. However, its fundamental principle is the same. First, a substrate is coated with a polymer layer or functional layer, and the stamp is pressed into the polymer layer. After curing, the imprint mold is released from the layer. As a result, the micro/nanopatterns are transferred on the layer.

The technology mainly has hot embossing lithography, UV lithography, and soft lithography. A variety of soft lithography techniques have been proposed [33], such as microcontact printing (μCP) [34], replica molding (REM) [35], microtransfer molding (μTM) [36], micromolding in capillaries (MIMIC) [37], solvent-assisted micromolding (SAMIM) [38], embossing, and injection. Depending on the kind of stamp used, hard nanoimprint lithography and soft nanoimprint lithography are defined. SiO_2, Ni, Si, Si_3N_4, and SiC molds are usually used for hard nanoimprint lithography. Polymer materials, such as PDMS, PMMA, PUA, PVA, PVC, PTFE, and ETFE, are main components for the fabrication of soft molds, which are used for soft nanoimprint lithography. According to curing approaches to spin-coated polymer, there are hot embossing lithography, UV nanoimprint lithography, and hot-UV nanoimprint lithography. Based on the imprinted area, full wafer nanoimprint lithography and step-and-roll nanoimprint lithography are defined. Roll-to-roll nanoimprint lithography process is suitable for mass fabrication and has application in state-of-the-art flexible nanodevices. The classification of nanoimprint lithography is indicated in Fig. 2.3 [39].

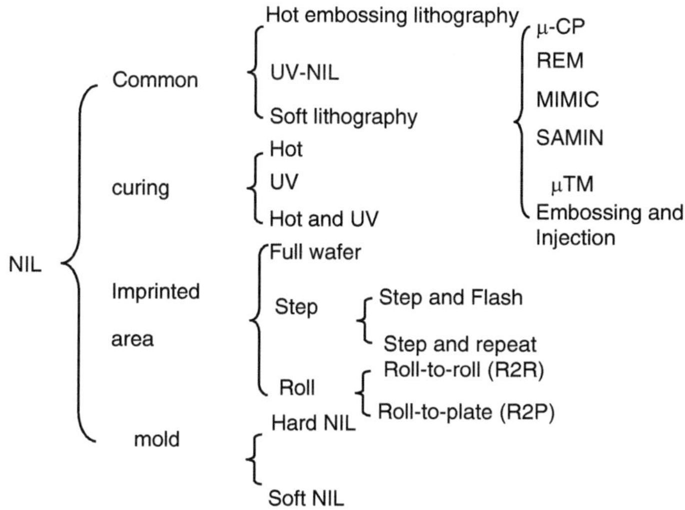

Fig. 2.3 Classification of nanoimprint lithography

2.3.4 Research Field of Nanoimprint Lithography

Nanoimprint lithography technology is a complicated subject including mechanical engineering, electron engineering, physics, and chemistry. It deals with state-of-the-art methods, process controlling, fabrication of equipment, and application of devices and system. Its aim is to develop fabrication equipment and process, which realize low cost, high efficiency, and mass product. Nanoimprint lithography involves two fundamental aspects: basic principles and process and application. The basic principles and process deals with process, mold fabrication, mold treatment, functional imprinting material (resists), tool, and imprinted pattern controlling, which concentrates on an achievement of a good fidelity of patterns. Its application is referred to nanoelectronic (high-density memory, etc.), nano-optoelectronics (light-emitting diodes, solar cell, etc.), nano-optic component (grating, TFT-LCD, plasma sensors, etc.), and biological fields (biochips, microfluidic devices, etc.). The whole framework of nanoimprint lithography is outlined and shown in Fig. 2.4.

2.4 Critical Issues in Nanoimprint Lithography

Nanoimprint lithography process can be divided into three steps: (1) mold fabrication and treatment, (2) imprint process, and (3) subsequent etching (if required). The involved processes are mold fabrication, photoresist, process control, and high-quality etching. To obtain small uniform nanopatterns, any process control in nanoimprint lithography is critical, and these become a series of related key technologies.

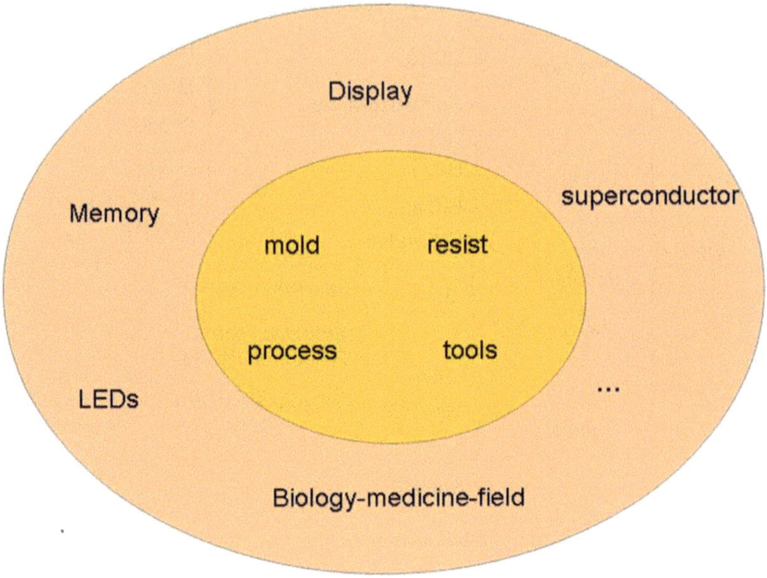

Fig. 2.4 Overview of nanoimprint lithography

2.4.1 Mold Fabrication

Mold is a functional unit for nanoimprint lithography. Mold fabrication (hard or soft mold) and mold treatment are the key technologies for the success of nanoimprint lithography.

Imprint mold (also called stamp, template) material used in hot nanoimprint lithography needs high hardness, compression strength, tensile strength, thermal expansion coefficient, good corrosion resistance, and other properties, to ensure that the template can wear, deformation, be an accurate and longer life. Silicon, quartz, silicon nitride, and diamond can be selected for mold material, and soft polymer (PDMS, etc.) is suitable for large-area imprint lithography. Usually, nanoscale structured mold is fabricated by electron beam lithography.

2.4.2 Imprint Resist

Photoresist used in nanoimprint lithography is different from that of conventional optical lithography. The resist has the following properties: inherent properties of resist (with a good substrate adhesion, shrinkage, and low viscosity), coating properties (controllable thickness of the 50–500 nm, smooth surface, degree of surface roughness is less than 5 nm), and good resistance to etching (at least 1:3 etch selectivity, relative to Si or SiO_2) [40–43].

As for thermal imprint resist, the viscosity of the imprinting resist becomes smaller, and it is easy to flow when heated temperature is above glass transition temperature. Some usually used thermal imprint resists are mr-I 8000, mr-I 9000, mr-L 6000, NXR-1000, Hybrane, PMMA, and SU8. However, the UV nanoimprint lithography process should be carried at room temperature. Resists with low viscosity can easily fill the cavity of the mold. The good fidelity of patterns can be obtained under UV curing. The UV-curable resists include mr-L600, NXR-2000, NXR-300, PAK-01, MMS4, WaterShedTM, etc.

In recent years, some functional materials can be used for resists to directly be imprinted. Some metal nanoparticles, metal oxides, and nanodots can be imprinted directly and act as a permanent component of devices or system.

2.4.3 Defects Control

In nanoimprint lithography process, due to the environment of experiment, materials, or process, for example, air bubbles, mold deformation, uneven coating, unparallel between substrate and mold, and so on, these will produce a variety of defects. How to control and avoid the emergence of various defects in imprint lithography process and improve imprint pattern quality are among the key issues.

2.4.4 Alignment and Overlay

With the increment of performance of electronic devices, the integrated chip is necessary. If nanoimprint lithography is adopted in the IC product line, overlay technology is also to be resolved. Currently, the overlay procedure aims at the field of nanoimprint, which is the most difficult and important issue [44]. The application of moiré fringe alignment method can achieve higher alignment accuracy, which reaches a precision of 20 nm [45].

2.4.5 3D Imprint Process

Three-dimensional (3D) structure in the microsystem field has a very important prospect. Small structure, lightweight, high sensitivity, low cost, and superior performance of device can be fabricated by 3D fabrication technology. It has promoted the development of product miniaturization and portability, increasing device and function of system density, information density, and interconnection density. And more importantly, it can greatly save materials and energy consumption.

A unique benefit of nanoimprint lithography is the ability to pattern 3D structures [46–48]. Multilayer interconnect structure can be fabricated by nanoimprint lithography, and its unique physical transfer process not only can reduce costs, with high resolution, but also significantly lessen process of building of complex three-dimensional structures.

2.4.6 Large-Area Imprinting and Throughput

Nanoimprint lithography is a low cost and mass production process. The large-area imprint lithography can improve throughput. Now, many approaches are proposed to realize the large-area imprinting. One of these approaches, roll-to-roll, is a continuous imprinting process and can be expected to meet the modern semiconductor manufacturing requirements [49].

2.5 Tools and Evaluation of Nanoimprint Lithography

2.5.1 Imprint Machine

There are several suppliers for nanoimprint lithography tools: EVG, Suss MicroTec, Molecular Imprints, Nanonex, and Obducat.

EVG provides nanoimprint lithography including single-step UV-imprinting systems and step-and-repeat large-area UV nanoimprint lithography and hot embossing system. The commercially available tools are EVG620 Automated NIL System, EVG6200 Infinity Automated NIL, IQ Aligner Automated UV-NIL System, EVG770 Automated NIL Stepper, EVG510HE Semi-automated Hot Embossing System, EVG520HE Semi-automated Hot Embossing System, and EVG750 Automated Hot Embossing System [50].

Suss MicroTec provides sub-50-nm resolution imprint system. Previously, MJB4 Mask Aligner can be added with a toolkit for nanoimprint lithography. Then, the second-generation nanoimprint lithography MA/BA6 enhanced the capability. The product can imprint with small rigid stamp and large-area soft stamp UV nanoimprint lithography. Based on MA/BA6, the MA/BA3 Mask Aligner can be optionally with various imprinting techniques for nanotechnology and UV replication. One technique is SUSS microlens imprint lithography (SMILE) using soft stamps to imprint high-quality microlens arrays in UV-curable polymers on 8-in. glass wafers [51].

Molecular imprints (MII) technique is based on its core technology, Jet and Flash™ Imprint Lithography (J-FIL™), which is a UV nanoimprint lithography process. Other than a conventional spin coating, imprint resist is to dispense by

Table 2.1 Nanoimprint lithography tool suppliers

Company	Location	Comments
EVG	Austria	Mask alignment and bonding tool, more than 100 tools installed worldwide
Molecular imprints (MII)	Texas, USA	Market and technology leader for high-resolution, low cost-of-ownership nanoimprint lithography systems and solutions in the hard disk drive and semiconductor industries
Nanonex	New Jersey, USA	Founded by Princeton University, supplier of tool, resist, and mold
Obducat	Sweden	Supplier of imprint tool. Besides, supplier of SEM and electron beam lithography
Suss MicroTec	Germany	Mask alignment and bonding tool. The first company to commercialize nanoimprint lithography (NIL) and electron beam recorder (EBR)

IntelliJet Drop Pattern Generator prior to the imprint step. The products are Perfecta TR1100, Perfecta MR5000, NuTera HD7000, Imprio 300, Imprio HD2200, and Imprio 1100 [52].

Nanonex provides nanoimprint lithography solution (machines, resists, and processes) based on its patented technologies. NX-B100 and NX-1000 are full-wafer thermal nanoimprintor. NX-2000 is a full-wafer UV imprintor. But NX-2500 can realize alignment on the full wafer (up to 8 in.). The most advanced imprinter is NX-3000, which is a step-and-repeat nanoimprinter with alignment [53].

Obducat provides some products of nanoimprint lithography and lithography solutions. The Eitre® Nano Imprint Lithography (NIL) series and the Sindre® series can meet the different requirements. Eitre® Nano Imprint Lithography (NIL) series are suitable for researches and have three types of Eitre3, Eitre6, and Eitre8. The Sindre® series are suitable for producing and have Sindre400/600/800, which can make full imprinted patterns on 4, 6, and 8 in., respectively [54].

Besides the five suppliers of nanoimprint lithography, other homemade systems are also used for research field. For example, Hitachi developed a thermal nanoimprint tool, which includes alignment, transfer, and press subprocess portions. It was reported that it has a capability of full imprinting up to 300-mm patterns [55]. Table 2.1 shows some suppliers of nanoimprint machine.

2.5.2 *Metrology for Nanoimprint Lithography*

Metrology tool for evaluation of nanoimprint patterns is very important to reveal the quality of structures replication from stamp into resist and pattern transfer from resist into substrate. Optical microscopy can be used for inspection of the whole structures of imprinted patterns and stamp. Scanning electron microscopes (SEM) and atomic force microscopes (AFM) can be used for metrology of submicrometer and nanometer features.

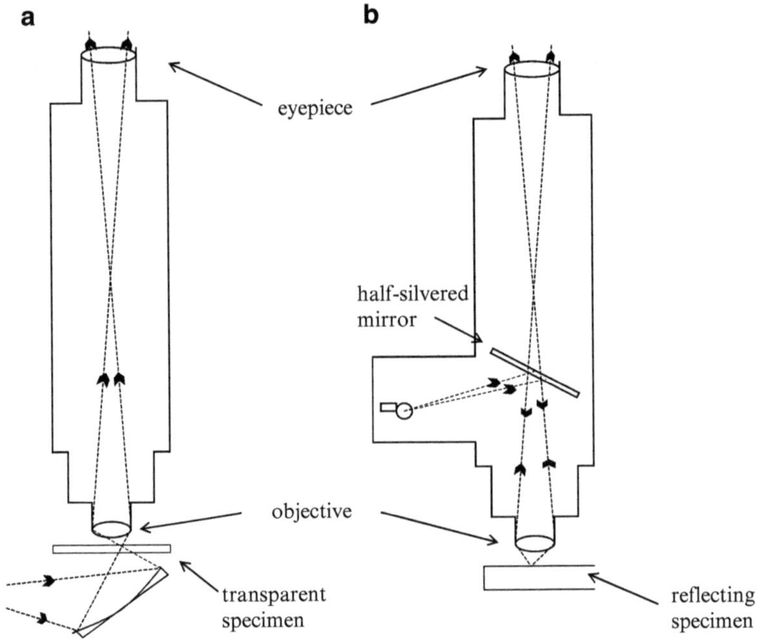

Fig. 2.5 Schematic diagrams (**a**) of a biological microscope, which images light transmitted through the specimen, and (**b**) of a metallurgical microscope, which uses light (often from a built-in illumination source) reflected from the specimen surface

2.5.2.1 Optical Microscopy

Optical microscopy is an optical instrument which uses visible light and lens to produce magnified images of micro-objects. Scanning electron microscope was primarily used for profiling the surface down to the nanometer scale further inspected by atomic force microscopy. A compound microscope contains two lenses: an objective and an eyepiece. The biological microscope and metallurgical microscope are two main basic forms, as shown in Fig. 2.5. A photo of microscope is shown in Fig. 2.6.

2.5.2.2 Atomic Force Microscopy

Atomic force microscopy (AFM) has been developed on the basic of scanning tunneling microscope (STM) invented by Gerd Binnig and Heinrich Rohrer. In 1986, AFM was invented by Binnig, Quate, and Gerber. It can further inspect insulators and is a novel technique for high-resolution imaging, as shown in Fig. 2.7. The force between the sample surface and a sharp probing tip can be sensed and

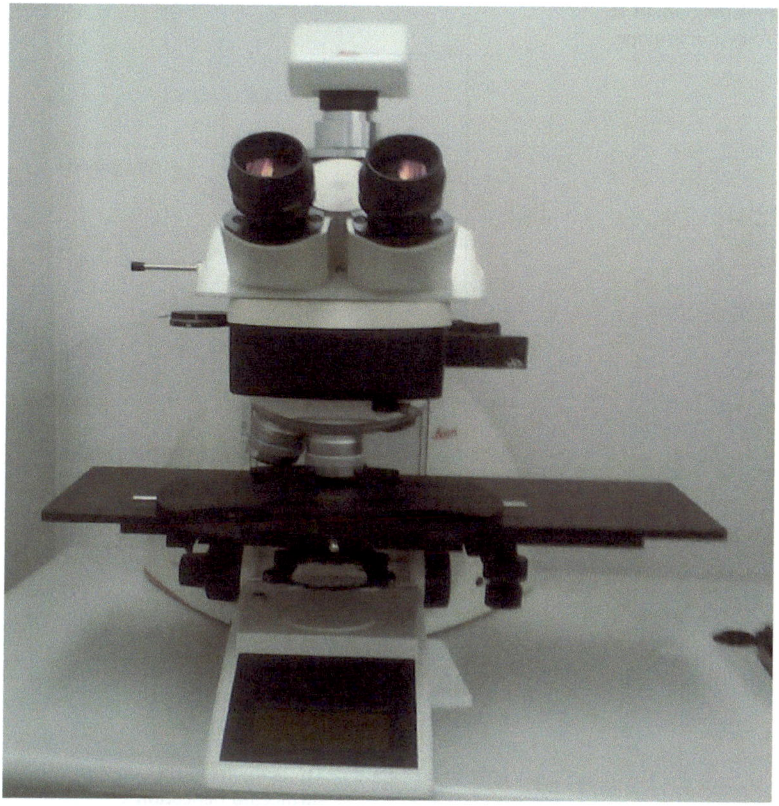

Fig. 2.6 A typical photo of microscopy (Leica DM4000)

converted to topographic information. A typical AFM instrument consists of a wafer stage and a measurement part including a cantilever, laser and photodiode, and other electronics for signal processing.

The force between the tip and the sample is shown Fig. 2.8. As the tip is approaching the sample's surface, the force is attractive. As it approaches closer, the force converts to be repulsive. The force between the sample and the tip can be calculated by measuring the deflection of AFM cantilever. The height of the tip is recorded and by a feedback system generates a topographic image of the sample surface.

AFM has three working modes: contact mode, noncontact mode, and tapping mode. In contact mode, a tip is attached to the end of a cantilever across the surface. The tip contacts the sample surface through the adsorbed fluid layer on the surface. In noncontact mode, the tip does not contact the sample surface but oscillates above the adsorbed fluid layer on the surface during scanning. In the tapping mode, the tip is periodically in contact with the sample surface [57].

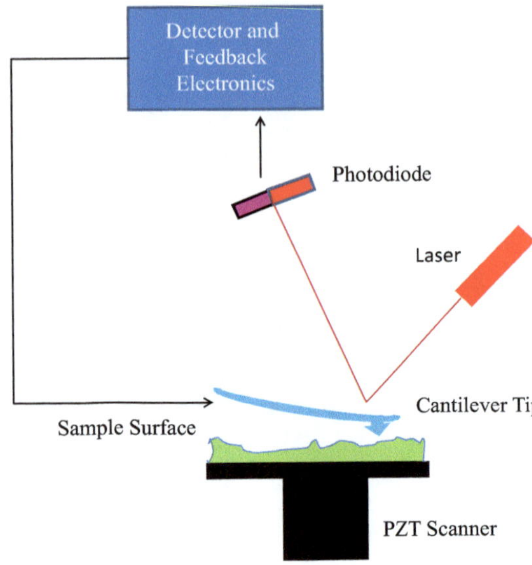

Fig. 2.7 A diagram of an atomic force microscope

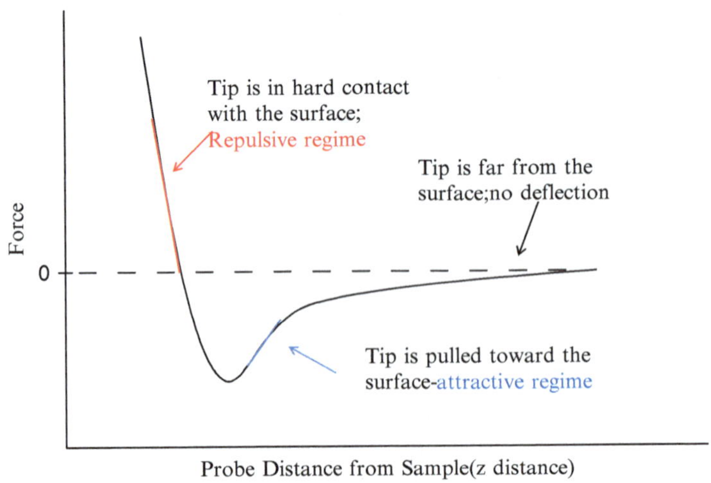

Fig. 2.8 Force as the function of distance between tip and sample

There are many types of cantilever and tip available commercially for various fields. The tip is very important for evaluating the properties of the sample. The tip materials have Si, SiO_2, Si_3N_4, single-walled carbon nanotube (CNT), etc. Besides measuring surface topography, AFM can be used to measure forces, which can characterize the adhesive property of the stamp. A digital instrument (Bruke) DI3100 is shown in Fig. 2.9. Figure 2.10 is an AFM image of soft imprinting stamp inspected by DI3100.

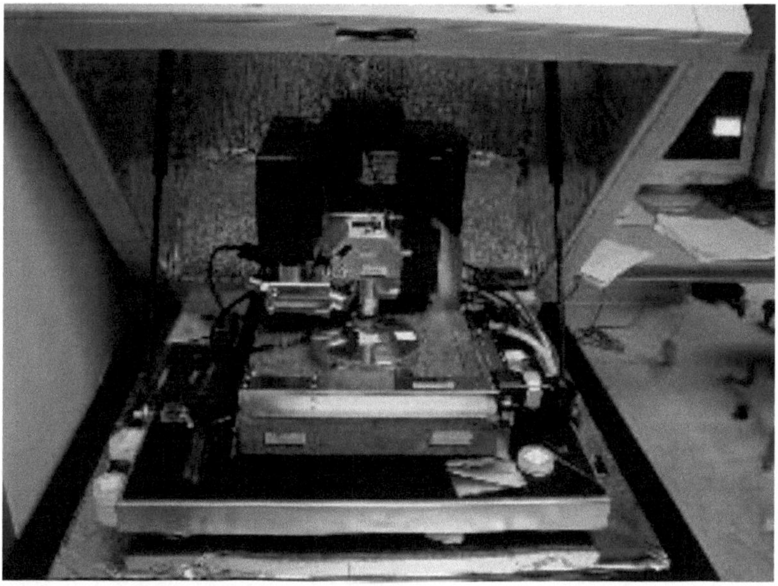

Fig. 2.9 Digital instruments (Bruke) DI3100 atomic force

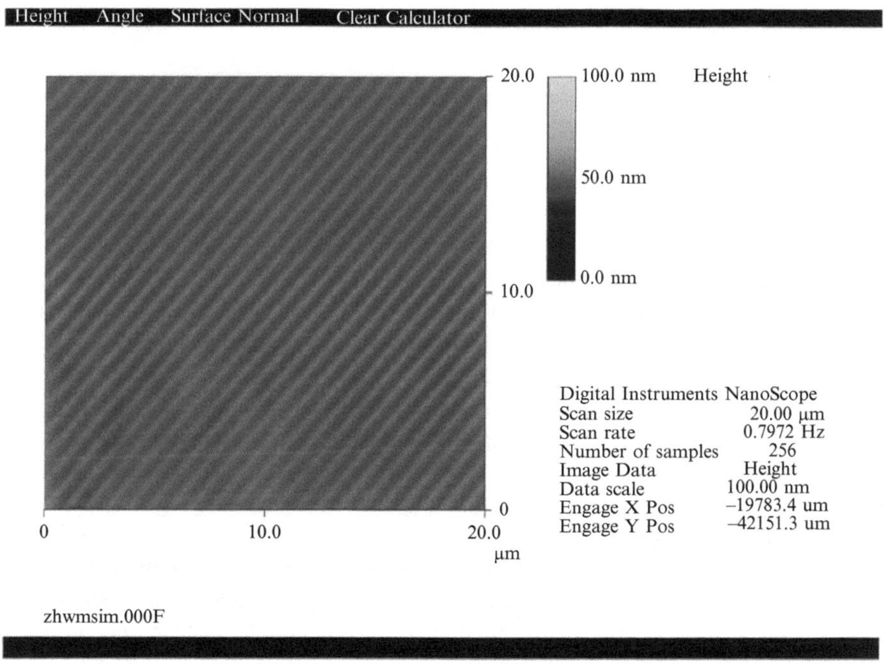

Fig. 2.10 AFM image of soft imprinting stamp

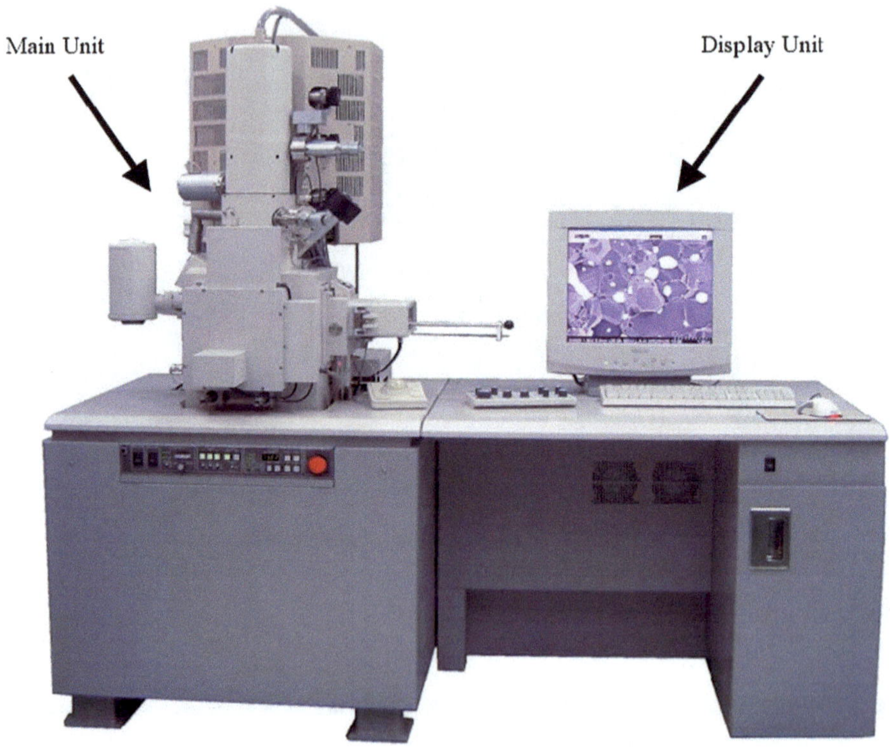

Fig. 2.11 Hitachi-S4800 field-emission scanning electron microscope

2.5.2.3 Scanning Electron Microscope

Scanning electron microscope (SEM) was first used by Max Knoll to get the microimage of a silicon steel in 1935. After the Second World War, SEM was further developed. Finally DuPont delivered the first SEM instrument. Steady commercial production has begun by Cambridge Instrument Company in 1965 [56]. Nowadays, SEM is widely used in various filed, such as biology, medicine, and metallurgy. SEM has some essential ingredients including electron, vacuum, image, and cooling system. Figure 2.11 shows one example of a modern instrument (Hitachi S-4800).

The schematic of a typical SEM is shown in Fig. 2.12. An electron beam is emitted from an electron gun, which typically has an energy ranging from 0.2 to 40 keV. The electron beam is focused by one or two condenser lenses to a spot of about 0.4–5 nm in diameter. It passes through pairs of scanning coils or pairs of deflector plates in the electron column (or the final lens) and scans over a rectangular area of the sample surface. The beam interacts with the sample surface and produces various types of signals including secondary electrons (SE), backscattered electrons (BSE), characteristic X-rays, light (cathodoluminescence), specimen current, and transmitted electrons, as shown in Fig. 2.13.

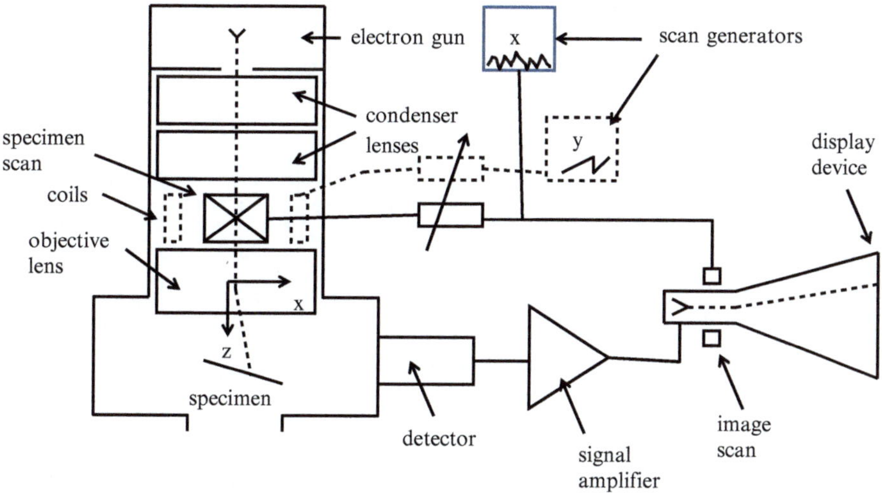

Fig. 2.12 Schematic diagram of a scanning electron

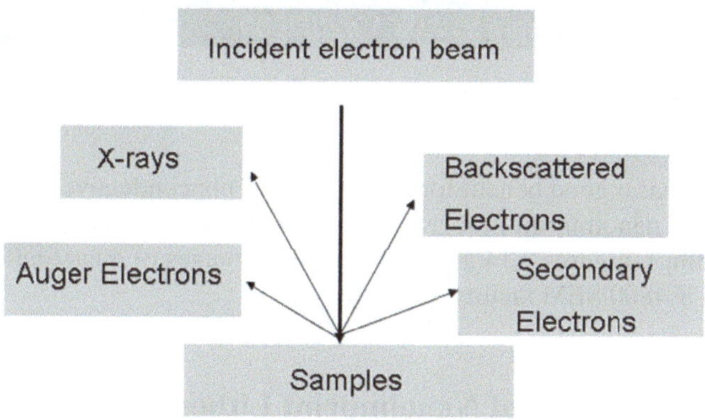

Fig. 2.13 Various types of signals

SE and BSE signals were detected to form the images of a sample surface. SE signal is produced only within small depth (<2 nm) below the surface. SE image is a property of the sample surface rather than any underlying structure. So it is said to display topographical contrast. BSE is a primary electron that has been ejected from a solid by scattering. An elastic scattering and inelastic scattering are composed of BSE signals. The elastic scattering is a scattering through an angle greater than 90° and only a small energy loss. The inelastic scattering causes both the losing of energy and direction of electron. The quantity of elastic scattering is more than that of inelastic scattering. BSE signals come from depth (5–200 nm) of a sample and can reveal the inner information of a sample.

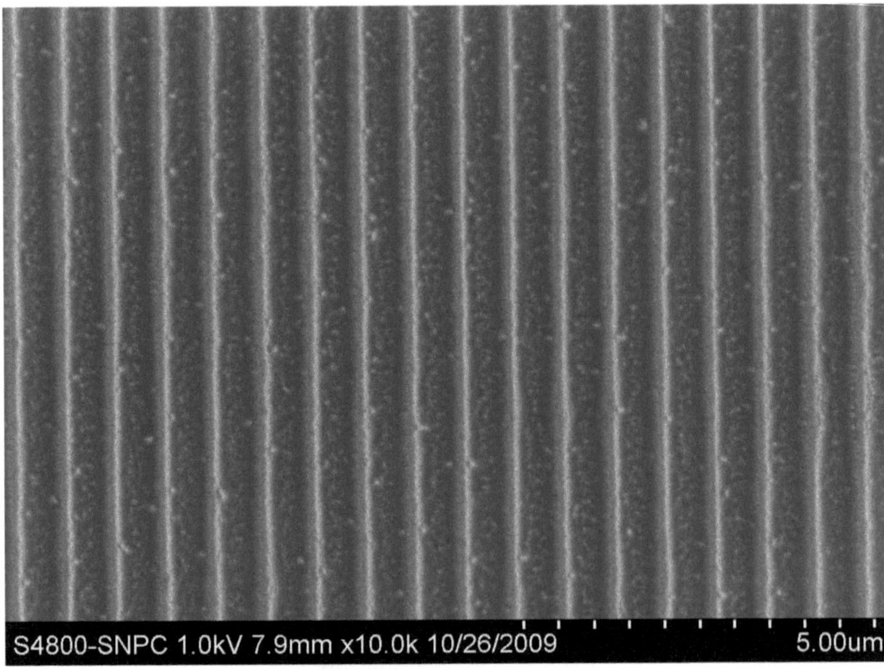

Fig. 2.14 A SEM image of quartz stamp (30° tilt)

Field SEM has a good benefit for observation of nonconductive sample and was widely used in nanoimprint lithography (quartz stamp, soft stamp, and imprinted resist patterning). Figures 2.14 and 2.15 are SEM images of quartz and soft stamp with Hitachi S-4800 SEM facility, respectively.

2.6 Research Status of Nanoimprint Lithography

Research on nanoimprint lithography has come into being the status of the USA, Europe, and Asia (Japan, Korea, China, and China Taiwan). Some groups that started researches were the universities and institutes, such as Wuppertal [58], Microresist [59], CNRS [60], Korea University [61], PSI [62], Michigan University [63], Princeton University [64], and University of Texas [65]. Also, there are some semiconductor companies, such as HP [66], Samsung [67], Toshiba [19], Motorola [68], Philips [69], LG [70], Canon [71], AMO [72], and IBM [73], involving in the research field of nanoimprint lithography. Some suppliers of the related nanoimprint lithography have been set up, as shown in Table 2.2.

In order to know the status of nanoimprint lithography, the published papers and patents are searched and analyzed. The obtained results are based on the database of Derwent Innovations Index. The number of the patents is 1,122 and is 3,869 for the published papers (as of January 2011).

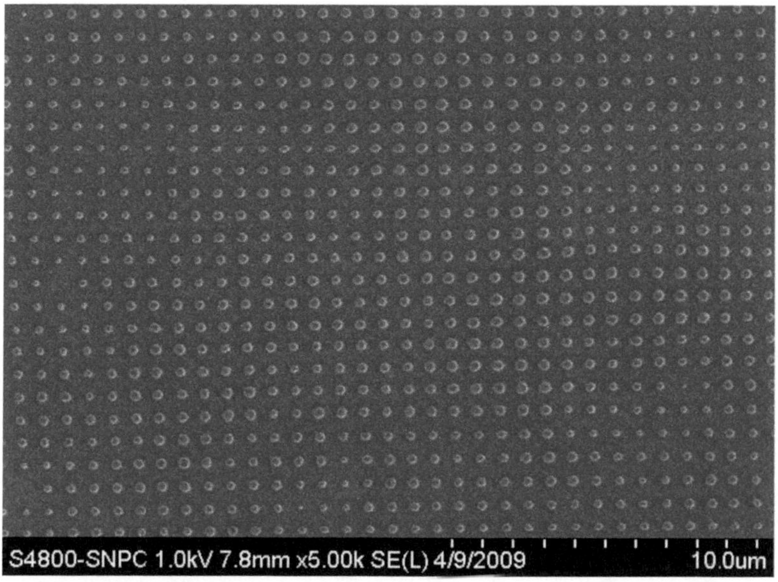

Fig. 2.15 A SEM image of soft stamp

Table 2.2 Information related to mold, resist, and tools

Mold	Hard and soft mold. Suppliers of the stamp (NILT and X-Lith)
Resist	AMONIL, PAK, mr-UV serious, PMMA, etc.
Suppliers of tools	EVG, Molecular imprints, Nanonex, Obducat, Suss MicroTec, Homemade

The number of patents has been increased continuously since the first patented paper was filed in 1998. Meanwhile, the number of published papers has climbed quickly and reaches the peak in 2007, then drops a little. It means that nanoimprint lithography has moved into maturity stage. The trend chart for the number of patents and papers related to nanoimprint lithography is shown in Fig. 2.16.

2.6.1 Publication Analysis

As far as paper quantity is concerned, the USA is listed as the top one, and the number of papers reaches 660; Japan and Korea are next to the USA. The documentation results are shown in Fig. 2.17.

The number of published articles is the first for physical subject and reaches 42.9%. Next two are nanoscience and nanotechnology, material, and engineering, respectively. Semiconductor device of applied physical field is a hot research topic in nanoimprint lithography, as shown in Fig. 2.18. The top ten of the number of published articles are shown in Table 2.3.

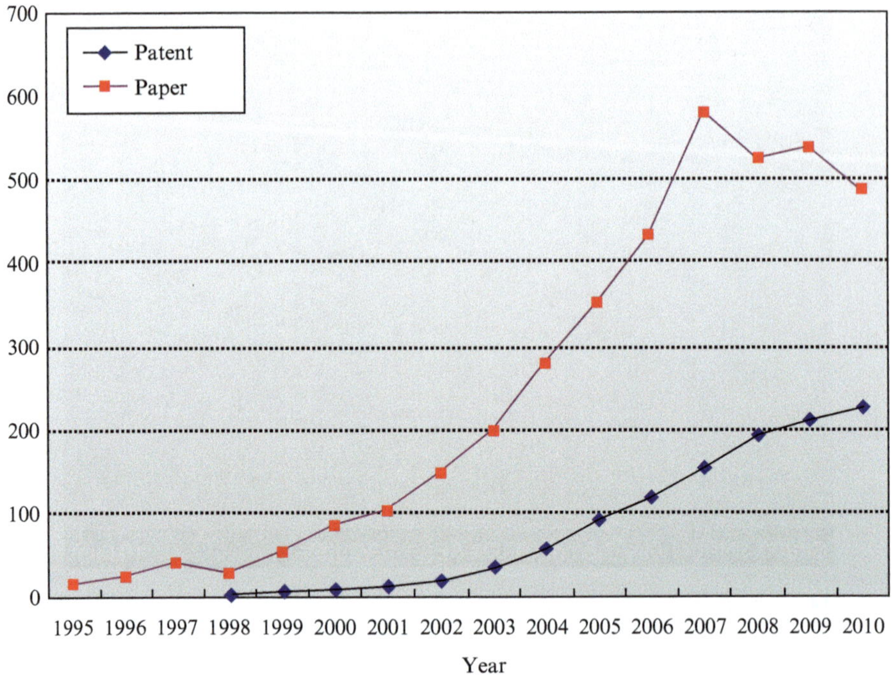

Fig. 2.16 Number of patents and papers related to nanoimprint lithography

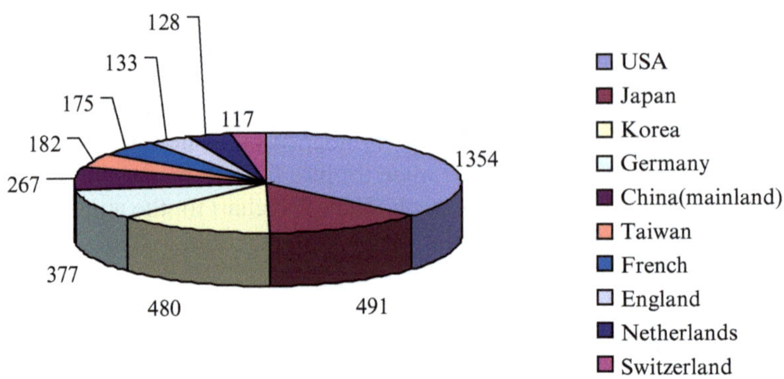

Fig. 2.17 Number of published articles related to country distribution

2.6.2 *Patents Analysis*

Patents are searched in databases of Derwent Innovations Index (DII), and 1,122 patent families have been obtained.

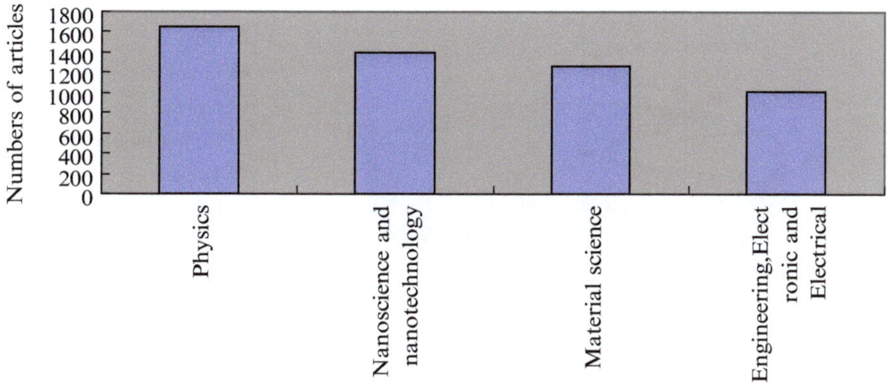

Fig. 2.18 The relation between subjects and the number of published articles

Table 2.3 Top ten lists in the numbers of published articles and company/institutes

Rank listing	Amount	Company/institute
1	93	University of Twente
2	89	Harvard University
3	86	National Institute of Advanced Industrial Science and Technology
4	82	Korea University
5	79	University of Michigan
6	77	Centre National de la Recherche Scientifique (National Center for Scientific Research)
7	77	University of Illinois
8	76	Princeton University
9	71	Korea Institute of Machinery Material
10	64	Seoul National University

Search strategy is as follows:

S1 = ts = (nano$imprint* or nano?Imprint* or ((step same flash) and (imprint* or impress* or lithograph*)) or "nano* press*" or nano*press* or nano*impress* or "nano* impress*")

S2 = ts = (nanopattern* or (nano* same pattern*) or (ultraviolet and lithograph*)) and ts = (imprint* or impress* or ("micro contact print*" or "microcontact print*"))

S3 = au = (chou stephen y or au = chou s y or au = c grant willson or au = willson c g or au = c g willson and ((ti = (imprint* or impress* or pattern* or lithograph*) or "micro contact print*" or "microcontact print*") or (ip = h011-021 or ip = g03f-007/00 or ip = b82b-003/00 or ip = b82b-001/00 or ip = g03c-005/00 or ip = g03f-001/00 or ip = g03f-009/00)))

S = s1 or s2 or s3

Search period (year):1998–2011

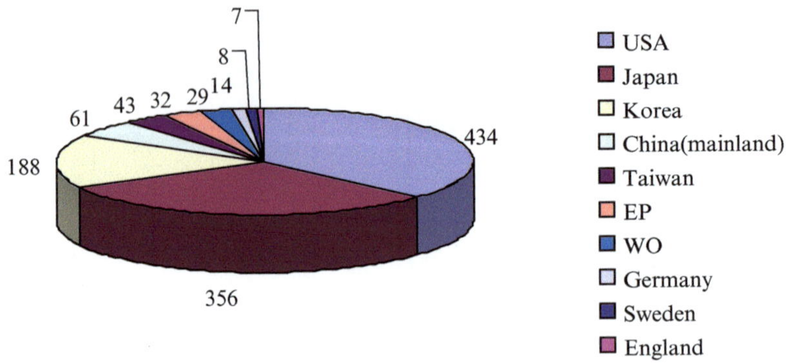

Fig. 2.19 Countries' and regions' map of nanoimprint lithography patent priority

Table 2.4 Top ten lists in patent quantity and company/institutes

Rank listing	Amount	Company/institute
1	44	Molecular Imprints, Inc.
2	41	Hewlett-Packard Dev Co., LP
3	36	Fujifilm Co., Ltd.
4	33	University of Texas System
5	32	Samsung Electronics Co., Ltd.
6	31	Canon KK
7	27	ASML Netherlands BV
8	24	LG Electronics, Inc.
9	23	Dai Nippon Printing Co., Ltd.
10	22	Toshiba KK

Companies make the top ten lists of patents, but university or research institute is responsible for the top ten of publication of papers. It has been proved that research institutes are paying attention to the fundamental research of nanoimprint lithography and companies more attach importance to its application and industrialization.

But for patent quantity, as shown in Fig. 2.19, Japan is close to the USA. Other countries in Asia, for example, China, Korea, and Taiwan region, have paid a role in nanoimprint lithography. Japan attaches more importance to patent than paper, meaning it thinks much of imprinting product.

The patent quantity is ranked except individual patentee. The patents for the company in the top ten are as shown in Table 2.4.

According to an estimation of international patents classification, the number of HO1L is 603, meaning that the hot topic of nanoimprint lithography is semiconductor devices. The number of B29C and G03F is the second and the third, respectively. No. IPC and the related field are shown in Table 2.5 and Fig. 2.20.

Table 2.5 No. IPC and the related field

Amount	No. IPC	Field
603	H01L	Semiconductor devices
391	B29C	Component parts, details accessories, auxiliary operations; molding; other shaping techniques
344	G03F	Photomechanical production of textured or patterned surfaces
138	B82B	Nanostructures formed by manipulation of individual atoms, molecules, or limited collections of atoms or molecules as discrete units; manufacture or treatment thereof

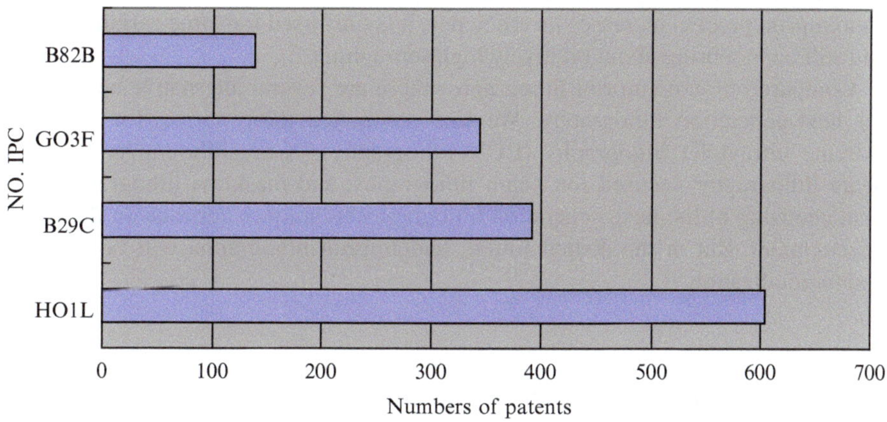

Fig. 2.20 The relation between no. of IPC and number of patents in the top four

2.7 Direction and Vision for Nanoimprint Lithography

2.7.1 Challenges for Nanoimprint Lithography

Although nanoimprint lithography has lots of advantages, it is still in its infancy and faced with many challenges to overcome. One is its own characteristics of the engine, the other is competition between the alternative lithography.

The stamp is a key ingredient for nanoimprint lithography. The defects of the mold must be decreased to zero. Otherwise, the defects will be transferred to imprinting resists. The 1× nature of nanoimprint lithography rather than 4× for photolithography needs a zero defect mold, which will be placed on mold fabrication. Therefore, the cost of mold inspection will increase. Besides, because of contact nature of nanoimprint lithography, stamp needs to have frequent in-fab inspection. This will add the total cost of nanoimprint lithography.

An important issue in semiconductor manufacturing is alignment (overlay). Nanoimprint machine lacks the expensive optics precise stages. The contact nature of nanoimprint lithography also increases alignment difficult, which affects the use of the technique in the fabrication of nanodevices.

Nanoimprint lithography is a direct contact process, and the adhesion between resists and mold has an impact on the pattern fidelity. How to modify the surface characteristics of mold and increase the imprinting times is an important topic.

The defect was regarded as the top issue for commercial application of nanoimprint lithography. The impact of defect on the product quality and yield is becoming more and more important. In nanoimprint lithography process, the contact nature increases the probability of defects.

To meet commercial manufacturing requirements, throughput is an important aspect for nanoimprint lithography. Throughput is affected by various factors, such as mold, mold modification, and imprinting process. Although roll-to-roll nanoimprint process increases its efficiency, it is only used for some certain product and still cannot bring about relatively high throughput.

Compared to nanoimprint lithography, there are several alternative approaches for next-generation lithography. We had mentioned these above, for example, extreme ultraviolet lithography (EUV lithography), X-ray lithography, electron beam lithography, focused ion beam lithography, and maskless lithography. The final choosing of the next-generation lithography techniques depends on their own development. But in the distant future, nanoimprint lithography will be used for commercialization.

2.7.2 *Vision for Nanoimprint Lithography*

Nanoimprint lithography has emerged as low-cost, high-resolution patterning technique that can be scaled up for mass production. It has widely used for the fabrication of biomedicine, high-density memory, photonic crystal, nanodevices, solar cell, sensor, etc., and has aroused immediate concern and widespread researching. If it is effectively put into industrial applications, the compatibility of NIL processes with current semiconductor processes must be addressed. Because it is in its infancy, one will have a decisive competitive advantage in the future if he occupies the priority of nanoimprint lithography.

Nanoimprint lithography is a comprehensive field and includes multidisciplines which need different professional researchers to develop this technique. For example, imprinting resists refers to chemical, material, plasma physics, etc. The company must involve in the research and development of nanoimprint lithography and speed up some further application of the techniques. EVG founded NILCom consortium, which aims to deliver complete process solutions for imprint lithography applications. The number of consortiums came from companies and research organizations [74].

Besides, some standards and terms are set up to regulate and guide the development of nanoimprint lithography.

Photolithography is a sustaining technology. Like I-line and DUV predecessors, EUVL, 193, and 157 nm (with or without immersion) are all extensions of optical lithography that apply light to pattern photoresist. Modern photolithography has a large market and renders nanoimprint lithography smaller market. Therefore, it usually aims at low-end market and cannot compete with current semiconductor process factory at all. But high-end product can also be found: One is the high-brightness LEDs, and the other is the patterned media recording hard disks. Besides, some important issues of nanoimprint lithography need to be overcome. It can find a market to have a distinct advantage over photolithography. At present, the technique is used from laboratory to factory manufacturing.

In the near future, nanoimprint lithography must preferably meet the needs of production because of the development of material and process. It will become the mainstream technique for batch fabrication of high-resolution patterns. This technology is particularly suitable for photon, organic electronics, and biomedicine field. Besides, it is suitable for some functional material that used for permanent components of devices or systems. We deeply believe that it has more and more application because of the further development of some novel materials and processes.

References

1. Stepanova M, Dew S (2012) Nanofabrication technique and principles. Springer, Heidelberg
2. Cui Z (2008) Nanofabrication: principles, capabilities and limits. Springer, Heidelberg
3. Quirk M, Serda J (2001) Semiconductor manufacturing technology. Prentice Hall, Upper Saddle River
4. The ITRS International Technology Working Groups, International Technology Roadmap for semiconductors. http://www.itrs.net/Links/2011ITRS/Home2011.htm. Accessed 12 Nov 2011
5. Chou SY, Krauss PR, Renstrom PJ (1995) Imprint of sub-25 nm vias and trenches in polymers. Appl Phys Lett 67:3114–3116
6. Hua F, Sun Y, Gaur A, Meitl MA, Bilhaut L, Rotkina L, Wang JF, Geil P, Shim M, Rogers JA (2004) Polymer imprint lithography with molecular-scale resolution. Nano Lett 4:2467–2471
7. Special Report (2003) 10 Emerging technologies that will change the world. Technol Rev. pp 33–49
8. Farber AS, Hibibrand J (1977) Method of preparing portions of a semiconductor wafer surface for further processing. US Patent 4,035,848
9. Fujimori S (2009) Fine pattern fabrication by the molder mask method (nanoimprint lithography) in the 1970s. Jpn J Appl Phys 48:06FH01–06FH07
10. Cobburm M, Johnson S et al (1999) Step and flash imprint lithography: an alternative approach to high resolution patterning. Proc SPIE 3676:379–389
11. Hua T, Andrew G, Chou SY (1998) Roller nanoimprint lithography. J Vac Sci Technol B 16:3926–39
12. Chow S, Keimel C et al (2002) Ultrafast and direct imprint of nanostructures in silicon. Nature 417:835–837
13. Bao LR, Cheng X, Huang XD, Guo LJ, Pang SW, Yee AF (2002) Nanoimprinting over topography and multilayer three-dimensional printing. J Vac Sci Technol B 20:2881–2886
14. Cheng X, Guo LJ (2004) One-step lithography for various size patterns with a hybrid mask-mold. Microelectron Eng 71:288–293
15. Cheng X, Guo LJ (2004) A combined-nanoimprint-and-photolithography patterning technique. Microelectron Eng 71:277–282
16. Ahn SH, Kim JS, Guo LJ (2007) Bilayer metal wire-grid polarizer fabricated by roll-to-roll nanoimprint lithography on flexible plastic substrate. J Vac Sci Technol B 25:2388–2391

17. Ahn SH, Guo LJ (2009) Large-area roll-to-roll and roll-to-plate nanoimprint lithography: a step toward high-throughput application of continuous nanoimprinting. ACS Nano 3:2304–2310
18. Tomi H (2011) Stamp fabrication by step and stamp nanoimprinting. Dissertation, VTT
19. Goto H (2007) Overview on nanoimprint technology – process, tools, applications and technical issues for industrialization. Key Eng Mater 345–346:1073–1077
20. Mitsuru H, Masahiko O (2008) Belt-shaped mold and nanoimprint system using the belt-shaped mold. US Patent 7,654,815
21. Chou SY, Krauss PR, Zhang W et al (1997) Sub-10 nm imprint lithography and application. J Vac Sci Technol B 15(6):2897–2904
22. Tan H, Gilbertson A, Chou SY (1998) Roller nanoimprint lithography. J Vac Sci Technol B 16(6):3926–3939
23. Lebib A, Chen Y, Bournexi J et al (2009) Nanoimprint lithography for a large area pattern replication. Microelectron Eng 46:319–322
24. Heidari B, Maximov I, Sarwe EL, Montelius L (2000) Nanoimprint lithography at the 6 in. wafer scale. J Vac Sci Technol B 18:3557–3560
25. Studer V, Pepin A, Chen Y (2001) Nanoembossing of thermoplastic polymers for microfluidic application. Appl Phys Lett 80:3614–3616
26. Beck M, Graczyk M, Maximov I et al (2002) Improving stamps for 10 nm level wafer scale nanoimprint lithography. Microelectron Eng 61–62:441–448
27. Gao H, Tan H, Zhang W (2006) Air cushion press for excellent uniformity, high yield, and fast nanoimprint across a 100 mm field. Nano Lett 6:2438–2441
28. Kehagias N, Zelsmann M, Torres CMS (2005) Polymers optical devices made by reverse and 3D nanoimprint lithography. Proc SPIE 5825:654–660
29. Haisma J (2006) Nanoimprint lithography combined with direct bonding: a possibility to construct quantum dots, wires, and planes in vertical cascade. Appl Phys Lett 89: 244105–244107
30. Wu W, Tong WM, Bartman J, Chen YF, Walmsley R, Yu ZN, Xia QF et al (2008) Sub-10 nm nanoimprint lithography by wafer bowing. Nano Lett 8:3865–3869
31. Jia R, Hornunga M, Verschuuren MA et al (2010) UV enhanced substrate conformal imprint lithography (UV-SCIL) technique for photonic crystals patterning in LED manufacturing. Microelectron Eng 87:963–967
32. Schift H (2008) Nanoimprint lithography: an old story in modern times? A review. J Vac Sci Technol B 26:458–480
33. Xia YN, Whiteside GM (1998) Soft lithography. Angew Chem Int Ed 37:550–575
34. Whiteside GM (1994) Microfabrication by microcontact printing of self-assembled monolayers. Adv Mater 6:600–604
35. Heidari B, Maximov I, Montelius L (2000) Nanoimprint lithography at the 6 in. wafer scale. J Vac Sci Technol B18:3557–3560
36. Zhao XM, Smith SP, Waldman SJ, Whiteside GM (1997) Demonstration of waveguide couplers fabricated using microtransfer molding. Appl Phys Lett 71:1017–1019
37. Kim E, Xia Y, Whitesides GM (1995) Polymer microstructures formed by molding in capillaries. Nature 376:581–584
38. King E, Xia Y, Zhao XM, Whitesides GM (2004) Solvent-assisted microcontact molding: a convenient method for fabricating three-dimensional structures on surfaces of polymers. Adv Mater 9:651–654
39. Zhou WM, Min GQ, Zhang J, Liu YB, Wang JH, Zhang YP, Sun F (2011) Nanoimprint lithography: a processing technique for nanofabrication advancement. Nano Micro Lett 3: 135–140
40. Hirai Y, Tanaka Y (2002) Application of nano-imprint lithography. J Photopolym Sci Technol 15:475–480
41. Kim EK, Stacey NA, Smith BJ, Dickey MD, Johnson SC, Trinque BC (2004) Vinyl ethers in ultraviolet curable formulations for step and flash imprint lithography. J Vac Sci Technol B 22:131–135
42. Cheng X, Guo LJ, Fu PF (2005) Room-temperature, low-pressure nanoimprinting based on cationic photopolymerization of novel epoxysilicone monomers. Adv Mater 17:1419–1424

43. Long BK, Keitz BK, Willson CG (2007) Materials for step and flash imprint lithography (S-FIL®). J Mater Chem 17:3565–3656
44. Zhang W, Chou SY (2001) Multilevel nanoimprint lithography with submicron alignment over 4 in. Si wafers. Appl Phys Lett 79:845–847
45. Muhlberger M (2007) A Moiré method for high accuracy alignment in nanoimprint lithography. Microelectron Eng 84:925–927
46. Sun HW, Liu JQ, Chen D (2009) Nanoimprint lithography for complex 3D micro-nano structures replication under low temperature. Proc SPIE 7402:74020U–74020U-8
47. Schleunitz A, Schift H (2010) Fabrication of 3D nanoimprint stamps with continuous reliefs using dose-modulated electron beam lithography and thermal reflow. J Micromech Microeng 20:201009–201014
48. Konijn M, Alkaisi MM, Blaikie RJ (2005) Nanoimprint lithography of sub-100 nm 3D structures. Microelectron Eng 78–79:653–658
49. McMackin I, Choi J, Schumaker P, Nguyen P (2004) Step and repeat UV nanoimprint lithography tools and processes. Proc SPIE 5374:222–231
50. Gourgon C, Perret C, Tallal J et al (2005) Uniformity across 200 mm silicon wafers printed by nanoimprint lithography. J Phys D: Appl Phys 38:70–75
51. Voelkel R, Duparre J, Wipperman F (2008) Technology trends of microlens imprint lithography and wafer level cameras (WLC). MOC'08, conference on micro-optics, Brussels, Belgium on 25 Sept–27 Sept 2008
52. McMackin I, Choi J, Schumaker P (2004) Step and repeat UV nanoimprint lithography tools and processes. Proc SPIE 537:222–231
53. Tan H, Kong L, Li M (2004) Current status of Nanonex nanoimprint solutions. Proc SPIE 5374:213–1382
54. Babak H, Maximov I, Montelius L (2000) Nanoimprint lithography at the 6 in. wafer scale. J Vac Sci Technol B 18:3557–3560
55. Ando T (2005) Development of nanoimprint lithography and applications. SPIE Optics and Photonics, San Diego
56. Egerton RF (2005) Physical principles of electron microscopy. Springer, Heidelberg
57. Veeco SPM Training Notebook. 004-130-000 (2003)
58. Zankovych S, Hoffmann T, Seekamp J, Bruch JU, Torres CMS (2001) Nanoimprint lithography: challenges and prospects. Nanotechnology 12:91–96
59. Pfeiffer K, Fink M, Ahrens G (2002) Polymer stamps for nanoimprinting. Microelectron Eng 61–62:393–398
60. Carcenac F, Vieu C, Lebib A (2000) Fabrication of high density nanostructures gratings (>500 Gbit/in^2) used as molds for nanoimprint lithography. Microelectron Eng 53:163–166
61. Jung GY, Wu W, Ganapathiappan S (2005) Issues on nanoimprint lithography with a single-layer resist structure. Appl Phys Mater Sci Process 81:1331–1335
62. Schift H, Saxer S, Park S (2005) Controlled co-evaporation of silanes for nanoimprint stamps. Nanotechnology 16:S171–S176
63. Ahn SH, Guo LJ (2008) High-speed roll-to-roll nanoimprint lithography on flexible plastic substrates. Adv Mater 20:2044–2049
64. Peng C, Liang XG, Chou SY (2009) A novel method for fabricating sub-16 nm footprint T-gate nanoimprint molds. Nanotechnology 20:185302–185304
65. Costner EA, Lin MW, Jen WL (2009) Nanoimprint lithography materials development for semiconductor device fabrication. Annu Rev Mater Res 39:155–180
66. Jung GY, Halperin JH, Wu W (2006) Circuit fabrication at 17 nm half-pitch by nanoimprint lithography. Nano Lett 6:351–354
67. Bessonov A, Kin JG, Seo JW (2010) Design of patterned surfaces with selective wetting using nanoimprint lithography. Macromol Chem Phys 211:2636–2641, Dauksher VJ, Le NV, Ainley ES, Nordquist KJ, Gehoski KA, Young SR, Baker JH, Convey D, Mangat PS (2006) Nano-imprint lithography: templates, imprinting and wafer pattern transfer. Microelectron Eng 83:929–932

68. Falco VD, Robert VDL, Marc V, Elmar P, Hans L (2010) Template masters for substrate conformal imprint lithography generated by charged particle nanopatterning techniques. Proc SPIE 7545:7545OS-1–7545OS-13
69. Ahn SW, Lee KD, Kim JS, Kim SH, Lee SH, Park JD, Yoon PW (2005) Fabrication of subwavelength aluminum wire grating using nanoimprint lithography and reactive ion etching. Microelectron Eng 78–79:314–318
70. Hideki I, Kazuyuki K, Eigo K, Kouji U (2007) Critical issues study of nano-imprint tool for semiconductor volume production. Proc SPIE 6517:65170M–65170P
71. Vratzov B, Fuchs A, Lemme M, Henschel W, Kurz H (2003) Large scale UV-based nanoimprint lithography. J Vac Sci Technol B 21:2760–2764
72. Houle FA, Guyer E, Miller DC (2007) Adhesion between template materials and UV-cured nanoimprint resists. J Vac Sci Technol B 25:1179–1181
73. Houle FA, Guyer E, Miller DC (2007) Adhesion between template materials and UV-cured nanoimprint resists. J Vac Sci Technol B 25:1179–1185
74. Tong WM, Hector SD, Jung GY, Wu W, Ellenson J, Kramer K, Hostetler T, Richards SK, Williams RS (2005) Nanoimprint lithography: the path toward high tech, low cost devices. Proc SPIE 5751:46–55

Chapter 3
Stamp Fabrication

3.1 Introduction

Nanoimprint lithography is a replication technique in essence, which can copy a master geometry. The master, also known as the template, the stamp, or the mold, is different from a mask for photolithography. The template material and geometrical parameters directly affect the template deformation and the quality of transfer pattern. Alignment accuracy is determined by the mark pattern on the master; also, the pattern transfer resolution is affected by the master geometry. Therefore, the production of high-quality, high-precision imprint template is a key issue. Besides, evaluation and repair for nanoimprint mold are increasingly becoming a focus. The aforementioned issue is a bottleneck for nanoimprint lithography process. Therefore, research on the stamp must be distinctly elucidated. There are three types of molds: hard mold, soft mold, and rigiflex mold. Silicon, quartz, or metals are used for hard mold, whereas polymers are typically used for soft and rigiflex molds. There are many methods for stamp fabrication, which have conventional and unconventional techniques. In the chapter, stamp materials, stamp fabrication, and evaluation of mold have been introduced.

3.2 Stamp Materials

Stamp quality is a key factor to determine the resolution of the pattern. Mechanical parameters of stamp material are hardness, thermal stability, thermal expansion coefficients and Poisson's ratio, roughness, Young's modulus, and resistance. These properties have to meet the need of nanoimprint lithography process. High hardness, low thermal expansion, and good resistance can guarantee a pattern with high fidelity.

Table 3.1 lists some properties of several different materials for stamps. Si and SiO_2 molds are two common materials for nanoimprint lithography because of their

W. Zhou, *Nanoimprint Lithography: An Enabling Process for Nanofabrication*,
DOI 10.1007/978-3-642-34428-2_3, © Springer-Verlag Berlin Heidelberg 2013

Table 3.1 Comparison of some different materials for stamps

	Yield strength (GPa)	Knoop hardness (GPa)	Elasticity modulus × 10^2 (GPa)	Density (g/cm^2)	Thermal conductivity (mW/cm × K)	Thermal expansion (10^{-6}/°C)
Diamond	53	68.6	10.35	3.5	20,000	1.0
SiC	21	24.3	7.0	3.2	3,500	3.3
Si_3N_4	14	34.2	3.85	3.1	190	0.8
SiO_2	8.4	8.0	0.73	2.5	14	0.55
Si	7.0	8.3	1.9	2.3	1,570	2.33

properties. SiO_2 mold is transparent to UV light and suitable for alignment in UV-NIL. Diamond mold has the highest hardness and has good resistance to acid/alkali corrosion. It can be rinsed time after time. Especially, it can be imprinted on metal substrate, such as Al and Cu.

Besides mechanical properties, anti-sticking property is also a key factor. In the nanoimprint lithography, contact area between mold and resist probably causes nanoimprint resist attaching on mold and a poor fidelity of pattern during demolding. So, the anti-sticking property of the mold is a key technology for nanoimprint lithography. The anti-sticking properties of the mold will be introduced in Chap. 4.

Soft stamp is popular for nanoimprint lithography. There are various types of material for soft stamp, and PDMS, PMMA, PUA, PVA, PTFE, and ETFE are the common stamp materials for NIL [1–5]. The soft stamp is created by replication technology on a master fabricated by photolithography. Once the master is available, thousands of soft molds are easily replicated.

3.3 Conventional Lithography for Stamp Fabrication

Nanoimprint mold can currently be performed by many methods, for example, electron beam lithography (EBL), focused ion beam (FIB), extreme ultraviolet lithography (EUVL), and X-ray lithography (XRL). However, at the finest resolution, the throughput is very slow and time-consuming. Some other techniques (AFM lithography, self- assembled, etc.) can also be used for imprinting mold. It has advantages of simplicity and low cost, but a large area and ordered pattering is very difficult to control. Some methods of mold fabrication are presented as follows.

3.3.1 Electron Beam Lithography

The high-resolution lithography technique originates from the scanning electron microscope technology and was developed in the twentieth century. In 1960, two scientists from Tubingen University first put forward an idea in which the patterning

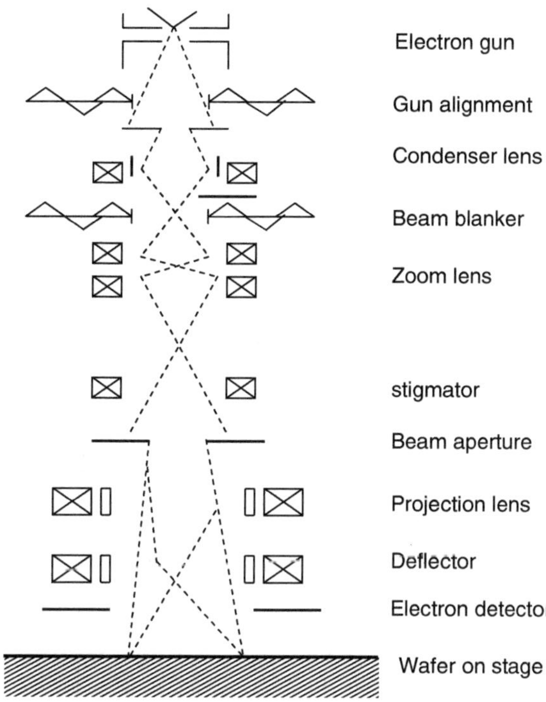

Fig. 3.1 Schematic of a modern electron optical column in electron beam lithography system

can be exposed by adjusting electron microscopy. Based on the technique, the first EBL machine was produced in Cambridge and used for commercial purpose in 1965. After 5 years, Thomson CSF Company from France produced a complete EBL machine composed of laser interference orientation system. From then on, the technique has become a practical state, and has gained attention from researchers around the world. Currently, the most advanced EBL system can focus into a spot of 2 nm, and the resolution of 8 nm can be easily obtained.

In the EBL process, a radiation-sensitive organic polymer thin film (also called resist) is used as a pattern definition medium. By spin coating, the resist was formed on the surface of the substrate and exposed by a focused electron beam. The exposed areas have different solubility in the developer solvent. Hence, the obtained pattern is used as a mask for transferring it to the substrate.

The main characteristics of EBL technology are (1) very high resolution, (2) very slow, and (3) expensive and rather complicated.

3.3.1.1 Electron Beam Lithography System

A typical EBL system consists of electron gun, lens, and deflector. But in fact, the EBL system is very complicated. Figure 3.1 shows a more complete column, as might be found on a modern dedicated EBL system.

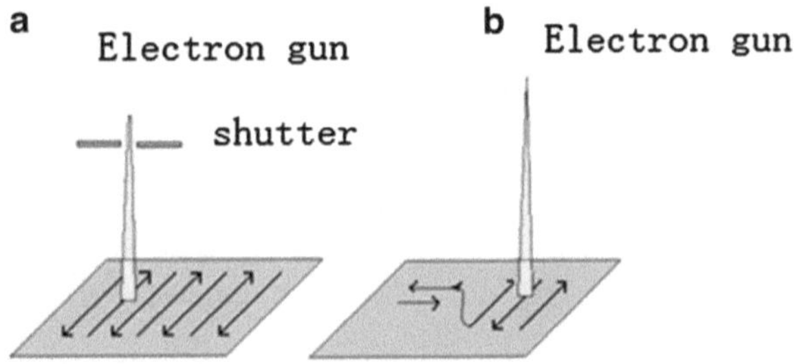

Fig. 3.2 Two types of scanning EBL system. (**a**) Raster scan. (**b**) Vector scan

According to the exposure method, EBL system is mainly divided into two types: scanning EBL system and projection EBL system. Scanning EBL system is classified into raster scan and vector scan. Raster scan is a method of scanning of the whole field depending on beam blanker to achieve patterning exposure. The raster scan method scans the entire field and blanks the beam where exposure is not needed, while the vector scan method scans only over the places where exposure is needed and is blanked as it moves between scanning localities [6], as shown in Fig. 3.2. Table 3.2 shows the properties of some EBL system.

3.3.1.2 SEM and Pattern Generation System

Compared to the EBL system, the SEM lithography system combined with Nanometer Pattern Generation System (NPGS) has become popular at research institutes around the world because of its low price. NPGS is powerful, versatile, and easy to use for doing advanced EBL. Because of its high resolution and low cost, SEM lithography is more widely used in the field of quantum device, fabrication of grating, and experiment process. The SEM lithography facility is installed in the physics laboratory, semiconductor laboratory, and microfabrication laboratory around the world. Figure 3.3 shows the photography of SEM lithography facility. Table 3.3 shows the properties of SEM lithography.

3.3.1.3 Mold Fabrication by Electron Beam Lithography

To prepare an imprinting mold by EBL, a resist was first spin-coated on the substrate; then after exposure and development, the designed nanostructure was patterned on the resist. Next, the patterning on the exposed resist could be transferred to the substrate (Si or SiO_2). There are two methods to transfer the pattern to the substrate: One is the process of reactive-ion etching to transfer the nanopatterning, and the other is the combination of lift-off process with reactive-ion etching to transfer the nanopatterning to the substrate. The schematic illustration for patterning transferring is shown in Fig. 3.4.

Table 3.2 The properties of some EBL systems

Manufacturer	JEOL			Vistec	Leica	Elionix	Crestec	NanoBeam	Raith
Model	JBX 5500	JBX 6300FS	JBX 9300FS	EBPG 5000	VB6-UHR	ELS750 0EX	CABL 9510	NBL-NB3	e-line
Electron gun	ZrO/W Thermal field	ZrO/W Thermal field	ZrO/W TFEG	ZrO/W Thermal field	TFE Thermal field	ZrO/W Thermal field	ZrO/W Thermal field	ZrO/W Thermal field	ZrO/W
Acceleration voltage	50/25 KV	100/50/25 KV	100/50 KV	100/50/25 KV	50–100 KV	5–50 KV	5–50 KV	30–100 KV	100 V–30 KV
Spot	Gaussian beam	Gaussian beam	Gaussian beam	Gaussian beam	Gaussian beam	Gaussian beam	Gaussian beam	Gaussian beam	Gaussian beam
Minimum spot size (nm)	3	2–3	4	5	<3	2	2	2–3	2
Deflection	Vector scan 18-bit DAC 12 MHz	Vector scan 12-bit DAC 12 MHz	Vector scan DAC 25 MHz	Vector scan 16 bit DAC 25 MHz	Vector scan 20 bit 50 MHz	Vector scan 18 bit 12 MHz	Vector scan 18 bit 10 MHz	Vector scan 20-bit 55 MHz	Vector scan 16-bit
Current density/cm^2	<20 nA	30 pA–20 nA	<4,000 nA	50 pA–200 nA		1 pA–50 nA	5 pA–100 nA	<50 nA	5 pA–20 nA
Resolution (nm)	10	8	20	35	8	10	10	20	20
Field stitching error (nm)	40	30	20	50	20	30	30	25	40
Sample size (mm)	100	200	300	150	200	200	100	200	100

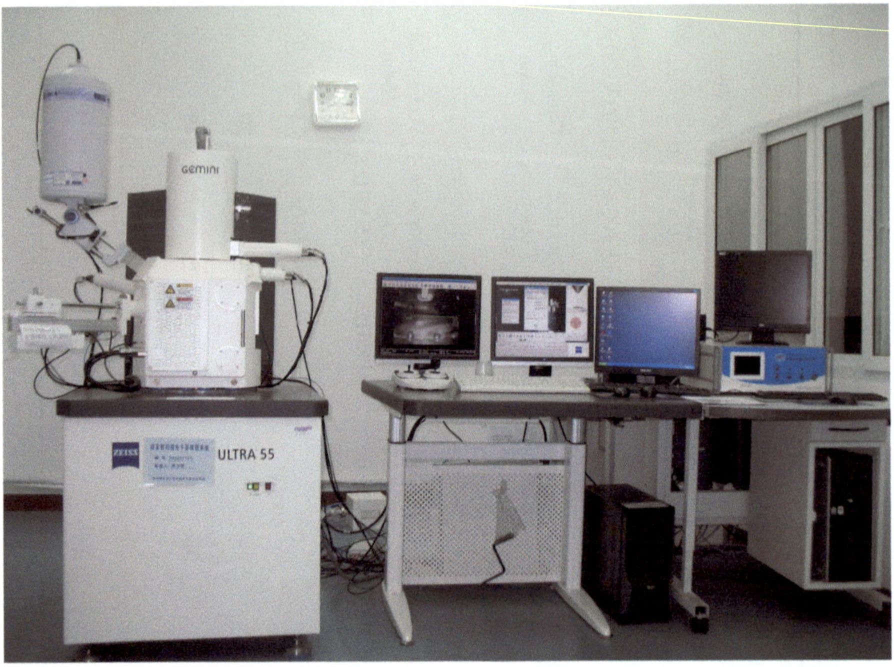

Fig. 3.3 A photo of SEM lithography facility

Table 3.3 The properties of SEM lithography

Manufacturer	Jc Nabity	Raith	Raith	Lecia
Model	NPGS	Elphy-plus	Raith 150	EBL nanowriter
Minimum spot size	Depend on SEM	Depend on SEM	2 nm	–
Resolution	Depend on SEM	Depend on SEM	5 nm	–
Scan field/um	<100	<100	1–800	–
Acceleration voltage/kV	0–40	0–40	0.2–30	10–100
Scan speed/MHz	0.1	2.6	10	1
Control computer	PC	PC	PC	PC

As for quartz material, a layer of metal film is deposited on the quartz surface to curb the effect of the charge before EBL exposure. Meanwhile, the high selectivity between metal and quartz during the etching process can be used to obtain the desired nanopatterning. The Cr or Ti film usually acts as the etching barrier. Figure 3.5 shows the process of the quartz mold of photonic crystal by EBL technique.

The recipe for quartz (SiO_2) mold is as follows:

(a) Cleaning of the quartz substrate.
(b) Sputtering of Ti metal film; the thickness is 40 nm. The wafer is spin-coated with the resist of hydrogen silsesquioxane (HSQ).

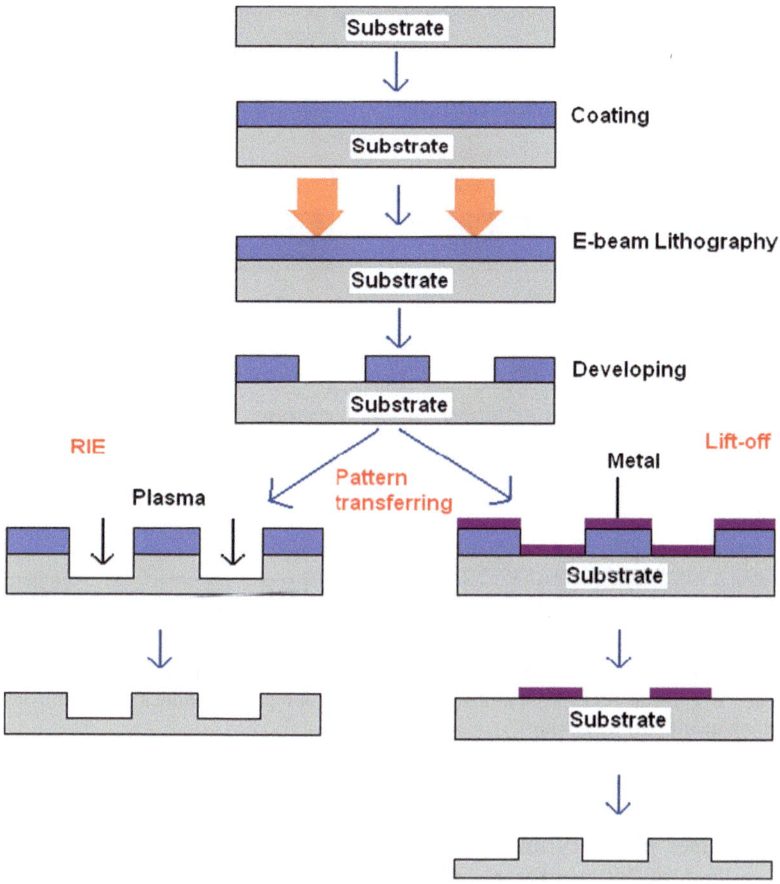

Fig. 3.4 Schematic illustration for patterning transferring

(c) Exposure by EBL system.
(d) Development.
(e) Etching of Ti film.
(f) Removing the residual resist.
(g) Reactive-ion etching of quartz material.
(h) Removing the Ti film and obtaining the quartz mold used for nanoimprint lithography.

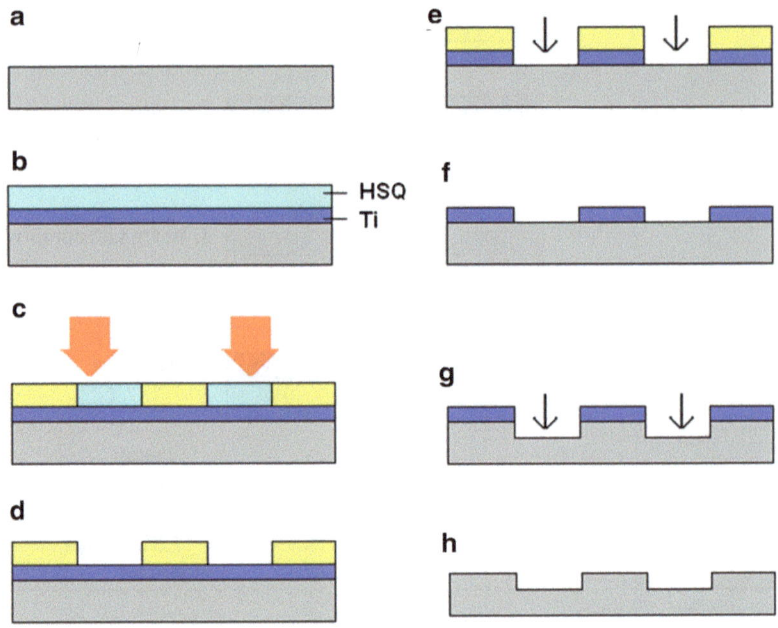

Fig. 3.5 The process of the quartz mold of photonic crystal by EBL technique. (**a**) Quartz substance. (**b**) Coating. (**c**) E-beam lithography. (**d**) Developing. (**e**) Etching Ti. (**f**) Removing HSQ. (**g**) Etching quartz. (**h**) Removing Ti

Hydrogen silsesquioxane (HSQ) is a negative resist that is capable of forming sub-5 nm lines in a very thin layer. The selectivity between metal and quartz is 1:6; the depth of 240 nm for quartz is obtained while etching Ti film of a 40-nm thickness.

Figure 3.6 shows SEM images of the quartz mold in our experiment. The hexagonal photonic crystal structure can be seen clearly, and its height, diameter, and pitch are 15, 100, and 200 nm, respectively.

3.3.2 Ion Beam Lithography

Ion beam lithography is similar to the optical lithography, but with ion source. Compared to parallel light, parallel ion beams can be applied for proximity lithography. The technique includes focused ion beam (FIB) lithography and ion projection lithography (IPL). Research on FIB technique appeared early, and a resolution of 10-nm nanopatterning can be produced by FIB technique.

Besides optical lithography, importantly, it can be used as a micro- and nano-machining tool. A FIB setup is an instrument that resembles a scanning electron microscope (SEM). The FIB system consists of metal ion source, lens, detector, and PC.

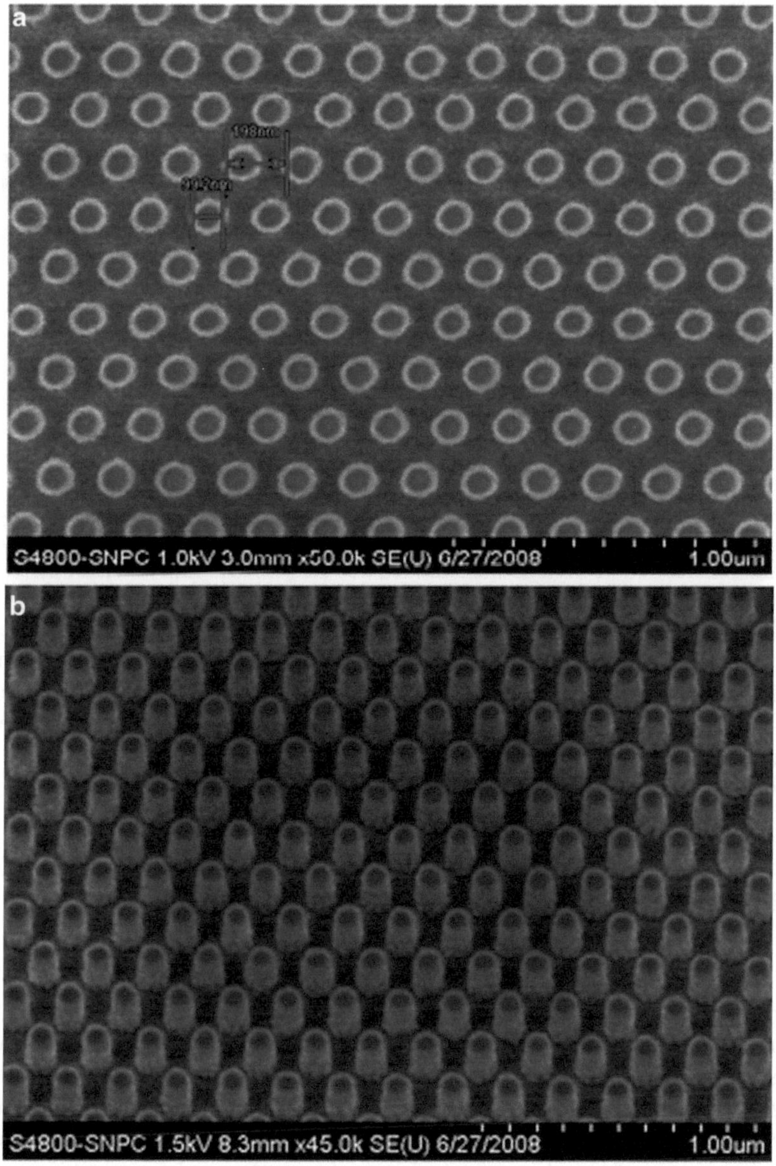

Fig. 3.6 SEM images of the quartz mold. (**a**) Plan. (**b**) 30° tilt

An ion source is at the heart of a FIB system. Usually, liquid metal ion source (LMIS) is a field ion emission source. Gallium is used as the emission material because of low melting point. At high field, metal atoms at the liquid surface escape from the metal surface in the form of field evaporation, resulting in ion emission. It can accurately mill the specimen down to a submicrometer or even a nanoscale.

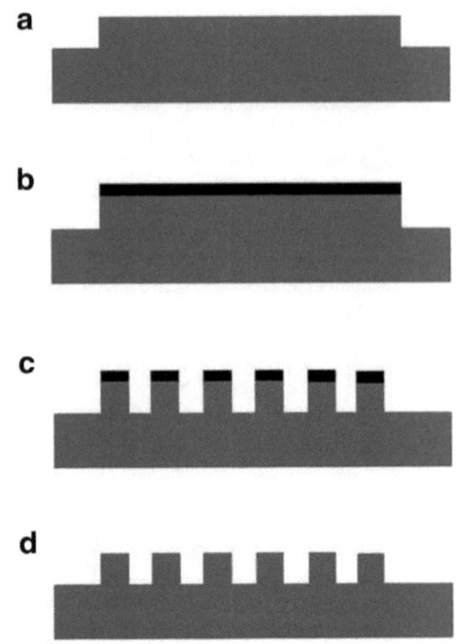

Fig. 3.7 The steps in the template fabrication process using FIB patterning (**a**) Form mesa and mold though grinding. (**b**) Coat mold with 15 nm Cr. (**c**) Structuring with FIB. (**d**) Cr etch and mold clean

FIB is a very strong tool to fabricate the mold. Besides, the mold with some defects can be repaired by FIB technique. A complex 3D structure such as "moth-eye" lens quartz mold is fabricated by the use of FIB milling [7]. The processing steps involved in the mold fabrication when FIB milling is applied are depicted schematically in Fig. 3.7. Figure 3.8 shows the SFIL replication of the "moth-eye" lens with FIB milled 3D structure mold.

3.3.3 Ultraviolet Lithography

Extreme ultraviolet lithography (EUVL) is a next-generation lithography technology using an extreme ultraviolet (EUV) wavelength, currently expected to be 13.5 nm. In fact, EUVL is a common optical lithography process and presented by a scientist of NNT Company in 1989. In EUVL, the wafer is spin-coated with ultraviolet-sensitive polymer named photoresist. With a mask with the designed patterns, the wafer is exposed to UV light and developed in a developer. Then, the remaining photoresist on the surface of the wafer can be used as an etching mask.

Schift et al. reported that producing large-area stamps with patterns of sub-50 nm half pitch for nanoimprint lithography is employed by extreme ultraviolet interference lithography. The resist patterns exposed by extreme ultraviolet interference lithography were transferred into underlying Si or SiO_2/Si substrates by reactive-ion etching. NIL stamps with feature sizes as small as 25 nm and patterned areas as large as 1×1 mm^2 were successfully produced. Figure 3.9 shows the fabrication of NIL stamps via extreme ultraviolet interference lithography [8].

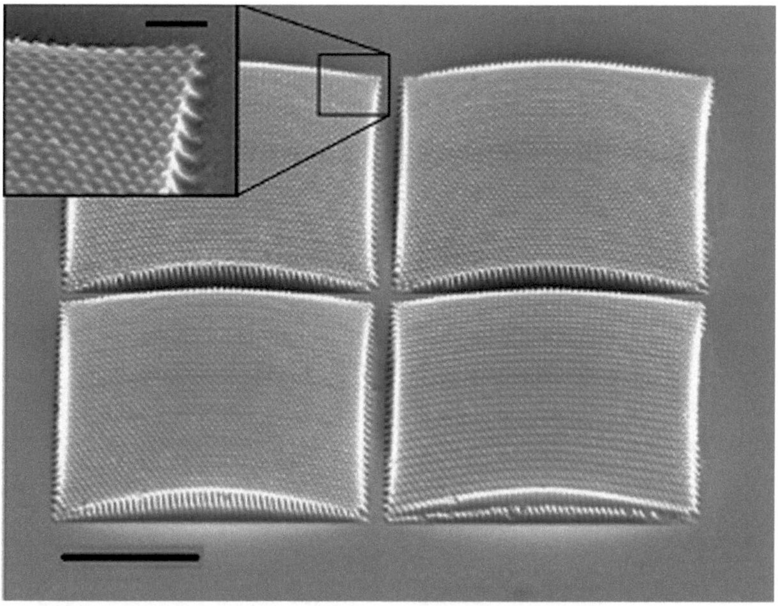

Fig. 3.8 SEM image of SFIL replication of structure with FIB fabricate mold (scale bar represents 4 lm) with inset showing 80-nm lenses on the surface of the larger lens (scale bar represents 500 nm) (Reprinted from Ref. [7], Copyright (2008), with permission from Elsevier)

3.3.4 X-Ray Lithography

An alternative to UVL is X-ray lithography (XRL), which was first developed by Henry Smith of MIT in the early 1970s. The principle of XRL originates from UVL. The technique has developed slowly because of the problem of the mask, photoresist, and optical source. In the1990s, some semiconductor companies and national laboratories were engaged in it, and some important research results were presented. Some countries with synchrotron radiation source, such as Europe, the USA, Japan, and China, have been carried out. The part listing of some synchrotron radiation sources is given in Table 3.4 [9].

Up to now, it is the most mature technology for all the next-generation optical lithography. Figure 3.10 shows the schematic diagram of a synchrotron radiation source. XRL can only be 1:1 proximity lithography and is a parallel process like UVL. XRL system includes X-ray source, X-ray mask, X-ray mask aligner, and X-ray resist. X-ray source is from point X-ray source or synchrotron radiation source. X-ray mask is the most critical part and composed of mask film, absorbing layer on the film and mask frame. X-ray mask aligner is like an optical mask aligner, except that it is structurally simpler. X-ray generates photoelectrons and Auger electrons which then interact with the resist molecules. The X-ray resists have PMMA,

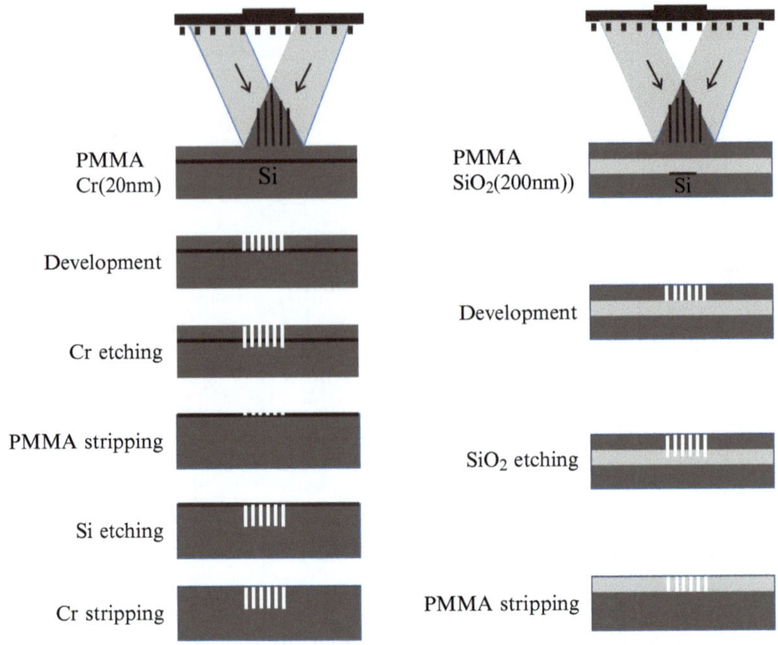

Fig. 3.9 Process schemes of transferring resist patterns exposed by EUV-IL into (**a**) of Si and (**b**) of SiO_2/Si substrates to be used as the stamp

ZEP52, AZPF514, and HSQ. Figure 3.11 shows SEM images of nanostructure with X-ray lithography in Shanghai Synchrotron Radiation Facility (SSRF).

3.3.5 Holographic Lithography

Holographic lithography (or interference lithography) is a maskless technique for patterning regular arrays of fine features, without the use of photomask or complex optical systems. An interference pattern between two or more light beams is set up, and its result reflects in the photoresist.

For two-beam interference, the fringe-to-fringe spacing or period is given by $(\lambda/2)/\sin(\theta/2)$, where λ is the wavelength and θ is the angle between the two interfering waves. The minimum period achievable is half the wavelength. By using three-beam interference, arrays with hexagonal pattern can be generated, while with four beams, arrays with rectangular pattern are generated. Hence, by superimposing different beam combinations, different patterns are made possible [10]. The periodic pattern ranging from one-dimensional to three-dimensional is easily made.

The benefit of using interference lithography is a quick generation of dense features over a wide range without a mask. Furthermore, the pattern can be

Table 3.4 Part listing of some synchrotron radiation sources

Country	Name
China	Beijing Synchrotron Radiation Facility (BSRF)
	National Synchrotron Radiation Laboratory (NSRL)
	Shanghai Synchrotron Radiation Facility (SSRF)
	National Synchrotron Radiation Research Center (NSRRC) (Taiwan)
Singapore	Singapore Synchrotron Light Source (SSLS)
Korea	Pohang Accelerator Laboratory
Japan	Hiroshima Synchrotron Radiation Center (HSRC)
	Institute of Free Electron Laser (iFEL)
	IR FEL Research Center (FEL-SUT)
	Nagoya University Small Synchrotron Radiation Facility (NSSR)
	Saga Light Source
	VSX Light Source
India	INDUS, Center for Advanced Technology
Danmark	ASTRID, Institute for Storage Ring Facilities (ISA)
French	European Synchrotron Radiation Facility (ESRF)
	Source Optimisée de Lumière d'Énergie Intermédiaire du LURE (Soleil)
Germany	ANKA, Forschungszentrum Karlsruhe
	Electron Stretcher Accelerator (ELSA)
	Dortmund Electron Test Accelerator (DELTA)
Italy	DAFNE Light, Laboratori Nazionali di Frascati (LNF)
Russian	Kurchatov Synchrotron Radiation Source (SIBIR-1, SIBIR-2)
	Siberian Synchrotron Radiation Centre (SSRC)
	Dubna Electron Synchrotron (DELSY)
Sweden	MAX-lab
Switzerland	Swiss Light Source (SLS), Paul Scherrer Institute (PSI)
England	Diamond Light Source, Didcot
Canada	Canadian Light Source (CLS)
USA	Advanced Light Source (ALS)
	Advanced Photon Source (APS)
	Cornell High Energy Synchrotron Source (CHESS)
	Duke Free Electron Laser Laboratory (DFELL)
	Stanford Picosecond FEL Center
	National Synchrotron Light Source (NSLS)
	Synchrotron Radiation Center (SRC)
	Jefferson Laboratory Free Electron Laser (Jlab)
	UCSB Center for Terahertz Science and Technology (CTST)
Australia	Australian Synchrotron (AS)

generated on different surfaces (metals, ceramics, and polymers), avoiding the requirement of roughness and morphology of substrate for optical lithography. Patterning arrayed features with submicron structure can be generated over large areas. The benefit of fabricating mold with interference lithography technology is low cost, time-saving, and a large-area patterning. Figure 3.12 shows a photo of nanoimprint mold using interference lithography, and the Si substrate is 4 in. and the nanopatterning is fabricated on the whole substrate surface.

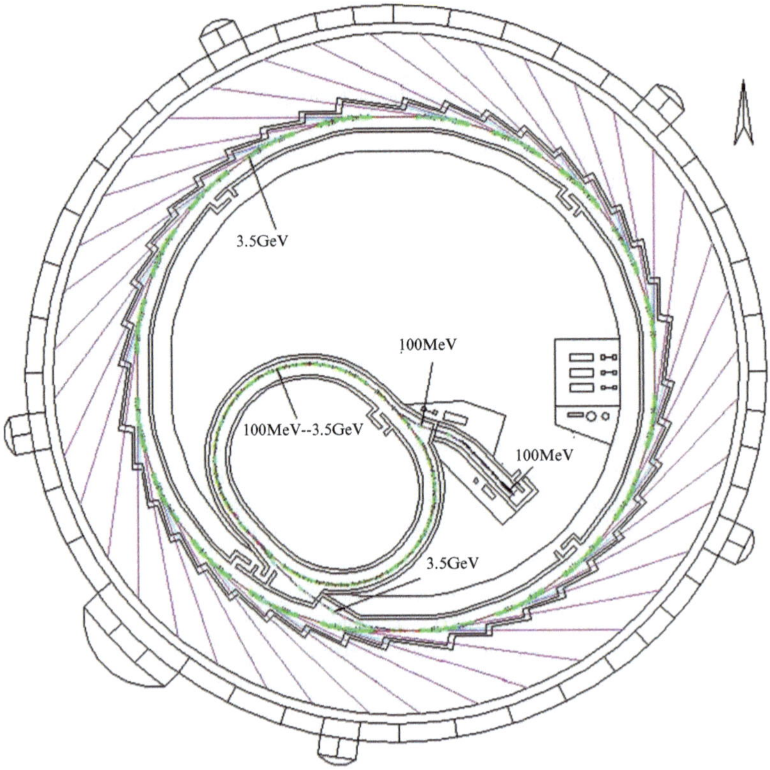

Fig. 3.10 The schematic diagram of a synchrotron radiation source

3.4 Unconventional Lithography for Stamp Fabrication

3.4.1 Edge Lithography

Edge lithography has been investigated as a method for NIL stamp fabrication. It can be referred to as selectively removing or depositing materials on the edges which are defined by conventional photolithography.

Take SiO_2 mold fabrication, for example, the detailed process is as follows and shown in Fig. 3.13 [11].

Silicon wafer surface is deposited with a layer of 15-nm-thick silicon-rich nitride (SiN_x), followed by a 40-nm conventional LPCVD TEOS (silicon oxide formed by decomposing tetraethyl orthosilicate):

(a) The pattern is transferred by photolithography technique into a TEOS layer using 1% HF.

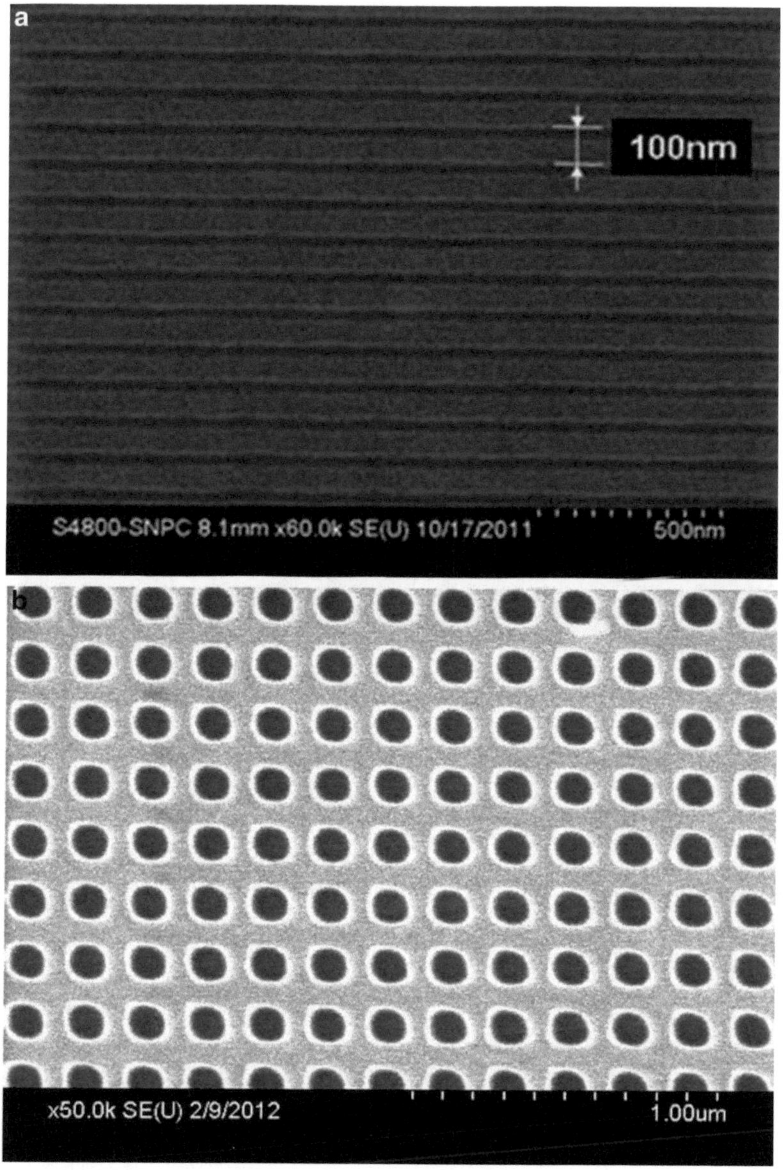

Fig. 3.11 SEM images of structure with X-ray lithography

(b) After removing the photoresist, 85% H_3PO_4 acid heated up to 180°C is used to pattern the SiN_x by using TEOS as the mask.
(c) The TEOS is removed in 1% HF. Then the patterned SiNx is transferred into the silicon substrate using OPD4262 as the etchant, which has a TMAH concentration of 2.5%.

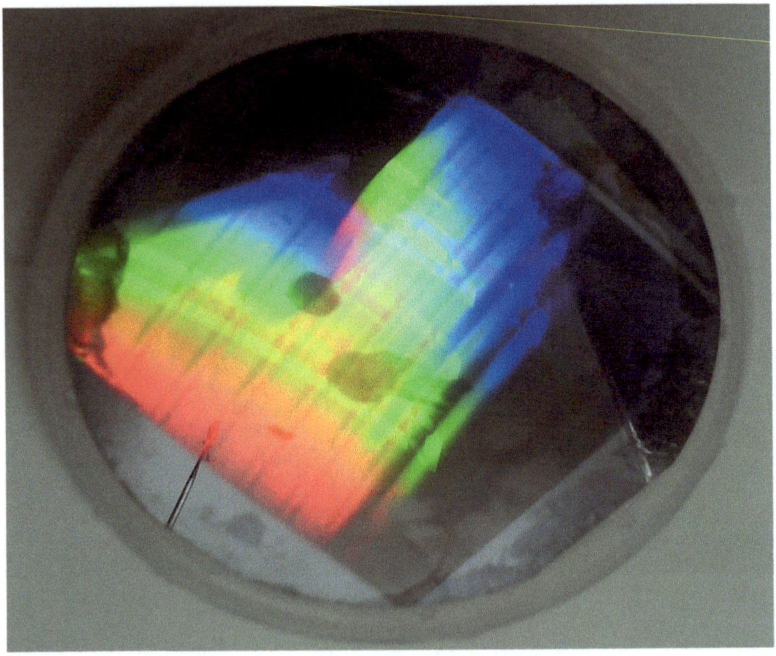

Fig. 3.12 A photograph of nanoimprint mold using interference lithography

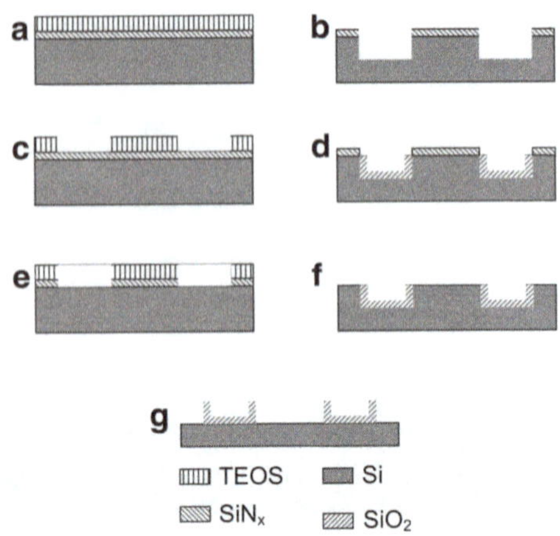

Fig. 3.13 SiO_2 nano-ridge mold fabrication process. (**a**) Si substrate coated with SiN_x and TEOS. (**b**) TEOS pattern. (**c**) SiN_x patten. (**d**) Silicon pattern. (**e**) Local oxidation of silicon. (**f**) Removing SiN_x. (**g**) Etching back into silicon, and form the SiO_2 nanomold

(d) Local oxidation of silicon process. The area which is not protected by SiN_x is dry oxidized at a temperature of 950°C.
(e) 180°C 85% H_3PO_4 acid is used to remove the SiN_x mask layer.
(f) Etching back into silicon using OPD4262 and forming the SiO_2 nanomold.

Size reduction lithography is similar to edge lithography, the dimension of pattern can be controlled to reduce by deposition, and a schematic process is shown in Fig. 3.14 [12–15].

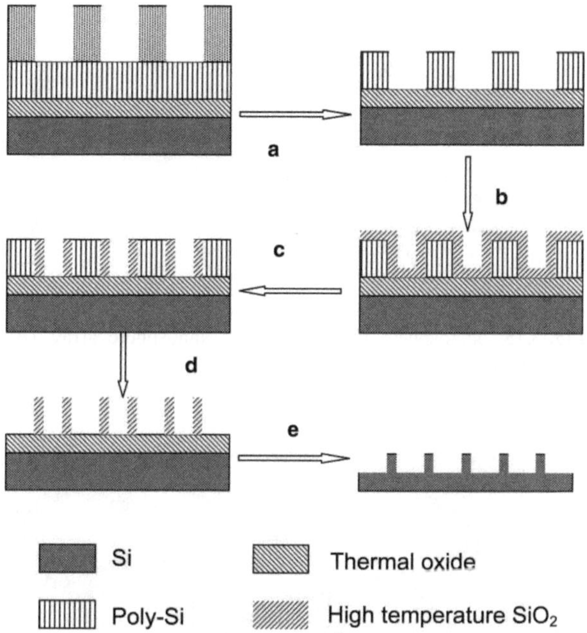

Fig. 3.14 Size reduction lithography process. (**a**) Etch poly-Si and remove photoresist. (**b**) Conformal deposition of silicon oxide. (**c**) Etch oxide and expose poly-Si. (**d**) Etch poly-Si. (**e**) Etch pattern into wafer and HF dip

3.4.2 Nanosphere Lithography

Nanosphere lithography (NSL) is an inexpensive, simple to implement, inherently parallel, high throughput, materials general nanofabrication technique capable of producing an unexpectedly large variety of nanoparticle structures and well-ordered 2D nanoparticle arrays [16]. The dimension of the array can be controlled by adjusting the diameter and gap of nanospheres. However, a uniform arrangement of nanospheres on the large area is very difficult to control.

Usually, polystyrene or silica nanosphere is used as a material tightly packed on a substrate surface as a masking layer. Take polystyrene nanosphere, for example, they were diluted in a solution of the surfactant Triton X-100/methanol (1:400 by volume). The surfactant was used to assist the solutions in wetting the substrate. Then, the solution of nanosphere was spin-coated on the substrate surface and generated hexagonally ordered nanostructure. Metal deposited by thermal evaporation on the surface can reach the substrate through the gaps. The polystyrene nanospheres were removed from the substrate by dissolving them in CH_2Cl_2 with the aid of sonication for 1–4 min [17]. The substrate surface is left with an array of dots, as shown in Fig. 3.15. Or the deposited metal is used as the etching mask. Figure 3.16 shows a SEM image of silicon nanopillar arrays by SNL.

As mentioned above, nanostructure stamp can be obtained by a mask of nanosphere or deposition mask, but the stamp with an ordered array is difficult to be realized on a large-area surface. In 2010, Jeong et al. reported that the nanospheres are arrayed on the substrate surface by tilted-drain methods, and the uniformity

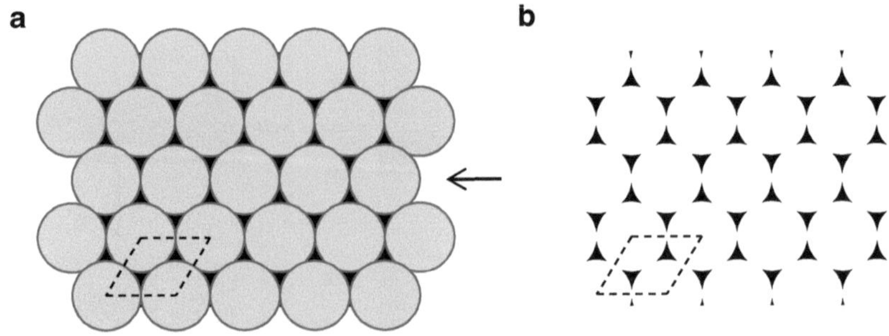

Fig. 3.15 Nanosphere lithography. (**a**) Nanosphere array. (**b**) Dot array is formed after metal deposition and removal of the nanospheres

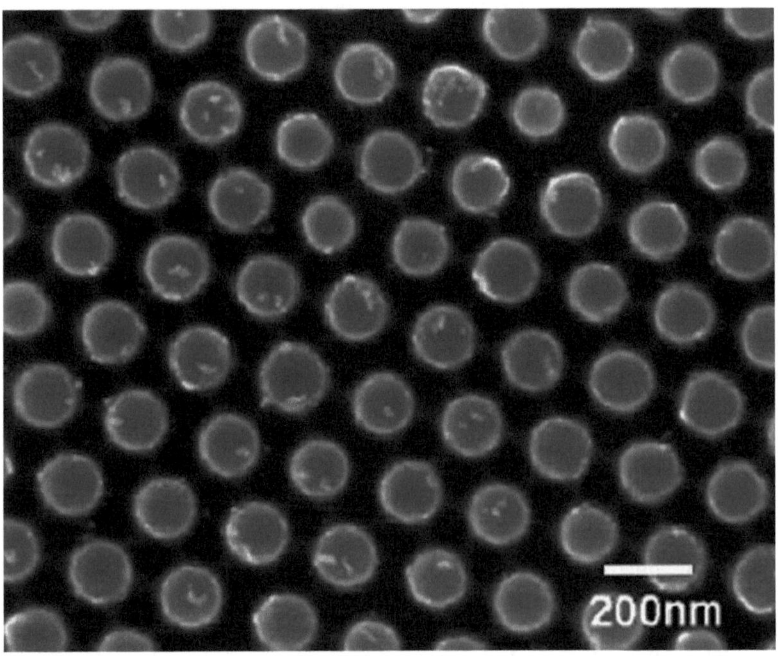

Fig. 3.16 SEM image of silicon nanopillar arrays by SNL. The mean FWHM of pillar is 160 nm (Reprinted from Ref. [16], Copyright (2007), with permission from Elsevier)

of nanospheres on the substrate surface is easily controlled. The simple and large-area arrangement patterning of nanospheres provides a mask for stamp fabrication. A schematic illustration of the coating process using the tilted-drain method of nanosphere is shown in Fig. 3.17 [18].

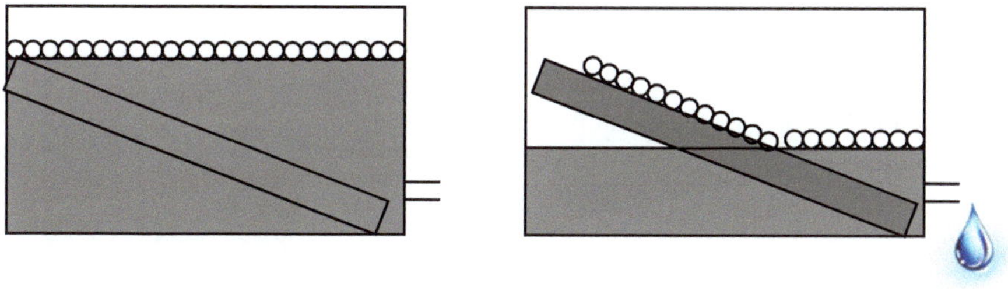

Fig. 3.17 Schematic illustration for the coating process using the tilted-drain method of nanosphere on Si wafer

3.4.3 Chemical Vapor Deposition

Chemical vapor deposition (CVD) is a chemical process used to produce high-purity, high-performance solid materials and often used in the semiconductor industry. In a typical CVD process, the substrate is exposed to one or more volatile precursors, which react and/or decompose on the substrate surface to produce the desired deposited production. Comparing to EBL, it can fabricate high aspect ratio patterning. CVD is classified into atmospheric pressure chemical vapor deposition (APCVD), low-pressure chemical vapor deposition (LPCVD), and plasma-enhanced chemical vapor deposition (PECVD).

Chow et al. reported that mold is fabricated by room-temperature conformal plasma-enhanced chemical vapor deposition. The process is as follows: A thin layer of a-Si is deposited onto the imprint resist; bonding of the a-Si layer to a second substrate by UV-curable adhesive; and separating the master mold from the duplicated mold. Schematic of the imprint mold replication process by CVD is shown in Fig. 3.18 [19].

3.4.4 Wet Etching

Wet etching is an etching process that uses liquid chemicals or etchants to remove materials from the substrate, which is often used in the semiconductor industry. The process may just consist of dissolution of the material to be removed in a liquid solvent, without changing the chemical nature of the dissolved material. It is an isotropy process. Although it is not suitable to fabricate smaller patterning with high vertical sidewall, wet etching is widely used in fabrication of an uncritical dimension of patterning for high selectivity, simple facility, and low cost and high capacity.

We have developed a method of a novel size reduction process using electron beam lithography (EBL) combined with wet etching technique, which is a possible solution for producing large-area and low-cost nanopattern stamp. According to

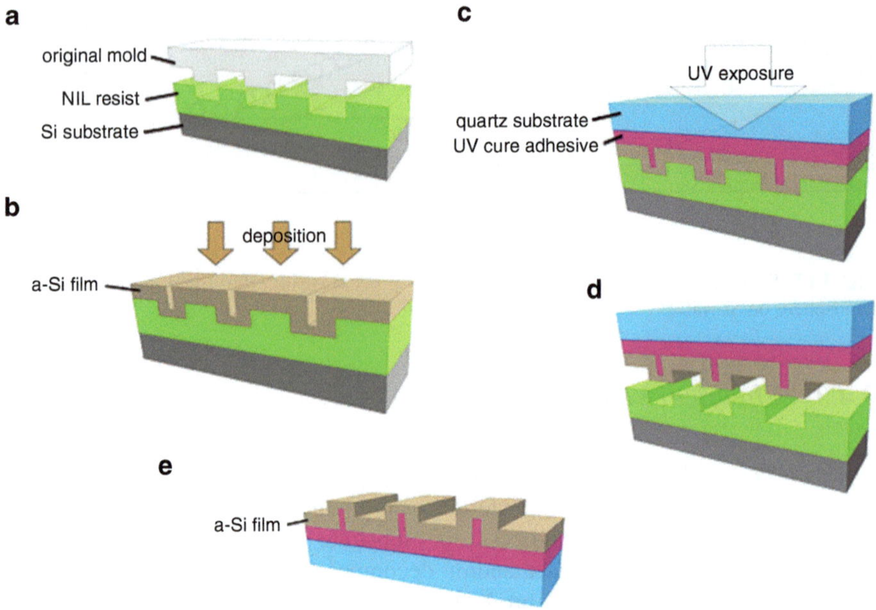

Fig. 3.18 Schematic of the imprint mold replication process by CVD. (**a**) NIL with master mold. (**b**) Conformal PECVD. (**c**) Bonding with UV curve adhesive. (**d**) Transfer to new substrate. (**e**) Complete mold replica (Reprinted from Ref. [19], Copyright (2007), American Institute of Physics)

the etching rate of SiO_2 materials, nanorods can be etched into nanocones, and dimension of stamp is reduced. The scheme of the size reduction process is shown in Fig. 3.19. Two steps are involved in the size reduction process:

1. Negative stamp fabrication with periodical pores by EBL
   The scheme process to fabricate periodical pores on the quartz surface by EBL and RIE process is shown in the (a–d) of Fig. 3.19. Two-inch SiO_2 wafer with a thickness of 1 mm was used. The surface was polished down to a root-mean-square (rms) roughness of less than 0.8 nm. In the first step, a 60-nm height chrome layer was deposited to suppress charging effects on the quartz surface during the EBL exposure, which also had the advantage to minimize the CD (critical dimension) losses during the pattern transfer. Primary patterns were defined by e-beam lithography (EBL) in hydrogen silsesquioxane (HSQ) resist. After the etching processes, the defined patterns in HSQ were transferred on the SiO_2 substrate. Figure 3.20 shows AFM image of negative stamp with 1-μm pitch. The diameter is 500 nm and the depth of the hole is 130 nm.
2. Negative stamp being changed into positive one by wet chemical etching
   The isotropic wet chemical etching process was the key step to reduce the pattern size and change the negative stamp into positive one. The developed 2-in. negative stamp was sawn into several 1×1 cm^2 parts and dipped in buffered

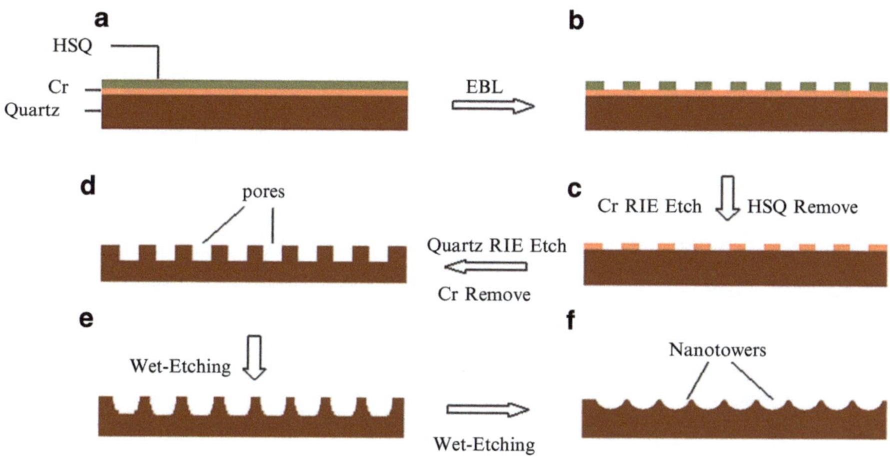

Fig. 3.19 AFM image of negative stamp by EBL

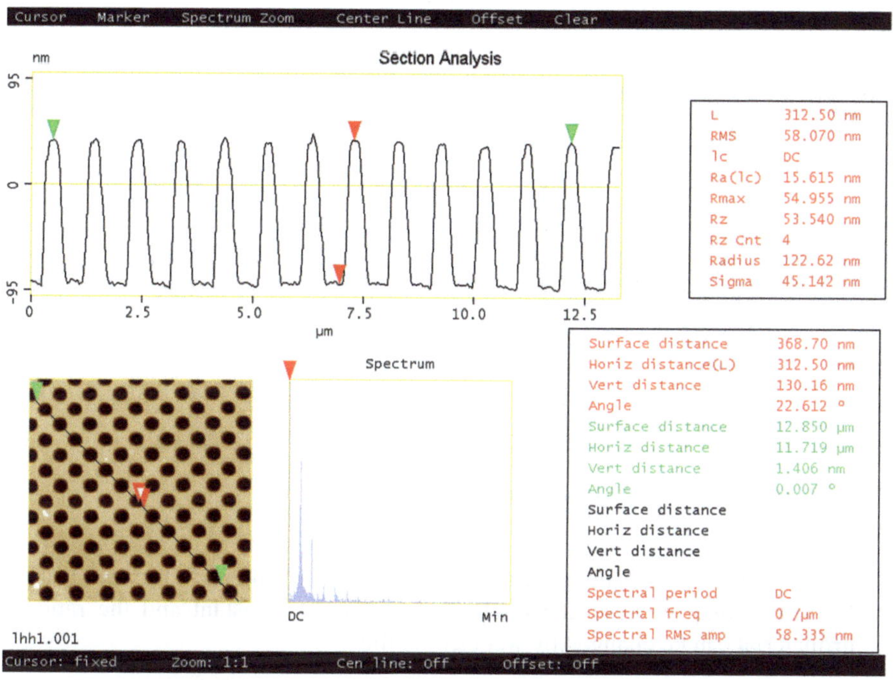

Fig. 3.20 The scheme of size reduction process for stamp fabrication. (**a**) quartz surface coated with Cr and HSQ. (**b**) EBL exposure and form HSQ pattern. (**c**) Cr pattern created by etching. (**d**) quartz pattern. (**e**) wet etching. (**f**) quartz surface with nanotowers pattern

oxide etch (BOE) solution (40% NH4F:49% HF = 5:1) at room temperature with different etching time. Micro-negative stamp is converted into nano-positive stamp, as shown in Fig 3.19e–f. If the etching time lasts, and finally photonic

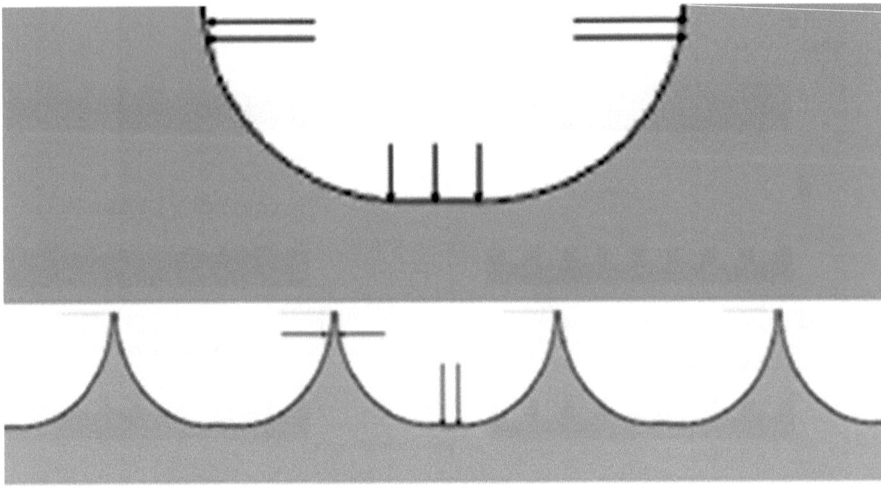

Fig. 3.21 The process of wet etching hole of SiO_2 stamp

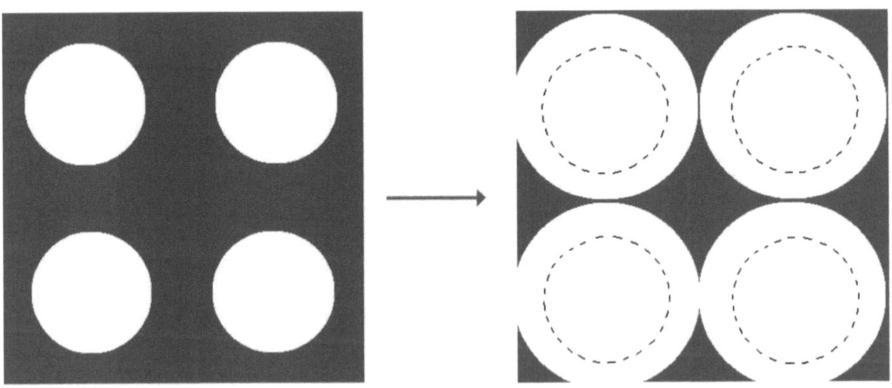

Fig. 3.22 The scheme of pores changes into protuberance

crystal stamp with less than critical dimension of 100 nm was fabricated. During the wet etching process, the pitch of the stamp is constant and the ratio of air/material can be changed by wet etching time.

Figure 3.21 shows the process of the wet etching hole of SiO_2 stamp. Because of isotropy for wet etching, buffered oxide etch (BOE) solution decreases the depth of SiO_2 holes and widens the diameter of the holes. As time passed, the interface of holes was broken and meshes have just become 2D periodic islands. In this period, the positive stamp has been changed successfully from the negative pattern, but the size of protuberances still keeps the same, as shown in Fig. 3.22.

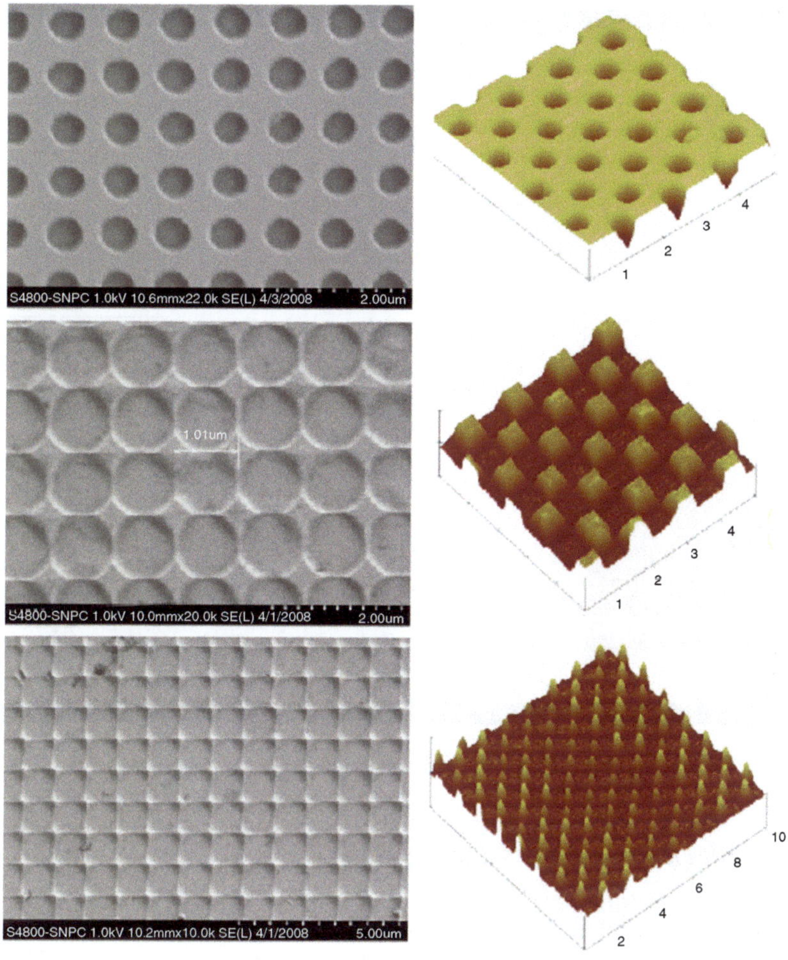

Fig. 3.23 SEM and AFM images of the changing process from negative stamp changes into positive one by wet chemical etching

Figure 3.23 shows the SEM and AFM images of the changing process from the negative stamp (pores) changes into positive one (protruding tips) by wet chemical etching. As can be seen, the holes are converted into pillars, the interface was broken, and meshes have just become 2D periodic islands. Keeping with the etching process, it can be found the diameter of pores is still enlarged following the former circular tracks, thus leading to the protuberances size smaller and smaller. Figure 3.24 shows an AFM image of photonic crystal SiO_2 stamp with 78-nm patterning.

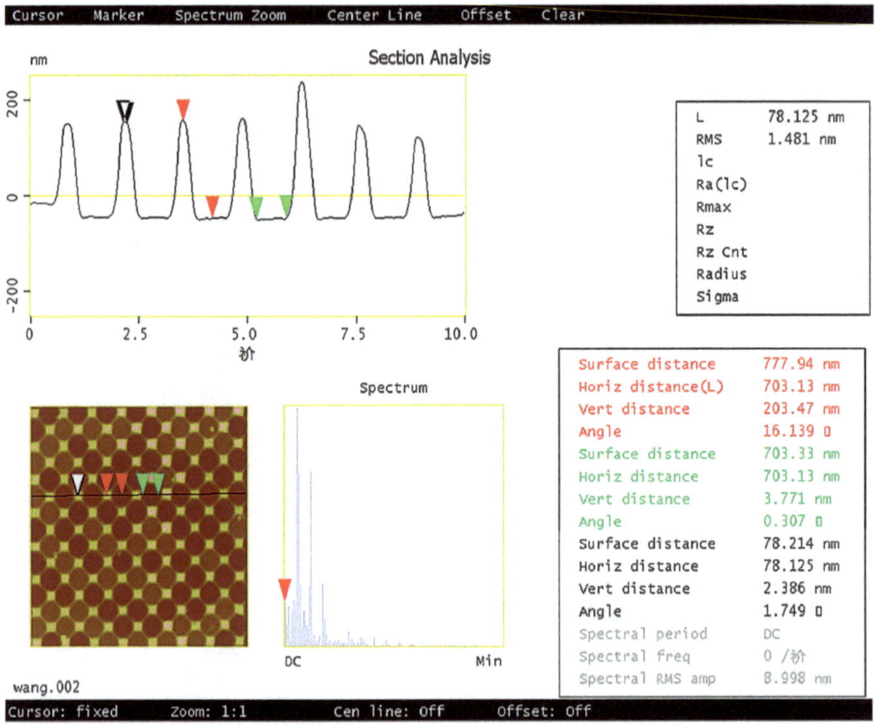

Fig. 3.24 AFM image of stamp fabricated by wet etching

3.4.5 Block Copolymers Lithography

A block copolymer is a polymer consisting of multiple sequences, or blocks, of the same monomer alternating in series with different monomer blocks. Diblocks, triblocks, tetrablocks, multiblocks, and so on can also be made. The chemical differences between the blocks can be used to generate a mask, allowing the block copolymer pattern to be transferred to the underlying substrate through plasma etching. Dots or wires can be created by block copolymer combined with plasma etching. The fabricated nanopatterning can act as a template.

Block copolymer (BCP) nanolithography is well suited for fabrication of nanopatterning, thanks to its fine resolution, cost-effectiveness, and scalability. The early investigation of block copolymers as a microlithographic photoresist can be dated back to late 1980s [20].

Since the magazine *Science* reported periodic arrays of dots in square centimeter substrate, research on the pattern formation by block copolymer has been the focus [21]. Figure 3.25 shows the procedure for pattern formation by block copolymer self-assembly and pattern transfer using plasma etching.

Fig. 3.25 The procedure for pattern formation by block copolymer self-assembly and pattern transfer using plasma etching. (**a**) Stamp. (**b**) Holes. (**c**) Dots

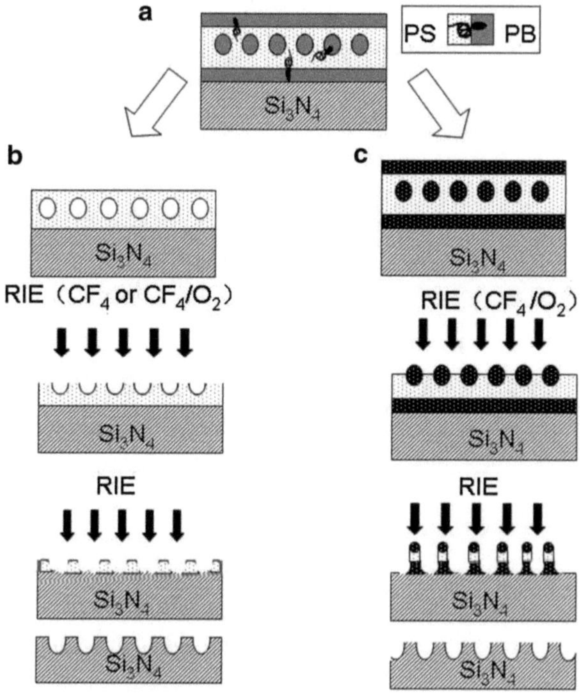

Ordered arrays of nanoscale PS-PMMA diblock copolymer can be formed over full 8-in. diameter silicon wafers and on various substrate materials. The polymer template pattern can be faithfully transferred using either RIE or metal lift-off [22].

Ti, W, Pt, Co, Ni, Ta, Au, and Al highly ordered nanowire arrays with widths down to 9 nm are created by transfer of high-quality linear block copolymer patterns [23].

3.4.6 Anodization Process for Anodic Aluminum Oxide

Anodic aluminum oxide (AAO) template is a promising and suitable option for nanoimprint lithography.

Since the porous alumina membrane (AAO) was fabricated in 1995, it has increasingly found its roles in preparing nanostructured materials and in the field of sensing and storage, such as aligned nanowires [24], aligned carbon nanotube [25] or metal and metal oxide nanotubes [26], dots arrays [27], and 2D photonic crystal [28]. Besides, AAO films have been acted as a mask for etching Si and GaAs substrates to gain nanopatterning surfaces [29–31].

AAO membrane can be made using a one-, two-, or three-step anodization process. Figure 3.26 presents a schematic of the two-step anodization process. In brief, the pre-handled Al sheet was placed in the solution of oxalic acid for 5–6 h

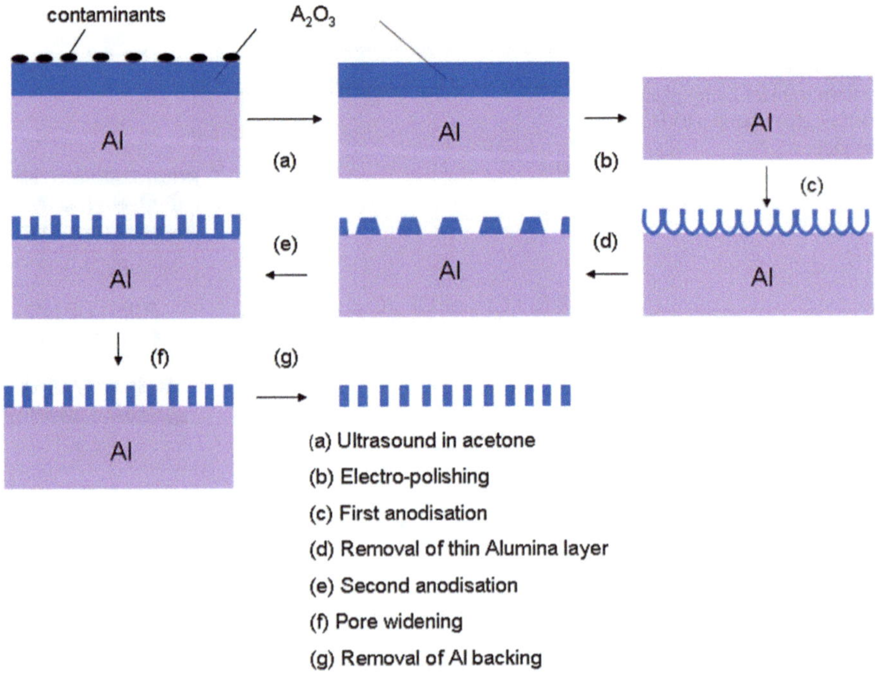

Fig. 3.26 Schematic of the two-step anodization process

Table 3.5 A typical two-step anodization parameters for AAO stamp forming electrolytes

Electrolyte	Concentration (M)	Potential (V)	First anodization (h)	Second anodization (h)
Sulfuric acid	0.3	24	5	3
Oxalic acid	0.3	30	8	5
Oxalic acid	0.3	60	5	3
Phosphoric acid	2.5	60	8	5

for a first anodization, followed by the 6% H_3PO_4 + 1.8% H_2CrO_4 aqueous solution for removing the oxide layer. Following the second anodization lasting for 10–12 h, the Al on the AAO membrane was eliminated by $CuCl_2$ solution. The removing of the barrier layer and the diameter of nanopore could be widened by 5% H_3PO_4 solution. Table 3.5 is a typical two-step anodization parameters for AAO stamp forming electrolytes.

The anodization setup is shown in Fig. 3.27. A Pt wire and a polished aluminum foil are used as the cathode and anode, respectively. A potential between two electrodes and the temperature of the electrolyte were kept constant to generate AAO master.

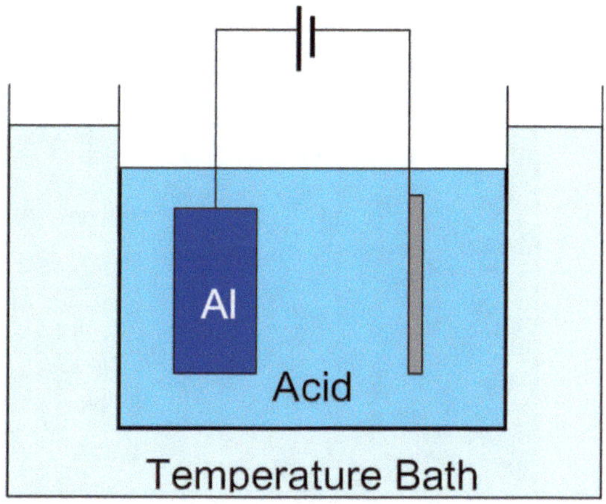

Fig. 3.27 Anodization setup

After two-step anodization process, a typical SEM image of AAO membranes is presented in Fig. 3.28. Masuda's group imprinted two-dimensional photonic crystals composed of polymer pillar arrays using ordered anodic porous alumina (AAO) molds. The period and diameter of the pillars were 500 and 200 nm and the height of the pillars is 1.5 um, as shown in Fig. 3.29 [32].

3.4.7 Electrodeposition

Electrodeposition (electrochemical deposition, electroplating) is to deposit (a dissolved or suspended substance) on an electrode by electrolysis. Usually an electrolytic cell (consisting of two electrodes, electrolyte, and external source of current) is used for electrodeposition. Figure 3.30 shows a fundamental theory of electrodeposition. The part of a stamp to be plated is the cathode of the circuit. The anode is made of the metal to be plated on the part. Both parts are immersed in a solution called an electrolyte containing one or more dissolved metal salts as well as ions. A dc supply provides a current to the anode, and the dissolved metal ions in the electrolyte solution are reduced and the metal leaves onto the cathode.

Ni is a common material for electrodeposition, which has the benefit of high strength, hardness, toughness, and resistance to corrosion. The holes of polymers resist stamp can be deposited with Ni film by electrodeposition, and some microstructures are also achieved with this method. Obducat, the world-leading supplier of lithography solutions, has developed a complete production process of Ni template. The resist patterning was imprinted by nanoimprint lithography with EBL-created stamp. Then, the seed layer is coated on the resist. Ni mold is finally fabricated by electrodeposition, which has both high strength and flexibility, and is suited for roll-to-roll nanoimprint lithography process. The replicated nickel stamp based on electroplating is shown in Fig. 3.31 [33].

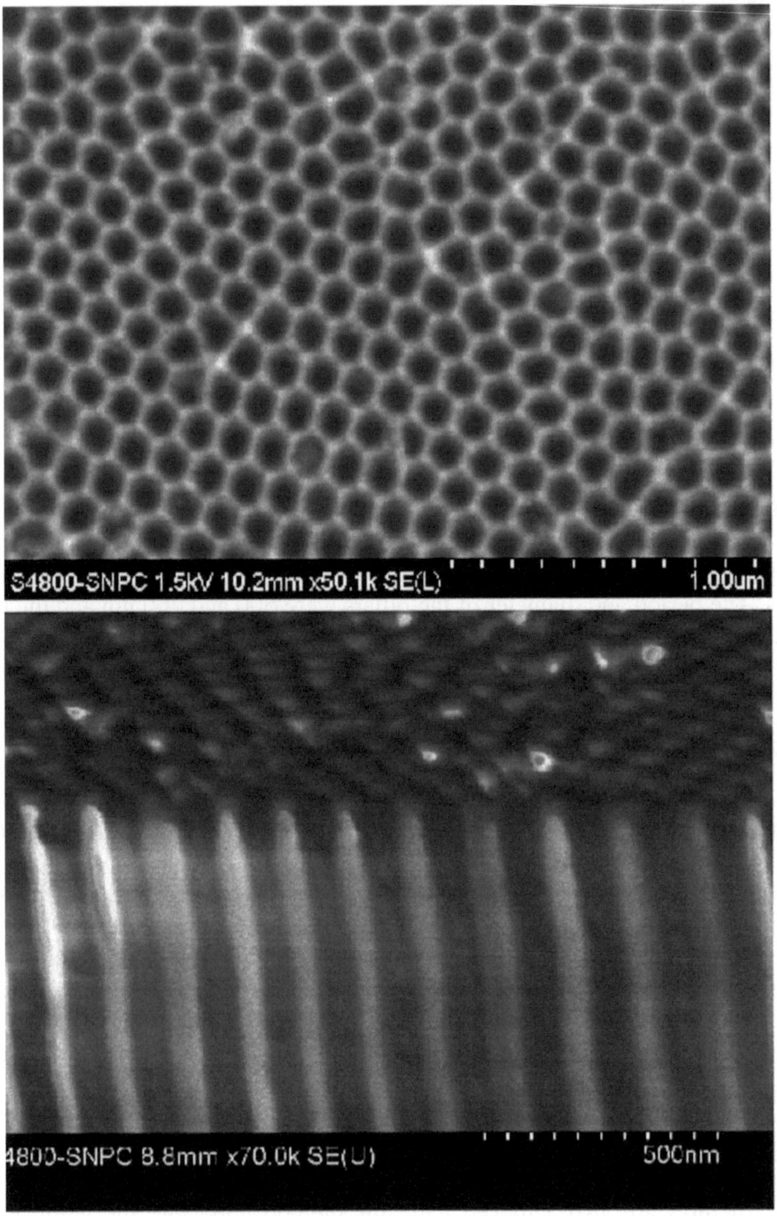

Fig. 3.28 A typical SEM images of AAO membranes

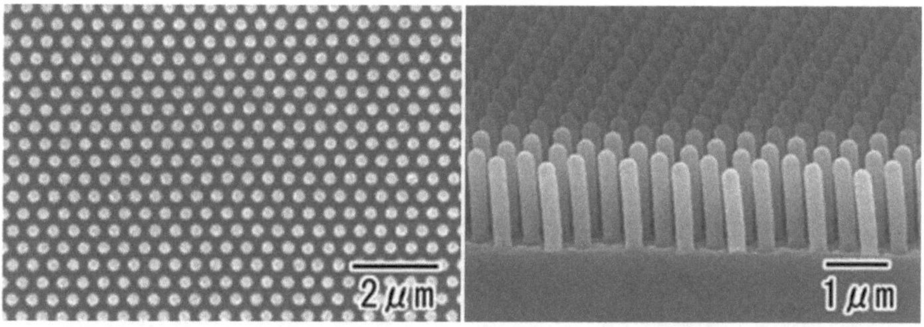

Fig. 3.29 SEM image of polymer pillars: surface and sectional view (Reprinted from Ref. [32], Copyright (2010), American Institute of Physics)

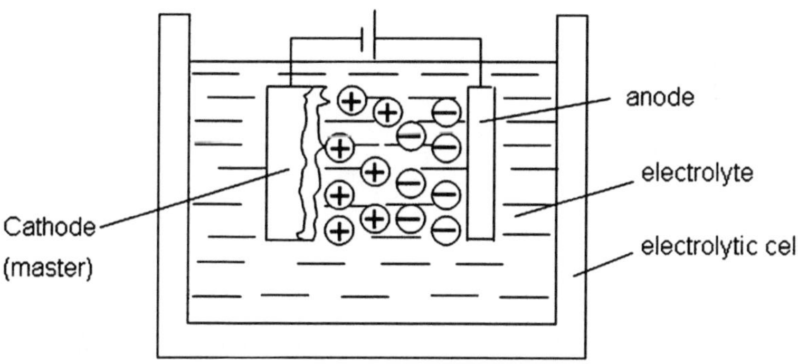

Fig. 3.30 Fundamental theory of electrodeposition

The nickel deposition rate on a flat surface is given by the equation [34]

$$\text{Rate} = \frac{a \cdot I}{d \cdot F \cdot n \cdot s}$$

The rate is the thickness of the nickel grown per second, where a is the atomic weight of Nickel (58.7 g/mole), I is the plating current, n is the number electron required to plate each ion ($n_{\text{Nickel}} = 2$) and s is the plating area, d is the density of the Nickel (8.9 g/cm^3), and F is Faraday's constant (96,500 c/mole).

The patented Intermediate Polymer Stamp (IPS) from Obducat enables contamination control and increases the stamp lifetime. In the process, the master stamp is replicated into a soft Intermediate Polymer Stamp that transfers the structures into the target substrate [35].

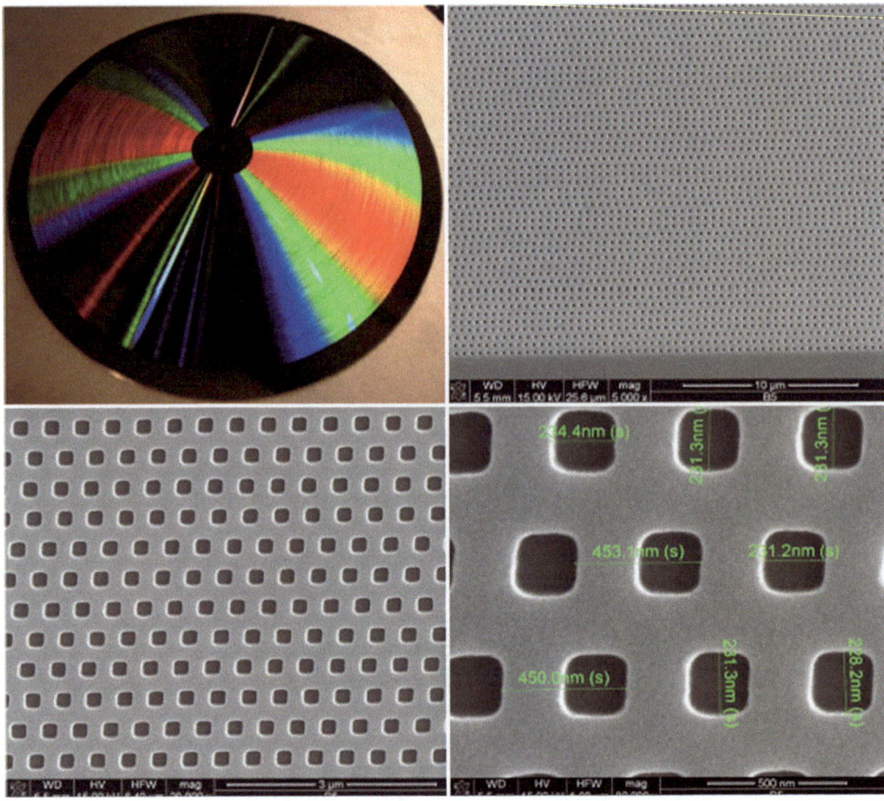

Fig. 3.31 A photograph and SEM images of electroformed Ni stamp (Reprinted from Ref. [33], Copyright (2012), with permission from Elsevier)

3.4.8 Liquid-Phase Deposition

Liquid-phase deposition (LPD) is a method that enables the controlled deposition of silicon dioxide and mixed silicon oxides onto a substrate from an aqueous solution at ambient temperatures and pressures. In comparison to CVD performed in a vacuum chamber at high temperature, it has a benefit of low temperature. Patterns with a uniform surface morphology can be obtained by the LPD process. Physical and chemical properties of SiO_2 by LPD are similar to CVD-SiO_2 [36].

The schematic process of LPD is shown in Fig. 3.32. The PR resist pattern is fabricated first by photolithography and plasma ashing. The ashed patterns are immersed in a solution, which is prepared by dissolving highly purified silica particles in hexafluorosilicic acid (H_2SiF_6) at 40°C to obtain a saturated solution of silica particles with a concentration of 3.4 mol/l using a magnetic stir bar. A boric acid (H_3BO_3) solution was continuously added for 12 h to the dissolving solution of silica particles. The LPD-SiO_2 mold is obtained after the lift-off process.

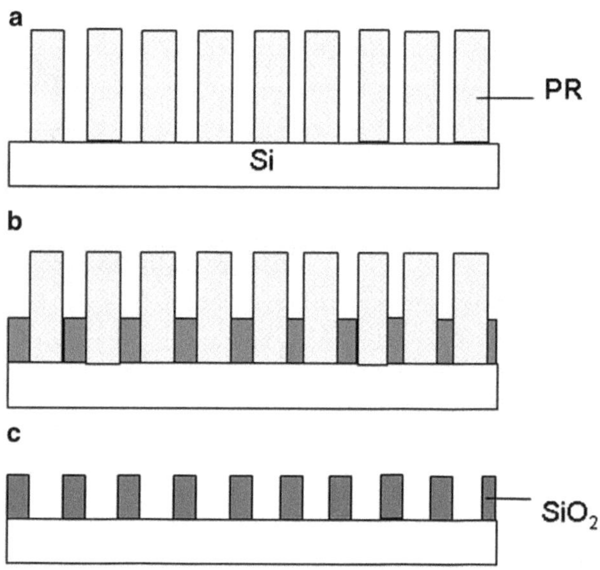

Fig. 3.32 A schematic diagram of PLD procedure. (**a**) Plasma ashing. (**b**) LPD process. (**c**) Lift-off process

3.4.9 Nanoimprint Lithography

Although imprint lithography can create nanopattern, it also produces stamps for NIL. First of all, the imprinted mold is provided. Transparent polymer stamps have been developed by thermal step and stamp nanoimprint lithography (SSIL). The benefit of the polymer stamp is without any other subsequent steps, for example, dry etching and anti-adhesive coating. The fluorinated material of the stamp provides good mold release with UV resist. The SSIL method offers a tool for fabricating polymer stamps for a simple and low-cost UV-mold fabrication [37].

3.4.10 AFM Lithography

Atomic force microscope (AFM) is usually used to characterize the surface structure of a sample. AFM lithography is a method in which a feature is drawn on a solid surface with an AFM tip. Scratching, oxidation, or reduction is widely used to modify the surface. It can create features less than 50 nm (as low as 10 nm) on silicon and other substrates (PMMA, self-assembled film, LB film). As for AFM scratching, the characteristic of the tip has an effect on the modified pattern. The method is an emerging technology, but makes it very slow (about 1 um/s) andvadjust it is limited

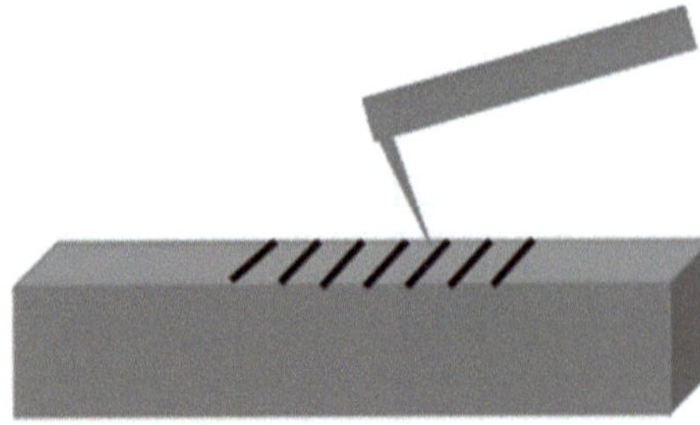

Fig. 3.33 Schematic of AFM lithography

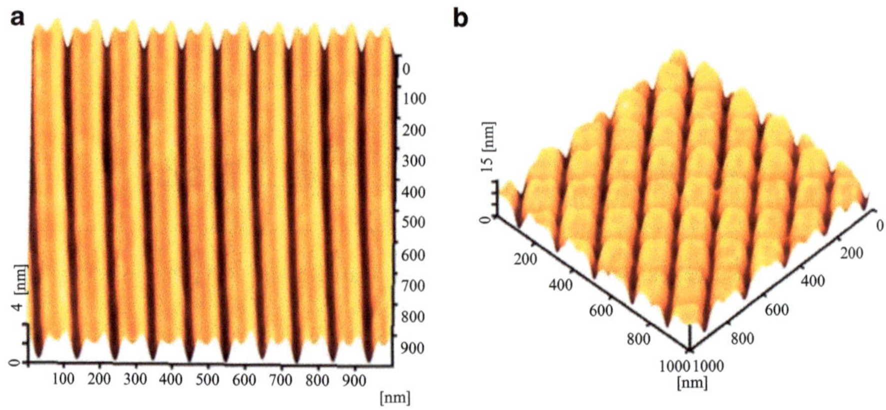

Fig. 3.34 Gratings and pillars pattern are fabricated by AFM lithography

to a further application for large-area stamp [38]. Figure 3.33 is a schematic of AFM lithography. The grating and pillars are created on the silicon substrate by AFM lithography, as shown in Fig. 3.34.

3.4.11 Replication of Master

Replication refers to a mold geometry that is copied by filling a material into it and is a recognized technology for mass production. In a replication process a master geometry is transferred to a substrate material, in this way copying the master geometry. This transfer can be induced by means of heat, force, chemical activation, or other energy or activation [39].

Soft lithography is a typical replication, which is invented by Prof. Whitesides from Harvard University [40]. The process can be separated into two steps: fabrication of a master by lithography or other methods and fabrication of elastomeric elements by casting and thermal curing of a liquid prepolymer on the master. A number of polymers can be used for molding, such as PDMS, PMMA, h-PDMS,

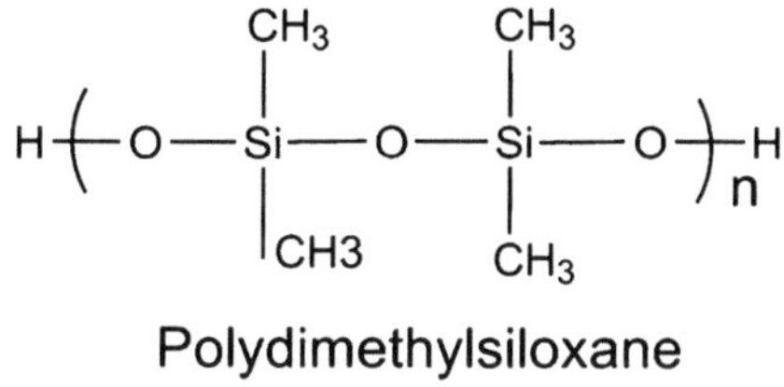

Fig. 3.35 Chemical formula of PDMS

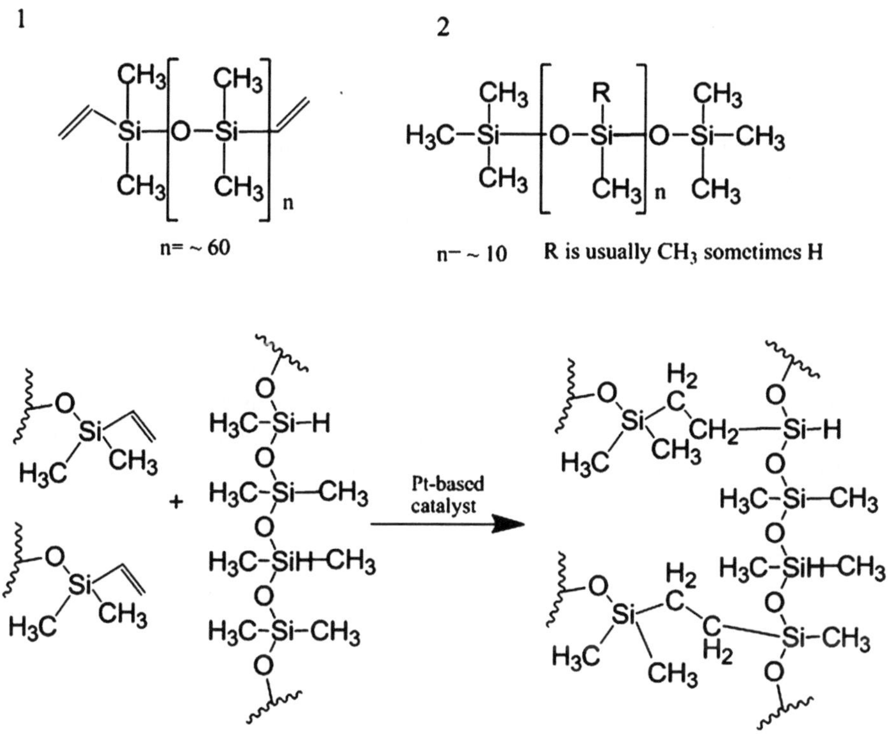

Fig. 3.36 Cross-linking reaction of PDMS precursors

PUA, PVA, PVC, PTFE, and ETFE. The most widely used elastomeric material for soft lithography is poly(dimethylsiloxane) (PDMS) polymer.

PDMS (Sylgard 184, Dow Corning) has a number of useful properties for replication as it is durable, inert to most of the materials being patterned or molded, and chemically resistant to many solvents. Moreover, the simplicity and the reproducibility of the reaction which transforms PDMS precursor chains into an elastomer are suited for low cost and adaptability to mass fabrication. A chemical structure of PDMS is given in Fig. 3.35.

PDMS is composed of two parts of kits. The first one contains vinyl-terminated polydimethylsiloxane chains of about 60 monomers long ($n = 60$). The second component contains a ready-to-use mixture of methylhydrosiloxane

Table 3.6 Cured PDMS (Sylgard 184) properties

Cured PDMS properties	
Stability temperatures	−50 to 200°C
Tensile strength	7.1 MPa
Young modulus	1–5 MPa
Surface energy	21.6 mJ/m^2
Modulus of elasticity	1.8 MPa
Shear modulus	≤1 GPa
Compressibility	2 N/mm^2
Optical properties	Optically transparent down to ~300 nm

and dimethylsiloxane copolymers of about ten monomers long ($n = 10$) and of a platinum-based catalyst. After mixing the two parts delivered by commercial available product, the linkages of Si–CH_2–CH_2–Si are created. A schematic process of the reaction that explains the conversion of the liquid prepolymer mixture into a solid elastomer is presented in Fig. 3.36. During cross-linking reaction, no gas or water release appear to be very convenient for replication.

PDMS stamp is a high-elasticity material and has the benefit of conformal contact with a substrate on large area. And it can be used in UV nanoimprint lithography because it is a transparent material within the spectrum of visible to ultraviolet light (800–300 nm). Ease of fabrication, low cost, and adaptability to mass production of PDMS stamp are very suitable for nanoimprint lithography. Some performances of cured PDMS (Sylgard 184) are given in Table 3.6.

The detailed process for PDMS stamp fabrication is as follows and shown in Fig. 3.37. The quality of replicated mold is related to pressure, cured time, materials, and selectivity of experiment parameter, which is important for the quality of the replicated mold.

1. Fabrication of a master stamp: the stamp can be created by conventional lithography or unconventional lithography. Usually, silicon master is used. The Si master needs surface treatment (clearing and anti-adhesive).
2. Poly(dimethylsiloxane) (PDMS, 184-PDMS) (Sylgard® 184, Dow Corning) in a commercial kit consists of a base and a curing agent. The two parts are mixed in 10:1 weight to weight ratio, and the mixture is pumped under vacuum for several min to allow the removal of any air bubble trapped during mixing. The mixture is then directly poured onto the master. Heating lasts for several hours and cures the prepolymer.
3. Peel off the PDMS mold from the stamp.

Despite of the advantages of PDMS stamp, the PDMS material also suffers from high compressibility which causes shallow relief features of a stamp to deform, buckle, or collapse; in addition, these relief structures tend to deform upon release from the master because of surface tension [41–44].

As shown in Fig. 3.38, the PDMS stamp can cause to pattern structure collapse if h/a is too large or to pattern sagging if w/a is too large. A composite stamp can

Fig. 3.37 Schematic procedure of PDMS fabrication

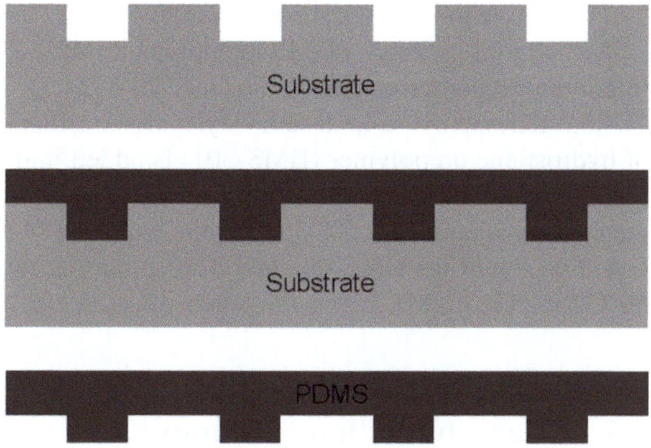

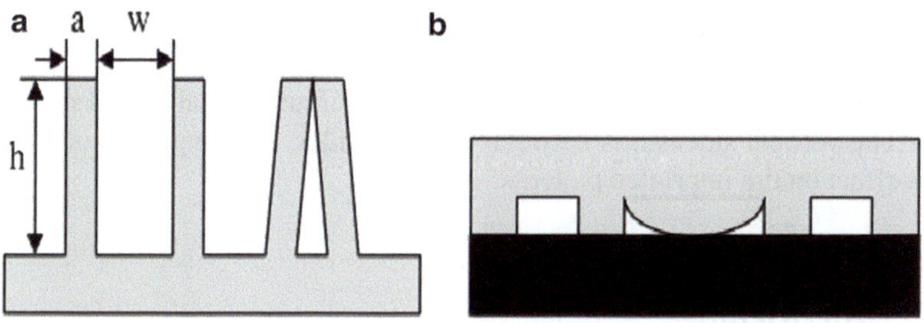

Fig. 3.38 Deformation of PDMS stamp. (**a**) Pattern collapse. (**b**) Pattern sagging

Fig. 3.39 Schematic of a composite mold consisting of h-PDMS and PDMS layer

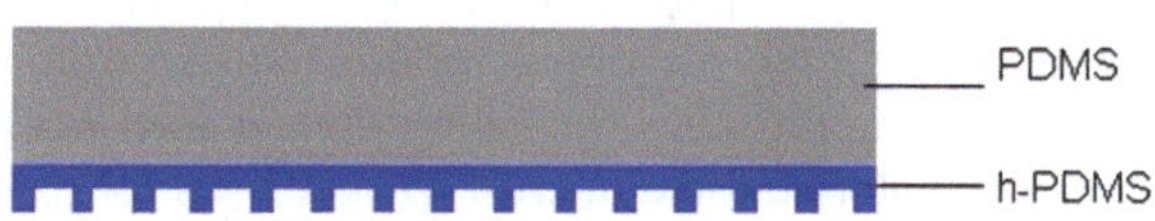

Table 3.7 Prepolymer of h-PDMS material

Vinyl PDMS prepolymer (VDT-731)	3.4 g
Platinum-divinyltetramethyldisiloxane	17.5 μl
2,4,6,8-tetramethyl −2,4,6,8-tetravinylcyclotetrasiloxane	5 μl
Hydrosilane prepolymer (HMS-301)	1 g

avoid the deformation of materials. A soft PDMS cushion layer and a hard PDMS imprint layer are designed. The high modulus PDMS guarantees the relief pattern, and the soft PDMS cushion layer adapts perfectly to the waviness of a substrate. The thickness of hard PDMS is usually several μm. The structure of a composite mold is given in Fig. 3.39. The h-PDMS is composed of polymer materials, which is presented in Table 3.7.

The bilayer mold fabrication is as follows:

A 3.4 g of a vinyl PDMS prepolymer (VDT-731), 18 ml of a Pt catalyst (platinum-divinyltetramethyldisiloxane, SIP6831.1), and one drop of a modulator (2,4,6,8-tetramethyl-2,4,6,8-tetravinylcyclotetrasiloxane, 87927) were mixed. 1 ml of hydrosilane prepolymer (HMS-301) is added into this mixture. Then the liquid prepolymer mixture is poured onto the master and spun for 1–2 min. After spinning, the mixture is cured at 60°C for 30 min, and the 184-PDMS prepolymer mixture is added on top of h-PDMS mixture. The composite mold (bilayer mold) is cured at 60°C for 24 h. Figure 3.39 is the schematic structure of a bilayer mold [45].

3.5 Characterization of Stamp

There are many methods to characterize the quality of an imprinting stamp. The nanopatterning on the stamp has an effect on the imprinted patterns. The tapered sidewall structure is the easiest demolding; the undercut sidewall of the stamp is the worst. The vertical sidewall is between them. Besides, the roughness of the stamp has an effect on the imprinted patterns.

3.5.1 Microscopy

Just mentioned in Chap. 2, the metrology tool for the evaluation of nanoimprint patterns is very important to reveal the quality of the structure replication from the stamp into the resist and pattern transfer from the resist into the substrate. Microscopy approaches, including optical microscopy, scanning electron microscope, and atomic force microscopy, are used for the testing of the stamp. Optical microscopy can be used to inspect the whole structures of the stamp. Scanning electron microscopes (SEM) and atomic force microscopes (AFM) can be used for the metrology of submicrometer and nanometer features. The related principles of these microscopy approached can be referred to the references [46–49].

3.5.2 CD-SAXS and SXR

Scientists from the national institute of standards and technology (NIST) developed critical dimensional small-angle X-ray scattering (CD-SAXS) and specular X-ray reflectivity (SXR) as feasible alternatives to quantify the pattern fidelity. It is a widely used method to quantify the thickness, density, and roughness of the nanostructures [50].

In order to evaluate the uniformity of nanopattern of the stamp given in Fig. 3.23, it is analyzed by measuring the side length of nanotip. First, SEM images of 36 sites

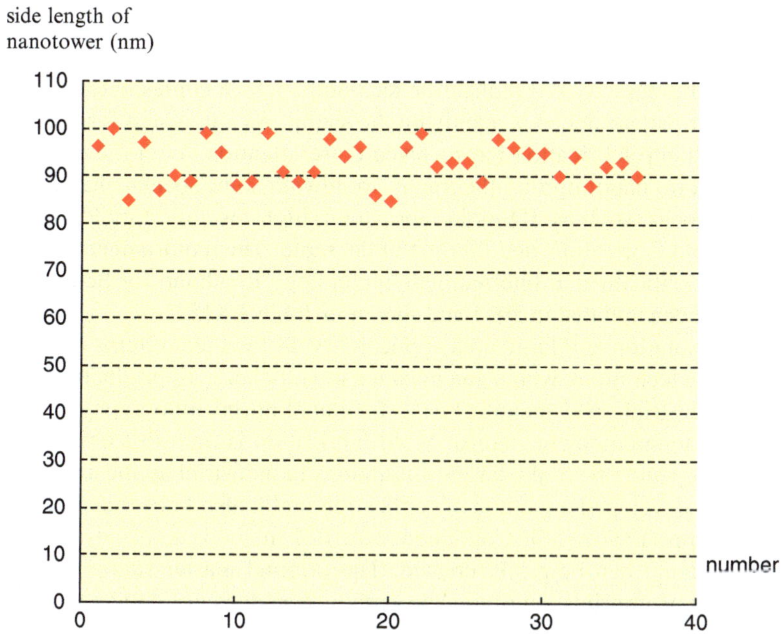

Fig. 3.40 Nanotip side size of each 6×6 array site across 1×1 cm^2

Fig. 3.41 Uniformity plot of nanotips across 1×1 cm^2 area

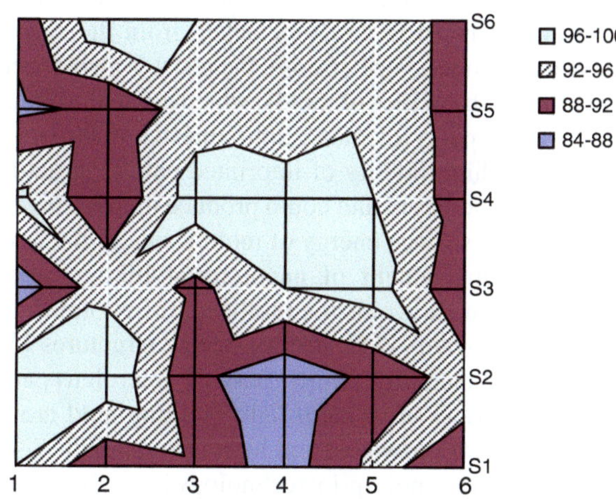

with 30,000 magnification were achieved by dividing the 1×1 cm^2 area sample into 6×6 array. For each site image, the ten nanotip top sizes are measured, and the average value is represented with this site size. Figure 3.40 presented the 36-site nanotip sizes. It has been demonstrated that the mean rhombic top size is 93.6 nm with a 3σ variation of 12.6 nm. The size distribution on the quartz substrate with 1×1 cm^2 area is plotted in Fig. 3.41 [51].

3.6 Critical Issues in Stamp Fabrication

Nanoimprint lithography is a replication technique, which copies master geometry. The quality of pattern depends mainly on the stamp. As a first step for nanoimprint lithography, stamp fabrication has received close attention. With the development of application for nanoimprint lithography, mold with some special characteristic is becoming a focus, such as 3D mold fabrication, high-resolution mold fabrication, large-area mold fabrication, and lifetime of the mold. The requirements of the mold fabrication specification in International Technology Roadmap for Semiconductor (ITRS) have been updated in 2011 and shown in Table 3.8 [52].

Feature resolution and large-area applicability are two main items of interest in any patterning technique, which can enhance the efficiency of production. In order to realize industrial-level manufacturing for nanoimprint lithography, one of the key issues is to supplement stamps. Mold fabrication is the most time- and cost-consuming step and one of the largest limitations in industrial application. Obducat company has developed a method of replication mold, also known as Intermediate Polymer Stamp (IPS®) [53]. Combined with IPS method, across 4-in. patterned area nickel stamp can be easily created. The original master stamp can be used for thousands of imprints, and replicated production does not damage the original master stamp. It is a simple and promising method for batch fabrication of working stamp.

A high-quality stamp cannot guarantee a good quality nanoimprint pattern unless the demolding process is also successful. During nanoimprint lithography, the nanoimprint resist is easily detached from the substrate using the stamp without surface treatment. Furthermore, the resist is also attached on the surface of the mold and filled with cavities of molds, which cause the damage of mold. In order to reduce the surface energy of mold, anti-adhesive treatment for mold is indispensable for obtaining a high fidelity of imprinted pattern structure [54]. Moreover, lifetime of the mold is increased and could produce thousands of imprints or more. Hence, how to reduce the surface energy of mold is one of the key challenges.

The unique benefit of nanoimprint lithography is the ability to pattern 3D structures, which avoids the multistep for photolithography. It creates 3D patterns, sub-40-nm T-gates [55], and air-bridge structures [56] in a single-step imprint in polymer and metal by lift-off. Besides, microlens pattern structures used in homogenizing a variety of modern light-emitting field are fabricated by 3D nanoimprint lithography with low cost and high throughput [57–60]. However, the key issue of developing 3D nanoimprint technology is the 3D stamp fabrication. It is still difficult to fabricate 3D mold.

Table 3.8 The International Technology Roadmap for Semiconductor (2011 update) expectations regarding the UV-NIL mold fabrication process

Year of Production	2011	2012	2013	2014	2015	2016	2017	2018	2019	2020	2021	2022	2023	2024	2025	2026
UV-NIL –special Mask Requirements																
Defect size impacting CD(nm) *x,y* [J]	2.2	2.0	1.8	1.7	1.5	1.4	1.3	1.2	1.1	1.0	0.9	0.8	0.8	0.8	0.8	0.8
Defect size impacting CD (nm) *x,y* [K]	4.4	4.0	3.7	3.4	3.1	2.8	2.6	2.4	2.2	2.0	1.8	1.6	1.6	1.6	1.6	1.6
Mask substrate flatness(nm peak-to-valley) [L]	180	153	126	110	88	72	56	45	36	29	24	21	21	21	21	21
Trench depth, mean (nm) [M]	70-76	62-64	54-56	48-50	42-44	38-40	34-36	30-32	27-28	24-24	21-22	19-20	17-18	14-16	14-16	14-16
Etch depth uniformity (nm) [N]	3.5-3.8	3.1-3.2	2.7-2.8	2.4-2.5	2.1-2.2	1.9-2.0	1.7-1.8	1.5-1.6	1.4-1.4	1.2-1.2	1.1-1.1	10.-1.0	0.9-0.9	0.7-0.8	0.7-0.8	0.7-0.8
Trench wall angle (degrees) [O]	87.9	88.1	88.3	88.5	88.7	88.8	88.9	89.1	89.2	89.2	89.3	89.4	90.4	91.4	92.4	93.5
Trench width roughness (nm, 3 sigma) [P]	2.8	2.5	2.2	2.0	1.8	1.6	1.4	1.3	1.1	1.0	0.9	0.8	0.7	0.6	0.6	0.5
Corner radius, bottom of feature (nm) [Q]	4.5	4	3.5	3.2	2.8	2.5	2.2	2	1.8	1.6	1.3	1.1	1.0	0.9	0.8	0.7
Corner radius, top of feature (nm) [R]	1.1	0.9	0.8	0.7	0.7	0.6	0.5	0.5	0.4	0.4	0.3	0.3	0.3	0.2	0.2	0.2
Trench bottom surface roughness (nm, three sigma) [S]	5.4	4.8	4.2	3.8	3.4	3	2.7	2.4	2.1	1.9	1.5	1.2	1.0	0.8	0.6	0.5
Template absorption [T] %	<2	<2	<2	<2	<2	<2	<2	<2	<2	<2	<2	<2	<2	<2	<2	<2
Near surface defect (nm) [U]	32	29	26	23	20	18	16	14	13	11	10	9	8	7	6	6
Dual Damascene overlay: metal/via on template (nm, 3 sigma) [X]	7.6	6.4	5.4	4.8	4.2	3.8	3.4	3.0	2.7	2.4	2.1	1.9	1.7	1.5	1.5	1.5

Source: ITRS 2011 update for lithography

Manufacturable solutions exist, and are being optimized (white)
Manufacturable solutions are known (yellow)
Manufacturable solutions are NOT known (red)

References

1. Viheriälä J, Viljanen M, Kontio J, Leinonen T, Tommila J, Dumitrescu M, Niemi T, Pessa M (2009) Soft stamp ultraviolet-nanoimprint lithography for fabrication of laser diodes. Proc SPIE 7271:72711O-1–72711O-10
2. Han T, Madden S, Bulla D, Debbarma S, Luther-Davies B (2011) Direct molding Chalcogenide glass waveguides using thermal nanoimprint lithography with a soft PDMS stamp. Conference on lasers and electro-optics Europe and 12th European quantum electronics conference, 22–26 May 2011
3. Park S, Choi K, Kim G, Lee J (2009) Nanoscale patterning with the double-layered soft cylindrical stamps by means of UV-nanoimprint lithography. Microelectron Eng 86:604–607
4. Chen Y, Roy E, Kanamori Y, Belotti M, Decanini D (2004) Soft nanoimprint lithography. Proc SPIE 5645:283–288
5. Plachetka U, Bender M, Fuchs A, Vratzov B, Glinsner T, Lindner F, Kurz H (2004) Wafer scale patterning by soft UV-nanoimprint lithography. Microelectron Eng 73–74:67–171
6. Hawkes PW (2006) Advances in imaging and electron physics, vol 140. Academic, San Diego
7. Kettle J, Hoyle RT, Dimov S, Perks RM (2008) Fabrication of complex 3D structures using step and flash imprint lithography (S-FIL). Microelectron Eng 85:853–855
8. Park S, Schift H, Solak HH, Gobrecht J (2004) Stamps for nanoimprint lithography by extreme ultraviolet interference lithography. J Vac Sci Technol B 22:3246–3250
9. Light Source Facilities. http://www.lightsources.org/cms/?pid=1000098. Accessed 20 Nov 2011
10. Luo Y (2007) Nanostructures design and fabrication for magnetic storage applications. Dissertation, North Carolina State University
11. Zhao YP, Berenschot E, Boer MD, Jansen H, Tas N, Huskensv J, Elwenspoek M (2008) Fabrication of a silicon oxide stamp by edge lithography reinforced with silicon nitride for nanoimprint lithography. J Micromech Microeng 18:064013–064018
12. Choi YK, Lee JS, Zhu J et al (2003) Sublithographic nanofabrication technology for nanocatalysts and DNA chips. J Vac Sci Technol B 21:2951–2955
13. Choi YK, Zhu J, Grunes J et al (2003) Fabrication of sub-10-nm silicon nanowire arrays by size reduction lithography. J Phys Chem B 107:3340–3343
14. Yan XM, Kwon S, Contreras AM et al (2005) Fabrication of large number density platinum nanowire arrays by size reduction lithography and nanoimprint lithography. Nano Lett 5: 745–748
15. Contreras AM, Yan XM, Kwon S et al (2006) Catalytic CO oxidation reaction studies on lithographically fabricated platinum nanowire arrays with different oxide supports. Catal Lett 111(1–2):5–13
16. Haynes CL, Van Duyne RP (2001) Nanosphere lithography: a versatile nanofabrication tool for studies of size-dependent. Nanopart Opt 105:5599–5611
17. Hulteen JC, Van Duyne RP (1994) Nanosphere lithography: a materials general fabrication process for periodic particle array surfaces. J Vac Sci Technol A 13:1553–1558
18. Jeong GH, Park JK, Lee KK et al (2010) Fabrication of low-cost mold and nanoimprint lithography using polystyrene nanosphere. Microelectron Eng 87:51–55
19. Murphy PF, Morton KJ, Fu ZL, Chou Y (2007) Nanoimprint mold fabrication and replication by room-temperature conformal chemical vapor deposition. Appl Phys Lett 90:203115–203117
20. Li MQ, Christopher K, Ober CK (2006) Block copolymer patterns and templates. Mater Today 9:3–39
21. Park M, Harrison C, Chaikin PM, Register RA, Adamson DH (1997) Block copolymer lithography: periodic arrays of 1011 holes in 1 square centimeter. Science 276:1401–1404
22. Guarini KW, Black CT, Milkove KR, Sandstrom RL (2001) Nanoscale patterning using self-assembled polymers for semiconductor applications. J Vac Sci Technol B 19:2874–2878

23. Jung YS, Lee JH, Lee JY, Ross CA (2010) Fabrication of diverse metallic nanowire arrays based on block copolymer self-assembly. Nano Lett 10:3722–3726
24. Liang Y, Zhen C, Zou D, Xu DJ (2004) Preparation of free-standing nanowire arrays on conductive substrates. J Am Chem Soc 126:16338–16339
25. Lee JS, Gu GH, Kim H, Jeong KS, Bae J, Suh JS (2001) Growth of carbon nanotubes on anodic aluminum oxide templates: fabrication of a tube-in-tube and linearly joined tube. Chem Mater 13:2387–2391
26. Dickey MD, Weiss EA, Smythe EJ, Chiechi RS, Capasso F, Whitesides GM (2008) Fabrication of arrays of metal and metal oxide nanotubes by shadow evaporation. ACS Nano 2:800–808
27. Deb P, Kim HY, Rawat V, Oliver M, Kim S, Marshall M, Stach E, Sands T (2005) Faceted and vertically aligned GaN nanorod arrays fabricated without catalysts or lithography. Nano Lett 5:1847–1851
28. Masuda H, Ohya M, Nishio K, Asoh H, Nakao M, Nohtomi M, Yokoo A, Tamamura T (2000) Photonic band gap in anodic porous alumina with extremely high aspect ratio formed in phosphoric acid solution. Jpn J Appl Phys 9:L1039–L1041
29. Jian YL, Hope C, Yin A, Xu J (2002) Two-dimensional lateral superlattices of nanostructures: nonlithographic formation by anodic membrane template. J Appl Phys 91:2544–2546
30. Crouse D, Lo YH, Miller AE, Crouse M (2000) Self-ordered pore structure of anodized aluminum on silicon and pattern transfer. Appl Phys Lett 76:49–51
31. Masuda H, Fukuda K (1995) Ordered metal nanohole arrays made by a two-step replication of honeycomb structures of anodic alumina. Science 268:1466–1468
32. Yanagishita T, Nishio K, Masudaa H (2010) Fabrication of two dimensional polymer photonic crystals by nanoimprinting using anodic porous alumina mold. J Vac Sci Technol B 28: 398–400
33. Zhou Y, Asbah I, Luo G, Eriksson T, Yamada S (2010) High volume manufacturing of nano imprint lithography produced devices: addressing the stamp supply challenge. Proc SPIE 7637:76371–76379
34. Durney LJ (1984) Electroplating engineering handbook, 4th edn. Van Nostrand Reinhold, New York
35. Eriksson T, Yamada S, Venkaatesh P (2011) High volume nanoimprint lithography on III/V substrates: imprint fidelity and stamp lifetime. Microelectron Eng 88:293–299
36. Hye Lee JH, Kimb KS, Lee BK, Kimb NH (2008) Performance and characteristics of SiO_2 imprint mould fabricated by liquid-phase deposition. Superlattices Microstruct 44:520–527
37. Haatainen T, Makela T, Ahopelto J, Kawaguchi Y (2009) Imprinted polymer stamps for UV-NIL. Microelectron Eng 86:2293–2296
38. Xie XN, Chung HJ, Sow CH, Wee ATS (2006) Nanoscale materials patterning and engineering by atomic force microscopy nanolithography. Mater Sci Eng R 54:1–48
39. Hansen HN, Hochen RJ, Tosello G (2011) Replication of micro and nano surface geometries. CIRP Ann Manuf Technol 60:695–714
40. Xia Y, Whitesides GM (1998) Soft lithography. Angew Chem Int Ed 37(5):551–575
41. Lee JN, Park C, Whitesides GM (2003) Solvent compatibility of poly(dimethylsiloxane)-based microfluidic devices. Anal Chem 2003(75):6544–6554
42. Hui CY, Jagota A, Lin YY, Kramer EJ (2002) Constraints on microcontact printing imposed by stamp deformation. Langmuir 18:1394–1407
43. Delamarche E, Schmid H, Michel B, Biebuyck H (1997) Stability of molded polydimethylsiloxane microstructures. Adv Mater 9:741–746
44. Bietsch A, Michel B (2000) Conformal contact and pattern stability of stamps used for soft lithography. J Appl Phys 88:4310–4319
45. Schmid H, Michel B (2000) Siloxane polymers for high – resolution, high-accuracy soft lithography. Macromol 33:3042–3049
46. Egerton RF (2005) Physical principles of electron microscopy. Springer, Heidelberg
47. Seizō M, Franz J, Giessibl RW (2009) Noncontact atomic force microscopy. Springer, Heidelberg

48. Ludwig R (1998) Scanning electron microscopy: physics of image formation and microanalysis. Springer, Heidelberg
49. Brian H, John JL (1993) Optical microscopy: emerging methods and applications. Academic, San Diego
50. Lee HJ, Soles CL, Ro HW, Jones RL, Lin EK, Karim A, Wu WL, Hines DR (2006) Characterizing nano-imprint pattern cross-section and fidelity from x-ray reflectivity. Proc SPIE 6151:61510N–61516N
51. Li XL, Wang QK, Zhang J, Zhou WM, Liu YB, Wan YZ, Niu XM (2009) Large area nanosize array stamp for UV-based nanoimprint lithography fabricated by size reduction process. Microelectron Eng 86:2015–2019
52. The ITRS International Technology Working Groups, International Technology Roadmap for semiconductors. http://www.itrs.net/Links/2011ITRS/Home2011.htm. Accessed 20 Nov 2011
53. Robert J, Hu LL, Eriksson T et al (2009) A non-destructive metrology solution for detailed measurements of imprint templates and media. Proc SPIE 7488:74881Z-1–74881Z-13
54. Zhou WM, Min GQ, Zhang J, Liu YB, Wang JH, Zhang YP, Sun F (2011) Nanoimprint lithography: a processing technique for nanofabrication advancement. Nano Micro Lett 3:135–140
55. Macintyre DS, Chen Y, Lim D, Thoms S (2001) Fabrication of T gate structures by nanoimprint lithography. J Vac Sci Technol B 19:2797–2800
56. Li MT, Chen L, Chou SY (2001) Direct three-dimensional patterning using nanoimprint lithography. Appl Phys Lett 78:3322–3324
57. Peng C, Liang XG, Fu ZL, Chou SY (2007) High fidelity fabrication of microlens arrays by nanoimprint using conformal mold duplication and low-pressure liquid material curing. J Vac Sci Technol B 25:410–414
58. Oh SS, Choi CG, Kim YS (2010) Fabrication of micro-lens arrays with moth-eye antireflective nanostructures using thermal imprinting process. Microelectron Eng 87:2328–2331
59. Thomas G, Gerald K (2010) Nanoimprint lithography: the technology makes its mark on CMOS image sensors and in the nano-world. Optik Photonik 2:42–45
60. Ong NS, Koh YH, Fu YQ (2002) Microlens array produced using hot embossing process. Microelectron Eng 60:365–379

Chapter 4
Stamp Surface Treatment

4.1 Introduction

Nanoimprint lithography is a simple process. In short, a substrate is coated with resin, and the stamp is pressed into the resin by mechanical deformation of imprint resist. The imprint resist is cured by heat or UV light during the imprinting process. Then, the mold is released from the substrate, referred to as demolding. Subsequent processes, such as plasma etching and lift-off, are applied for the pattern transferring [1]. The schematic of the nanoimprint lithography processes is given in Fig. 4.1.

Mold release in nanoimprint technology has been regarded as a significant obstacle to achieve reliable, high-quality patterning at the nanoscale. Owing to the high surface area between the stamp and the imprinted resist, the adhesion is often too high for the mold release, resulting in the destruction of patterns and even the contamination.

Adhesion between resist and stamp can be controlled to allow proper release. During demolding, the mechanical forces between the resist and the mold and the resist and the substrate affect the quality of pattern structures. If adhesion force between the resist and the mold is less than that between the resist and the substrate, ease of demolding can guarantee. However, the stamp has a complex pattern of fragile nanostructures, which affects the demolding process and the fidelity of pattern structures. It is to say adhesion between the resist and the stamp is one of the key factors which has profoundly affected the imprinted nanopatterns.

Adhesive force between the resist and the stamp is determined by the surface energy of their counterparts. There are three ways to reduce the stamp and resist adhesive forces: by using a self-assembled monolayer of molecules on the stamp surface, by depositing a polymer on the stamp surface or introducing anti-sticking molecules in the imprint resist material, or depositing metal layer. This makes the stamp surface more hydrophobic.

W. Zhou, *Nanoimprint Lithography: An Enabling Process for Nanofabrication*,
DOI 10.1007/978-3-642-34428-2_4, © Springer-Verlag Berlin Heidelberg 2013

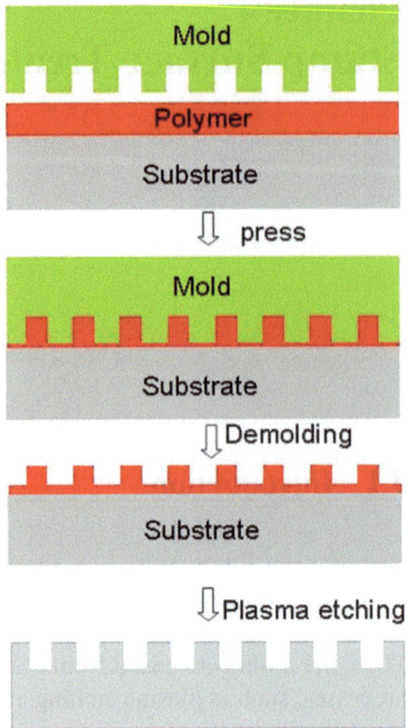

Fig. 4.1 Schematic process of nanoimprint lithography

Anti-adhesive layer can prevent stamp mold fouling, preserve the fidelity of the imprint pattern, and improve stamp lifetime. In this chapter, the issues related to adhesive are introduced, and characterization methods of the anti-sticking layer are presented. Several kinds of modified mold methods are described later.

4.2 Adhesion and Friction

After nanoimprint lithography, the stamp and the resist are attached. There is a risk of both the stamp and the imprint patterns being damaged in the separation process and possibly destroy the master and the imprinted nanostructures. To avoid the destruction of pattern and the stamp, research on adhesion is significantly necessary. The demolding force is simplified and shown in Fig. 4.2 [2].

The sidewalls of a nanopattern mold cause friction with resist during demolding. Besides, adhesion occurs between the upper and bottom surfaces of the nanopattern stamp and resist. The adhesion forces can be divided into three categories: mechanical, physical, and chemical force [3]. The mechanical adhesion is responsible for a large area when two substrates are in contact. It can be classified into three types: surface roughness, dovetail structures, and friction interlocking, as shown in Fig. 4.3.

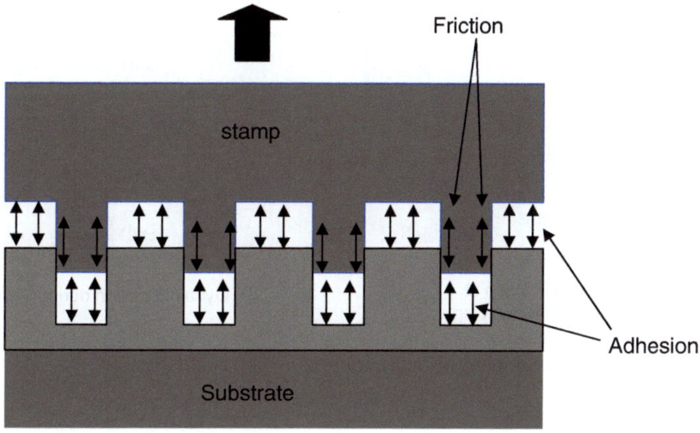

Fig. 4.2 Schematic of demolding process

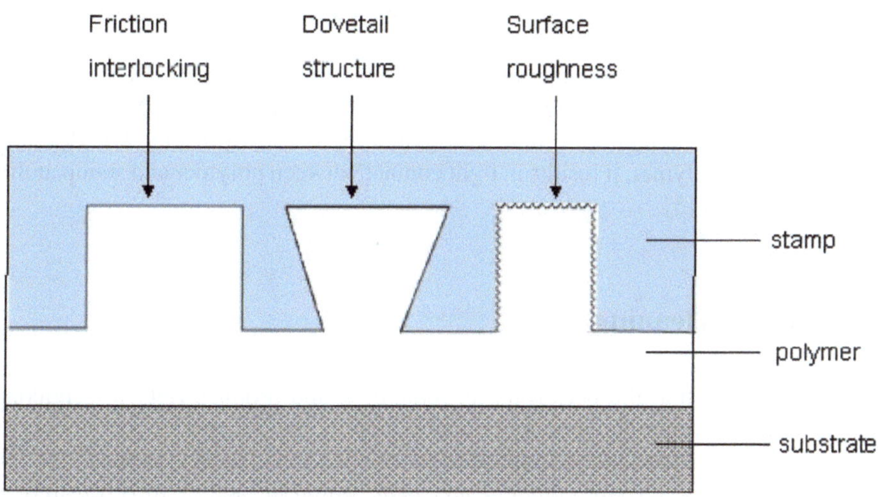

Fig. 4.3 Different contributions to the mechanical adhesion

Physical adhesion is a result of bonding between the two surfaces such as van der Waals bonds and hydrogen bridges. Chemical adhesion is due to chemical bonds resulting from ionic, atomic, or metallic forces.

Adhesion is the tendency of dissimilar particles and/or surfaces to cling to one another. The resist and the stamp are bonded together after UV or thermal cured. The adhesion energy is given by [4]

$$W_{\mathrm{AB}} = U_{\mathrm{A}} + U_{\mathrm{B}} - U_{\mathrm{AB}},$$

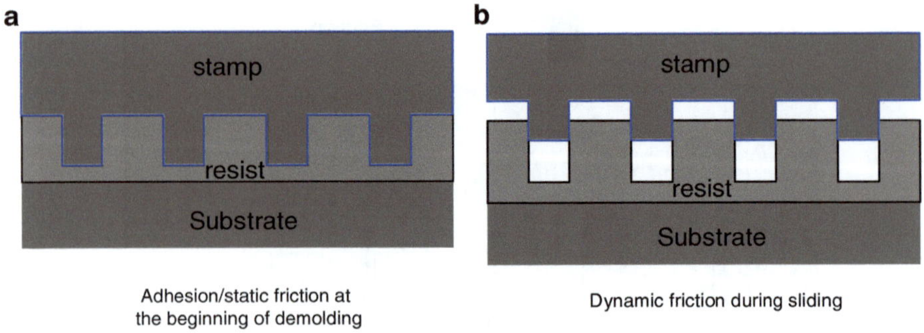

Fig. 4.4 Friction during demolding. (**a**) Static friction. (**b**) Dynamic friction

where U_A is the surface energy of a polymer (resist) , U_B is the surface energy of the solid (stamp), and U_{AB} is the interfacial energy between the polymer and solid.

Simplify the equation, $W_{AB} = 2C\sqrt{U_A U_B}$, C is a constant.

From the equation, it is noted that for a given imprint resist, the adhesion energy is dependent of the surface energy of the stamp.

Friction between resist (polymer) and mold is considered during demolding of micro/nanostructures. The friction force is in static friction at the beginning of demolding and dynamic friction during sliding, as shown in Fig. 4.4. Because of shrinkage of polymer, it results in tight contact between polymer and stamp at the vertical sidewalls [5].

4.3 Stamp Cleaning

It is well known that the performance, reliability, and product yield of semiconductor microelectronic devices are critically affected by the presence of chemical contaminants and particulate impurities on the wafer or device surface. As for nanoimprint lithography, the contaminants on the stamp surface are truly transferred onto the surface of the resist, which has a significant impact on the fidelity of the pattern. Therefore, the cleaning of the hard stamp is necessary and a key step for nanoimprint lithography.

Quartz or Si stamp can be cleaned by a semiconductor cleaning process, such as RCA cleaning process [6].

The process consists of two consecutively applied hot solutions known as "RCA standard clean," SC-1 and SC-2, featuring pure and volatile reagents. These solutions have been widely used in their original or modified form for over 40 years in the fabrication of Si semiconductor devices. The SC-1 solution for the first processing step consists of a mixture of NH_4OH (ammonium hydroxide), H_2O_2, and H_2O, also known as "APM" for "ammonia/peroxide mixture." It can be removed

Table 4.1 RCA chemistries showing typical dilution and temperature ranges used for stamp cleaning

RCA solution	Chemicals and conditions	Contaminant removal
SPM	H_2SO_4 and H_2O_2, 2:1–4:1, 90–140°C	Organics
SC-1	NH_4OH:H_2O_2:H_2O, 1:1:5–1:1:50, 40–75°C	Organics, some metals, particles, surface conditioning of Si along with DHF (dilute HF)
SC-2	HCl:H_2O_2:H_2O, 1:1:6–1:1:50, 40–75°C	Metals: alkaline and trace metals

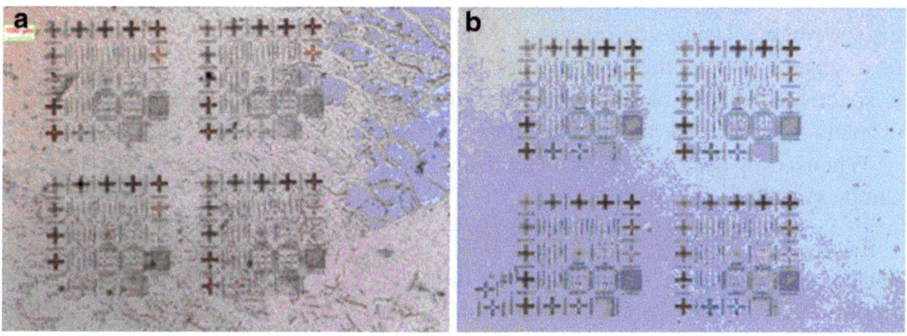

Fig. 4.5 Microscopy of a stamp. (**a**) Before cleaning. (**b**) After cleaning

from Si and quartz surface organic contaminants and some periodic group IB and IIB metals such as Cu, Au, Ag, Zn, and Cd. The SC-2 solution for the second processing step consists of a mixture of HCl (hydrochloric acid), H_2O_2, and H_2O, also known as "HPM" for "hydrochloric/peroxide mixture." It can remove from the stamp surface alkali residues and any residual trace metals (such as Au and Ag), as well as metal hydroxides, including $Al(OH)_3$, $Fe(OH)_3$, $Mg(OH)_2$, and $Zn(OH)_2$. Table 4.1 is RCA chemistries showing typical dilution and temperature ranges used for stamp cleaning. Removal of gross organic materials from Si or SiO_2 wafers such as hardened photoresist polymer patterns can be accomplished by using mixtures of 98-wt.% H_2SO_4 (sulfuric acid) and 30-wt.% H_2O_2. Volume ratios of 4:1–2:1 are used at a temperature of 100–130 for 10–15 min. These "sulfuric peroxide mixtures" (SPM) are also known as "piranha etch." Attention, SPM is extremely dangerous to handle in the lab. When using it, the operators need to be protected with goggles, face shields, and plastic gloves.

Figure 4.5 indicates the microscopy image of SiO_2 stamp before and after cleaning. Before cleaning, the stamp is attached with a layer imprint resist because it is not coated with an adhesive layer. After repeated imprinting operations, the stamp surface is attached with more and more resist, which affects the fidelity of imprint pattern. After cleaning, the mold surface is clean and without any contaminants.

4.4 Assessment of Anti-adhesive Layers

An anti-adhesive layer quality of a stamp is shown to be related with the fidelity of imprinted patterns. Contact angle measurement, XPS, FTIR, and AFM are used to demonstrate the quality of the anti-adhesive layer.

4.4.1 Wettability

The surface energy of the stamp or the resist is usually tested by contact angle measurements. Using goniometry is a simple and fast way to measure the property of a surface. When the droplets are free from force field space, its appearance is spherical because of the interfacial tension. But in fact, the appearance of the droplets is determined by a force balance between adhesive and cohesive forces.

When a drop of liquid drops on a surface, the contact angle, θ_c, defined in Fig. 4.6, can be measured [7] as:

$$\Upsilon_{LG} \cos \theta_c = \Upsilon_{SG} - \Upsilon_{SL},$$

where surface tension Υ denotes solid, liquid, and gas as S, L, and G, respectively. Defining the angle θ_c at which the liquid contacts the surface. We can calculate the wetting angle from the surface tensions under equilibrium conductions.

If the contact angle $\theta_c < 90°$, the liquid is said to wet the surface, and if $>90°$, the liquid is said to dewet the surface. The contact angle can be obtained from Young's equation. Figure 4.7 shows a range of different wetting conditions [8].

Some material surface energy for nanoimprint lithography is shown in Table 4.2.

4.4.2 Chemical Analysis

The release layer could be analyzed with chemical instrument. Usually, the adhesive layers can be done with attenuated total reflection Fourier transform infrared spectroscopy (ATR-FTIR) [9, 10] and X-ray photoelectron spectroscopy (XPS) [11–15].

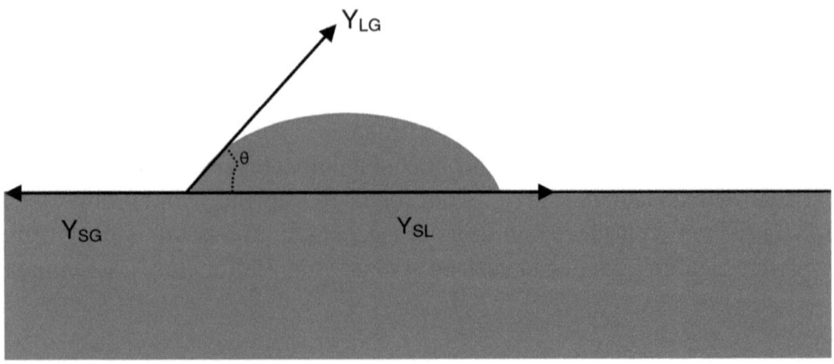

Fig. 4.6 Contact angle of a liquid droplet and a solid surface

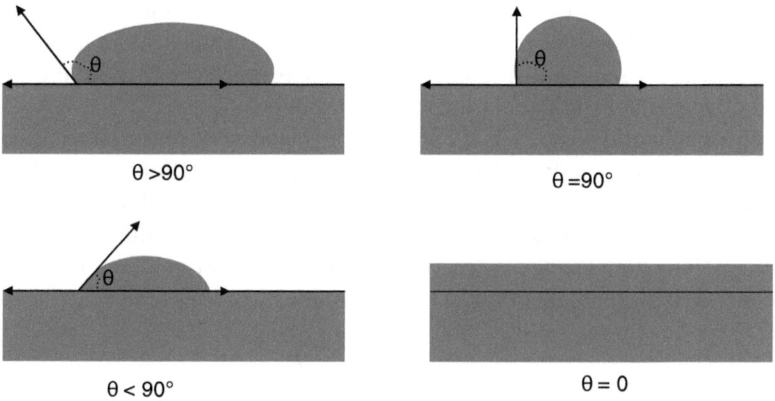

Fig. 4.7 Schematic diagram of wetting angles for various wetting properties

Table 4.2 Values of surface energy for different materials

Material	Surface energy (mJ/m^2)
PMMA	41.1
PS	40.7
PTFE	15.6
CF_2 and $-CF_3$	15
Silicon surface 20	26

Attenuated total reflection (ATR) spectrometry is a well-known method and suitable for the analysis of thin films deposited on a substrate. With the development of infrared microspectrometry, Fourier transform infrared (FTIR) spectrometry has made great improvements. ATR-FTIR technique provides a new method to evaluate the chemical analysis (specific functional group and chemical structures) of the deposited film on the stamp.

X-ray photoelectron spectroscopy (XPS) or electron spectroscopy for chemical analysis (ESCA) is a quantitative spectroscopic technique, which can measure the elemental composition, empirical formula, chemical state, and electronic state of the elements existing within a material. By irradiating a material with a beam of X-rays while simultaneously measuring the kinetic energy and number of electrons that escape from the top 1–10 nm of the material, the spectrum will be recorded. From the recorded spectrum, a specific element in the sample surface has the presence of a specific peak at a particular energy. Moreover, the intensity of the peaks is related to the concentration of the element of sample surface. Therefore, the XPS technique also provides a quantitative analysis of the surface composition.

4.4.3 Film Thickness

If thickness of a release layer is too large, the critical dimension of the stamp deviates from the actual value. Therefore, the coating on the stamp must be thick.

In the spin-coating process, the speed and concentration of release agent have an impact on the thickness of the adhesive layer. The film thickness can be directly measured by cross-sectional scanning electron microscopy (SEM). But this method is destructive and not suitable for the modified stamp.

Spectroscopic ellipsometry is a nondestructive and very sensitive measurement technique that uses an optical technique for the investigation of the dielectric properties (complex refractive index or dielectric function) of thin films, which is based on the change of polarization in light reflected or transmitted from a sample [16]. The polarization change is described by an amplitude ratio, tan (Ψ), and phase difference, Δ, between light oriented in the ρ- and s-directions relative to the sample surface:

$$\rho = \tan(\Psi)e^{i\Delta} = \frac{R_\rho}{R_s},$$

where R_ρ and R_s are the complex Fresnel coefficients for ρ- and s-polarizations, respectively. Ellipsometry measurements are used to determine optical constants and film thickness both for single- and multilayer coatings.

4.4.4 Morphology and Adhesion

Various phenomena of the modified stamp, such as morphology, roughness, adhesion, friction, and indentation, can be characterized by atomic force microscopy (AFM).

Force versus distance curves can describe the interactions. The corresponding curves for the tip/sample can be obtained with AFM and shown in Fig. 4.8. AFM can measure the adhesion force (or pull-off force) between the surface and the tip.

As the tip comes close to the sample, various attractive forces reach out and "grab" the tip. This is evidenced at point 2 in the graph. Point 2 sometimes called the "jump-to-contact" point is usually due to electrostatic attraction and/or surface tension (capillary) forces. As the tip is pressed into the surface, the cantilever bends upward. Cantilever relaxes downward until tip forces are in equilibrium with surface forces ($3 \rightarrow 4$). As the cantilever is pulled away from the sample, the force is increasing ($4 \rightarrow 5$). Eventually, the sample and the tip are separated ($5 \rightarrow 6$). Therefore, it is possible to measure the attractive forces of tip-sample (adhesive layer) interactions by knowing the spring constant of the cantilever [17].

4.4.5 Thermostability and Chemical Stability

The modified stamp must satisfy some thermal stability requirements. It has been shown that various testing methods can be used depending on the stamp application.

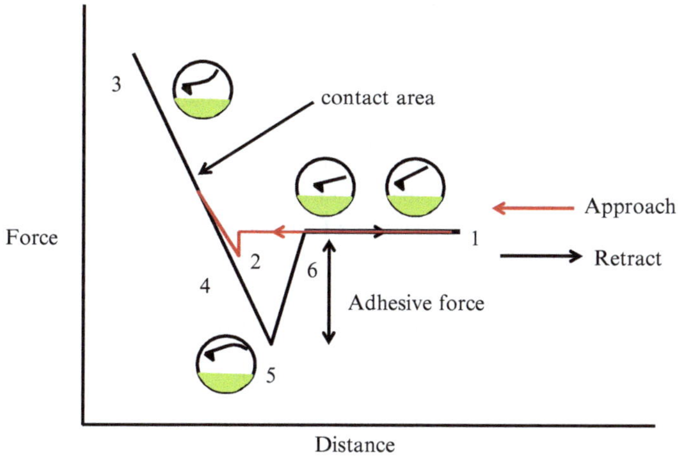

Fig. 4.8 Measurement principle of local adhesive force measurement with AFM

Table 4.3 Techniques for characterizing SAMs on the stamp

Property of SAMs	Technique
Composition	XPS, FTIR
Wettability	Contact angle
Thickness	Ellipsometry, SEM, AFM
Structure	STM, AFM, XRD
Defect	STM, AFM, wet etching
Adhesion	AFM

The thermal stability of the adhesive layer is investigated by the annealing of the stamp at different temperatures as a function of time. The stability of the adhesive layer can also be tested by plasma or chemical resistance. Besides, it should be evaluated by repeated imprinting operations and then measure the film property.

4.4.6 Roughness

After modification, the stamp surface should be smooth, which can decrease friction between stamp and resist. If a value of friction shearing force is more than that of tensile strength, the polymer will fracture. Therefore, the stamp roughness is significant for the demolding process, and its roughness can be investigated by AFM.

In all, the structures and properties of SAMs have been investigated using a number of techniques, summarized in Table 4.3.

4.5 Fabrication Methods of Anti-adhesive Layers

Release layer coated on the stamp is to decrease the surface energy, and ease of demolding is realized. Meanwhile, it can prevent mold contamination and decrease mold cleaning frequency and improve stamp lifetime. Therefore, how to decrease the mold surface energy is a key issue of nanoimprint lithography process. There are some methods to decrease surface energy of the stamp. A self-assembled monolayer on Si or quartz by chemical vapor deposition is often used. Or polymerized PTFE film is deposited by plasma. Recently, a popular material used in NIL is diamond-like carbon (DLC), a dense metastable form of amorphous carbon or hydrogenated amorphous carbon.

4.5.1 Plasma Polymerization

Plasma polymerization is widely used to indicate the process of forming high-molecular-weight products in electrical discharges. Over the past 50 years, plasma has become a useful method for the modification and deposition of various materials [18]. Since plasma polymerization was first observed in 1874, it has developed quickly. By plasma polymerization, plasma polymer films can be produced with an excellent coating adhesion on almost all substrates. The plasma polymer films have an excellent chemical, mechanical, and thermal stability. Plasma polymerization refers to formation of polymeric materials under the influence of plasma and known as "glow discharge polymerization." Various plasma polymer deposition methods include DC, AF-magnetron, and RF deposition. Plasma is a direct consequence of the ionization of the gases present in the reactor, and fragmentation leading to polymerization is a secondary process [19]. In the reactor, high-energy electrons collide with hydrocarbon molecules to produce CF_2. The CF_2 bonds in the depositing film determine the surface energy and anti-adhesive properties of the mold.

In the experiment, two types of gases (CHF_3 and SF_6) have been used to investigate the anti-adhesive behavior. CHF_3 or SF_6 can be decomposed into fluoride material on the stamp under the plasma. Figure 4.9 presents the dependence of SF_6 flux on the wetting angle. The wetting angle as a function of CHF_3 is given in Fig. 4.10. The wetting angle is relatively low if one gas is used.

To evaluate the dependence of both gases on wetting angle, the ratio of SF_6 and CHF_3 is changed, and the result is presented in Fig. 4.11. When the ratio is between 1:1 and 1:3, the contact angle is in maximum.

The working pressure is a key factor for reactive ion etching. When increasing the working pressure, the reaction speed of gas will increase and the contact angle will also increase. The maximum contact angle is obtained when the working pressure is 28 m Torr (Fig. 4.12).

PTFE-based materials are the most popular way to reduce the problem of adhesion between surfaces. PTFE-like films are achieved by deposition of ion

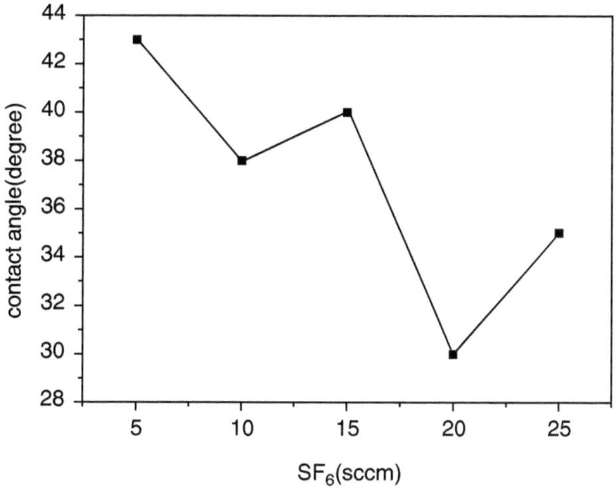

Fig. 4.9 Wetting angles as a function of SF_6 flux

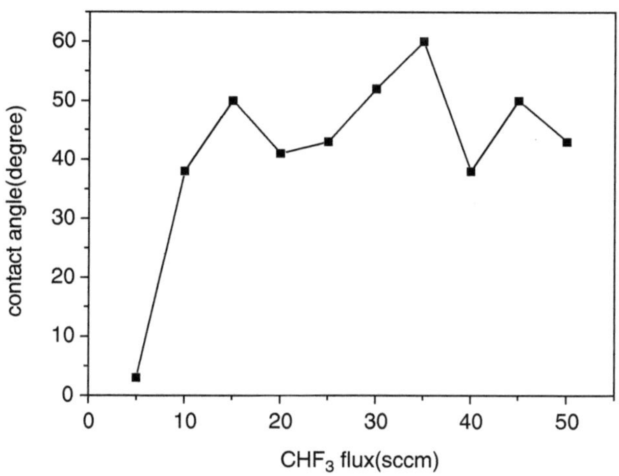

Fig. 4.10 Wetting angles as a function of CHF_3 flux

sputtering [20]. To generate PTFE-like film, a Teflon target and CHF_3 are used as the sputtering gas. High radio frequency (RF) is used to create the plasma. PTFE-like film on the Ni substrate is chemically homogeneous. For comparison, the plasma polymerized film is prepared on the other Ni substrate. Figure 4.13 shows XPS spectra of the two as-deposited PTFE-like films. Six major peaks, CF_3(293.2 eV), CF_2(291.4 eV), CF(289.0 eV), CCF(286.9 eV), CCC(284.8 eV), and C–Ni(283.6 eV) were detected. A high percentage of CF_2 bonds in the film implies a low surface energy and consequently brings about excellent anti-adhesive properties.

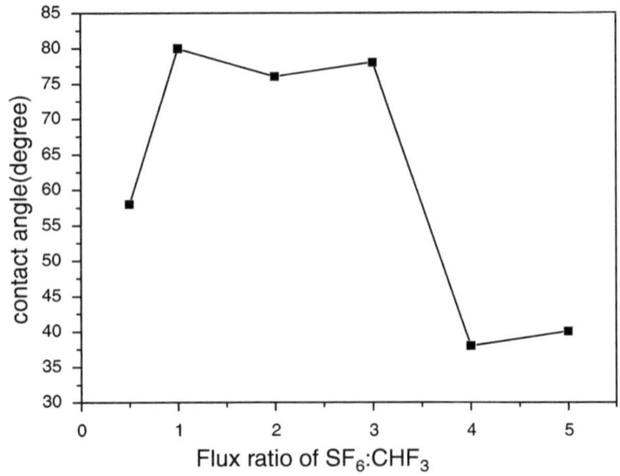

Fig. 4.11 Wetting angles as a function of SF_6/CHF_3 flux

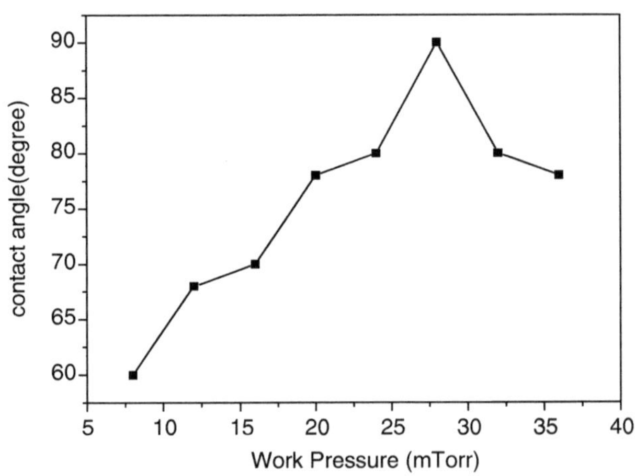

Fig. 4.12 Wetting angles as a function of working pressure

The excellent anti-adhesive property of the films was tested by hot imprint lithography with polycarbonate (PC) and poly(methyl methacrylate) (PMMA). After 50 imprints, the anti-adhesive layer remained constant. But for increasing imprint times, the percentage of CF_2 bonds decreased with the number of imprints, and fluorine molecules were transferred into the resist.

It should be noted that the deposited films were generated on micrometer-patterned nickel stamps. It is not suitable for the surface of the nanoscale features of mold to deposit polymeric structure. Meanwhile, the polymer layer is adsorbed weakly on the stamp; the deposited anti-sticking layer must be renewed after some few imprints. Only by optimized imprint parameter, hundreds of embossing are possible with one anti-adhesive PTFE-like film.

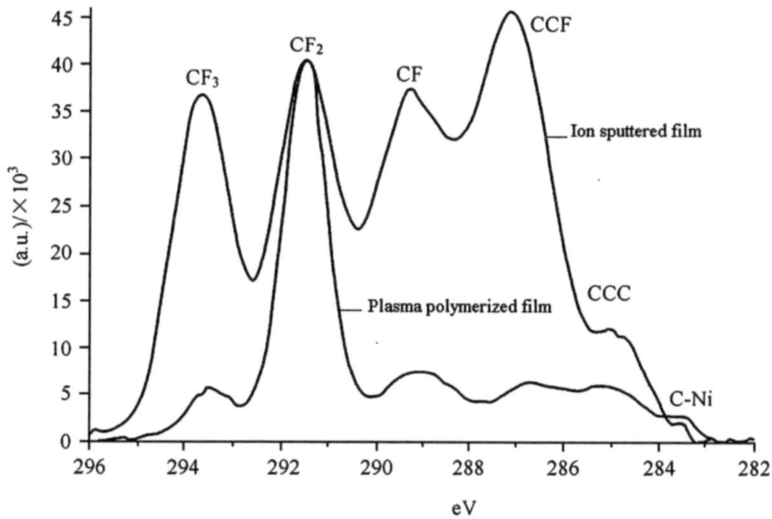

Fig. 4.13 A comparison of the two as-deposited PTFE-like films (bulk measurement) (Reprinted from Ref. [20], Copyright (1999), with permission from Elsevier)

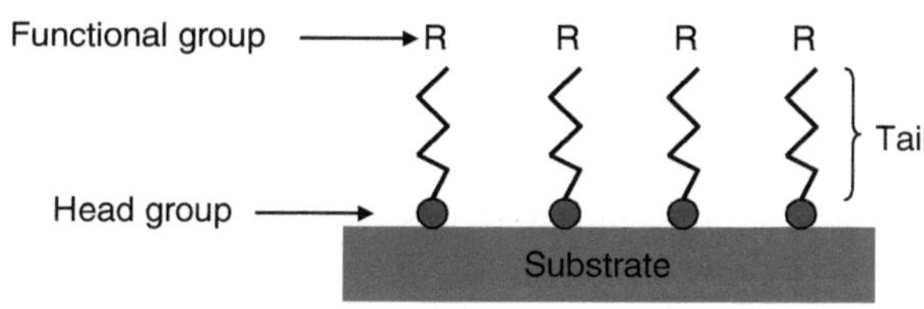

Fig. 4.14 Schematic illustration of SAM layer

4.5.2 *Self-Assembly Monolayer*

In 1946, Zisman published the preparation of a monomolecular layer by adsorption (self-assembly) of a surfactant onto a clean metal surface [21]. After some years, the field of self-assembled monolayers (SAMs) has witnessed tremendous growth in application of engineering. Lots of self-assembly systems have been investigated, but monolayers of alkanethiolates on gold are probably the most studied self-assembled monolayers to date.

SAMs are created by the chemisorption of hydrophilic "head groups" onto a substrate from either the vapor or liquid phase followed by a slow 2D organization of hydrophobic "tail groups." The hydrophilic "head groups" assemble together on the substrate, while the hydrophobic tail groups assemble far from the substrate. Areas of close-packed molecules nucleate and grow until the surface of the substrate is covered in a single monolayer and hold the assembly together, as shown in Fig. 4.14 [22].

Table 4.4 Chemical systems of adsorbates and substrates that form SAMs on some substrates

Surface	Substrate	Adsorbate(s)
Metal	Au	R–SH, R–SS–R, R–S–R, R–NH_2, R–NC, R–Se, R–Te
	Ag	R–COOH, R–SH
	Pt	R–NC, R–SH
	Pd	R–SH
	Cu	R–SH
	Si	R–Si, RCC–Si
	Hg	R–SH
Semiconductor	GaAs (III–V)	R–SH
	InP (III–V)	R–SH
	CdSe (II–VI)	R–SH
	ZnSe (II–VI)	R–SH
Oxide	Al_2O_3	R–COOH
	TiO_2	R–COOH, R–PO_3H
	ITO	R–COOH, R–SH, R–Si(x)$_3$
	SiO_2	R–Si(x)$_3$

Table 4.5 Some typical precursors for SAMs

Name	Chemical composition	Boiling point (pressure)
Dichlorodimethylsilane (DDMS)	$C_2H_6Cl_2Si$	70°C (760 Torr)
Octadecylmercaptan (ODT)	$C_{18}H_{38}S$	210°C (11 mTorr)
Trichlorooctadecylsilane (OTS)	$C_{18}H_{37}Cl_3Si$	220°C (10 mTorr)
Undecyltrichlorosilane (DTS)	$C_{11}H_{23}Cl_3Si$	155°C (15 Torr)
10-Undecenyltrichlorosilane (UTCS)	$C_{11}H_{21}Cl_3Si$	100°C (0.8 Torr)
Perfluorodecyltrichlorosilane (FDTS)	$C_{10}H_4C_{13}F_{17}Si$	216°C (760 Torr)
Perfluoroctyltrichlorosilane (FOTS)	$C_8H_4C_{13}F_{13}Si$	192°C (760 Torr)
Perfluorodecanoic acid (PFDA)	$CF_3(CF_2)_8COOH$	218°C (740 Torr)

The presence of SAMs molecules chemically attached to the substrate renders the properties of the modified interface (i.e., adhesion, wetting, chemistry, conductivity) to be completely different from those of bare substrate. Many substrates, including metals, semiconductors, and oxides, have been shown to be a self-assemble on such substrate. Chemical systems of adsorbates and substrates that form SAMs on some substrates are listed in Table 4.4.

A variety of experimental methods have been used to probe the quality and chemical nature of SAMs, including macroscopic AFM and STM, contact angle goniometry, XPS, FTIR, and electrochemistry.

In nanoimprint lithography, a fluorinated self-assembled monolayer (F-SAM) is one of the most mainly used in an anti-sticking layer. In terms of F-SAM, its release agent is generated by chemisorption onto a mold of quartz or silicon from either vapor or liquid phase deposition. During the process, the hydrolysis effect results in the formation of one HCl for each Si–Cl bond. If the silicon mold is used as a substrate, they should be first oxidized in order to form a chlorosilane SAM [23]. Some typical precursors for SAMs are given in Table 4.5.

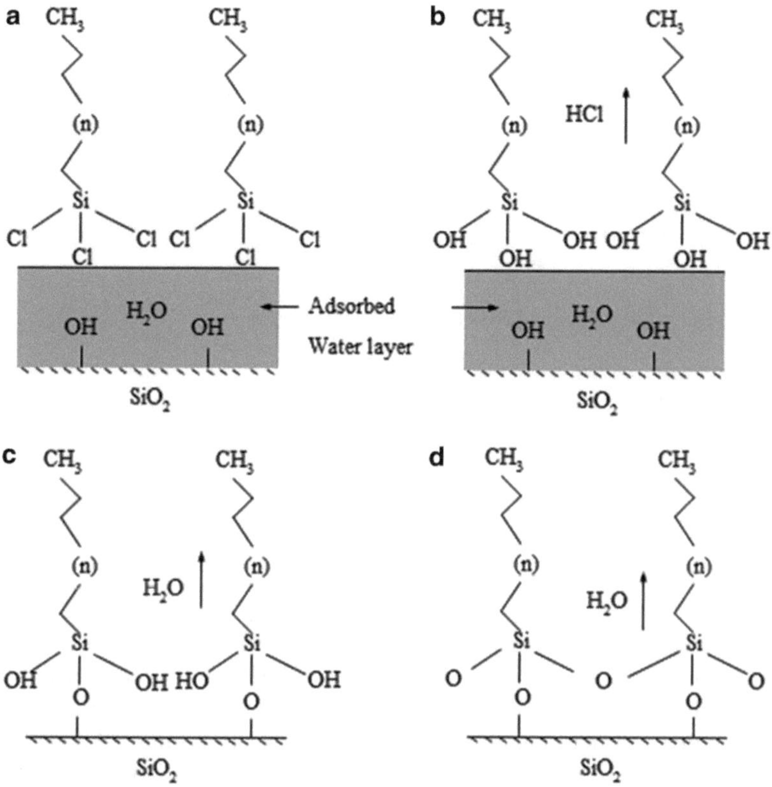

Fig. 4.15 Schematic of the SAM formation mechanism on fused silica surface (**a**) Physisorption. (**b**) Hydrolysis. (**c**) Covalent grafting to the substrate. (**d**) In-plane reticulation

The SAM is grafted on the fused silica or Si stamp, and a likely formation mechanism of SAM could be proposed in Fig. 4.15. Silanes with short chain lengths and high vapor pressures are easily deposited on the silica through vapor phase. (1) First, the fused silica mold is treated with piranha solution and rinsed with DI water to form large number of hydroxyl (^{-}OH) groups attached on mold surface as shown in Fig. 4.15a. (2) The silane head groups hydrolyze and form trisilanols (Fig. 4.15b). (3) The $Si(OH)_3$ groups form covalent bonds with the hydroxyl groups on SiO_2 surface, which is grafted to the silica surface (Fig. 4.15c). (4) Condensation reaction goes on between neighboring molecules and the Si–O–Si covalent bonds can take place (Fig. 4.15d). Finally, the networked coating is its ease of formation on the mold surface to lower the surface energy.

Liquid phase deposition (LPD) method refers to the master dipped into a solution of the mold release polymer dissolved in chloroform. The system is placed in a tank at 65°C and lasts for 1 h. In the process, the percentage of water in the solution

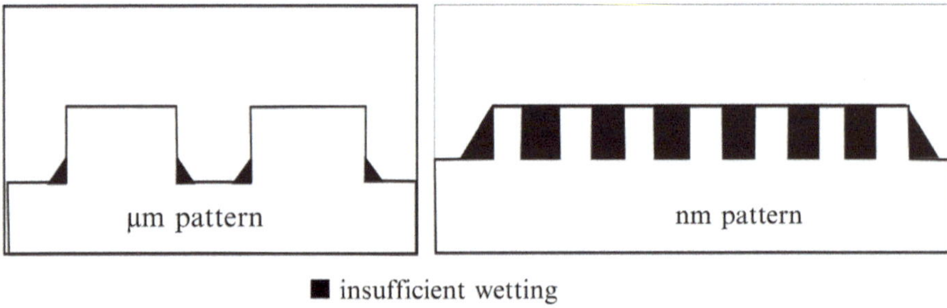

Fig. 4.16 Wetting defects during LPD of the anti-sticking layer

Fig. 4.17 Schematic of self-assembled monolayer on mold

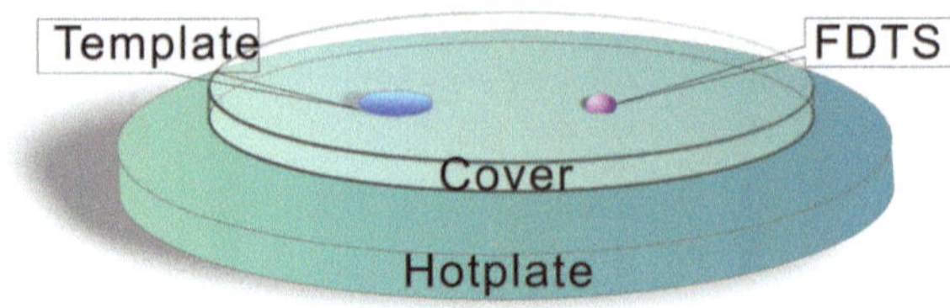

must be controlled. Less percentage of water will render an incomplete aggregation; otherwise, the aggregated silane in the solution will deposit the mold surface.

LPD method is not suitable for nanostructure molds for the following reasons: (1) the polymeric structure makes a monolayer deposition impossible and (2) the polymer is only adsorbed weakly on the mold. This brings about both significant changes in feature size and successive anti-sticking layer destruction by layer-by-layer removal during each imprint step [24].

Figure 4.16 shows wetting defects occurring during LPD of the anti-sticking layer. The release agent in a solvent does not have sufficient wetting properties on a nanopattern surface (black area). Larger microstructure can be covered by LPD since wetting defects are not as serious on the surface of the mold.

Vapor phase deposition (VDP) method is taken for prior consideration when the mold is the nanometer size range. The monolayer by VDP has a more uniform surface than that from LPD. In our research, a commercial available 1H, 1H, 2H, 2H-perfluorodecyltrichlorosilane (FDTS) is used for release agent. The schematic of vapor phase deposition was presented in Fig. 4.17. Before vapor deposition, the fused silica was first soaked in a piranha solution at 90°C for 30 min and thoroughly rinsed with deionized (DI) water and dried with N_2. The glove box was flushed with N_2 after the mold and FDTS were put in. After it is full with N_2 (ca. 2 h), a drop of FDTS is syringed on the petri dish and the coating is placed on the petri dish. After 3 h, the hot plate is cooled to room temperature and the excess is washed by anhydrous hexane to prevent FDTS polymerization.

A water droplet on the blank and SAM-coated quarter mold is shown in Fig. 4.18. The water contact angle for blank mold is 33.01°, but the markedly increased values

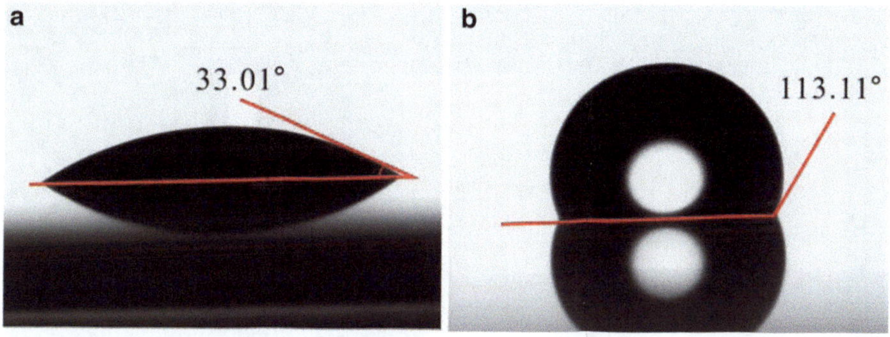

Fig. 4.18 Photos of water droplet on (**a**) of blank mold surface and (**b**) of SAM-coated mold surface

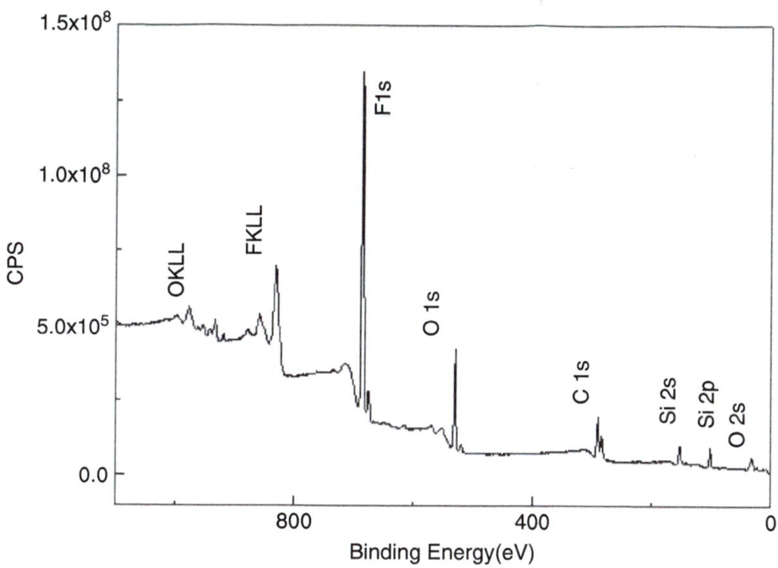

Fig. 4.19 XPS survey spectra of the FDTS-deposited quartz mold

of water contact angle for the FDTS-coated stamp are up to 113.11°, indicating that the anti-adhesive film is highly hydrophobic and suitable for nanoimprint lithography.

An XPS spectra of SAM of the mold is shown in Fig. 4.19. It can be seen that the F/C ratio of mass is 3.8; the C (1 s) spectra mainly consist of CF_3, CF_2, and CF, which are centered at binding energies of 295 eV, 292 eV, and 291 eV, respectively. The C (1 s) spectra indicated clearly that a thin organic layer successfully was coated on the surface of the fused silica mold.

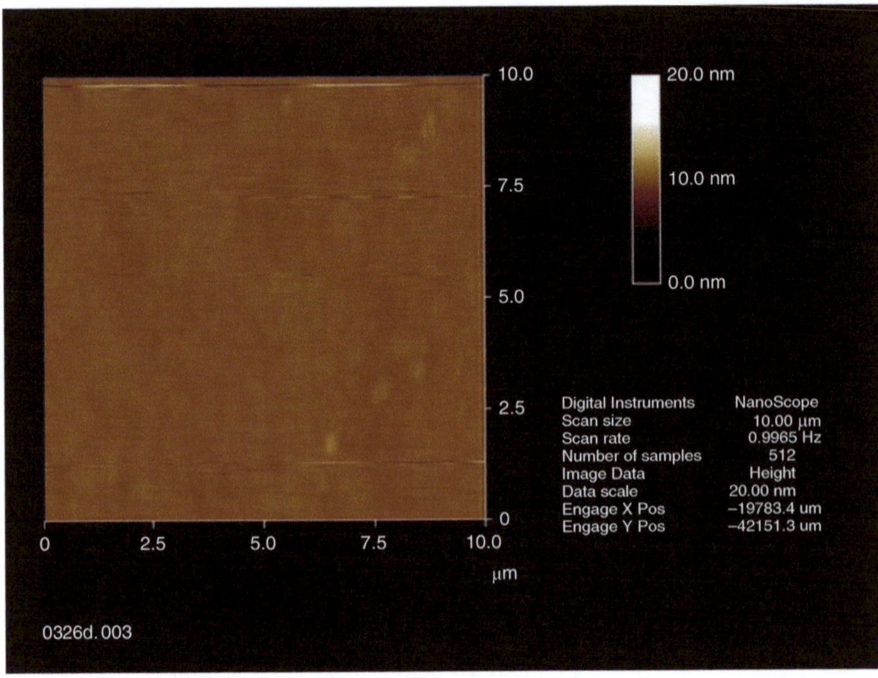

Fig. 4.20 AFM images of the FDTS-coated fused silica surface

The roughness of SAM obtained from AFM image analysis from Fig. 4.20 is 1.06 nm, which is close to the blank fused silica. It is shown that a dense and uniform nanofilm was deposited on the fused silica.

The local adhesive force was characterized by pull-off forces by an AFM. Figure 4.21 shows local adhesive force characterization on the non-treated quartz surface and template surface handled with a SAM anti-sticking layers. Seen from Fig. 4.21, the adhesive forces of the mold surface covered with fluorosilanes were drastically reduced compared to that of the non-treated quartz surface.

The values of adhesive forces are determined by their modification conditions. The detailed characteristics of the surface of the stamp can be further evaluated with a frictional force microscopy (FFM) and a pulsed force mode AFM (PFM-AFM).

Although SAM can reduce the force of demolding, it has been investigated that release characteristics of the mold deteriorate with repeated imprint process. Thickness and amount of the release layer on the mold surface were analyzed by X-ray reflectivity (XRR) and XPS measurements, respectively. Table 4.6 summarizes thickness and density of the release layer determined by XRR [25].

The thickness was found to decrease with increasing imprint times; however, density was seen to be almost the same. It can be concluded that the release layer deteriorates with repeated imprint number.

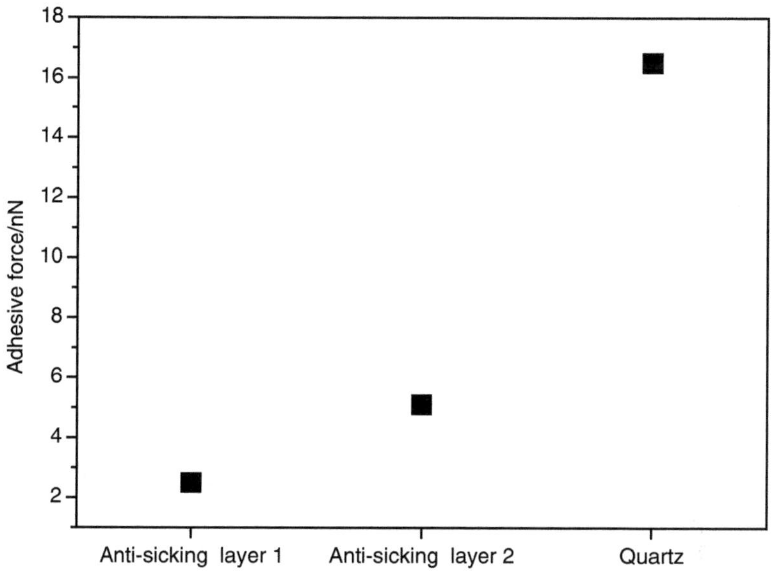

Fig. 4.21 The measured adhesive forces on the surfaces of quartz and the template modified with fluorosilane anti-sticking layers

Table 4.6 Changes in thickness and density of Optool DSX release layer on the stamp surface with repeated imprint times by XRR

Imprint times	Thickness (nm)	Density (g/cm^3)
0	2.2	1.7
25	1.9	1.6
50	1.5	1.7
75	1.3	0.9[a]
150	1.0	1.6

[a]The density of release layer is after 75 imprint operations. The value may be caused by measurement error

The change in C1s band spectra of the mold surface was measured by XPS with repeated imprint operations. And it was found that the intensities of -OCF_2- and -CF_2- were very weak after 150 times, which suggests the decreasing of the release layer with repeated imprint operations. The two possible reasons are: one is radicals generated by photoinitiator; the other is O_2 generated by UV irradiation [25].

The released layer is evaluated by heat durability. Researchers used F3-SAM, F13-SAM, and F17-SAM formed by using (3,3,3-trifluoropropy)trichlorosilane (F3-TCS), (tridecafluoro-1,1,2,2-tetrahydrooctyl)trichlorosilane(F13-TCS), and (he ptadecafluoro-1,1,2,2-tetrahydrodecyl)trichlorosilane(F17-TCS), respectively, as the anti-adhesive layers. The annealing temperature dependence of contact angles of F3-SAM, F13-SAM, and F17-SAM is shown in Fig. 4.22 [25].

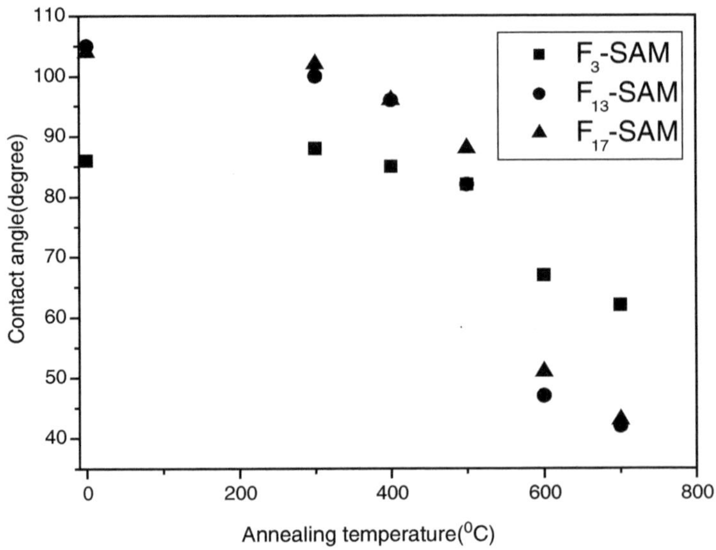

Fig. 4.22 Annealing temperature dependence of contact angle of F3-SAM, F13-SAM, and F17-SAM

The SAM films were annealed at 400°, 500°, and 600°, respectively. It is noted that the adhesion forces of F3-SAM, F13-SAM, and F17-SAM rapidly increased at 500°. It means that these SAMs were decomposed at this temperature [26].

4.5.3 Depositing Diamond-Like Carbon Film

Diamond-like carbon (DLC) is a kind of amorphous carbon materials that displays some of the typical properties of diamond, which contains significant amounts of sp^3-hybridized carbon atoms. DLC has shown to be a high value of density, hardness, electrical resistance, chemical inertness, and IR transparency and lack detectable long-range structural order [27].

Many methods have been developed for the deposition of DLC coatings; they include:

- Direct ion beam deposition
- Pulsed laser ablation
- Filtered cathodic arc deposition
- Ion beam conversion of condensed precursor
- Magnetron sputtering
- RF plasma-activated chemical vapor deposition
- Plasma source ion implantation and deposition

Compared to the property of the common anti-adhesive layer, DLC is a promising mold coating material for nanoimprint lithography because of its superior mechanical properties and a wide range of DLC deposition methods existing. Moreover, it has a relatively low surface energy compared to quartz and glass. However, because the contact angle of DLC is approximately 60° [28], an anti-sticking layer of DLC may still not meet the need of fine imprint patterns and therefore may still require an additional surface treatment for fine imprinting.

Hence, fluorinated DLC (FDLC) has been used in anti-adhesive layer for nanoimprint lithography. FDLC film has both relatively lower surface energy and a high hardness. The contact angle of water on the FDLC film surface is 103°, a 30° higher than that of the DLC surface in the same preparation condition.

DLC material can also be used for imprinting stamp. Schvarzman deposited 100–200 nm DLC film on Si substrate by PECVD. Combined with EBL and RIE process, the DLC nanopattern was created. Then, the stamp of DLC material was exposed to fluorocarbon plasma to reduce the surface energy. The direct fabricated DLC stamp is suitable for nanoimprint lithography. With this mold, sub-50-nm features have been imprinted [29].

4.5.4 Depositing Metal Nanoparticles

Some kinds of materials such as metal oxide form the anti-sticking layer by deposition. The technique is used in anti-adhesive layer of large feature of imprinting mold. Some metals, such as Cr, Al, and Ni for hot embossing, behave like non-wettability and chemical inertness. The Cr, Al, and Ni are used for anti-adhesive material in the hot embossing process, and the experiment shows that Ni is the most excellent anti-sticking material.

TiO_2 is a very strong and nonstick coating which is commercially used in some industries. To coat the mold uniformly in a very thin layer, it would require Ti evaporation so that vertical surfaces could be coated, then oxidate the surface [30].

4.5.5 Spin Coating

Spin coating is a method of forming uniform thin films on flat substrates. Certainly, the technique is also used to form a uniform release film on the stamp surface. It is very simple but highly reliable and reproducible. Spin coating can be roughly divided into four stages, that is, deposition, spin-up, spin-off, and solvent evaporation. As for film-forming process, viscosity and spin speed have a primary impact on the thickness [31].

Faircloth reported that a 2-mg/ml solution of the mold release polymer dissolved in chloroform is drawn into a pipette and enough solution is expelled onto the imprinter to form a meniscus. The stamp is immediately spun at 6,000 rpm for 30 s [32].

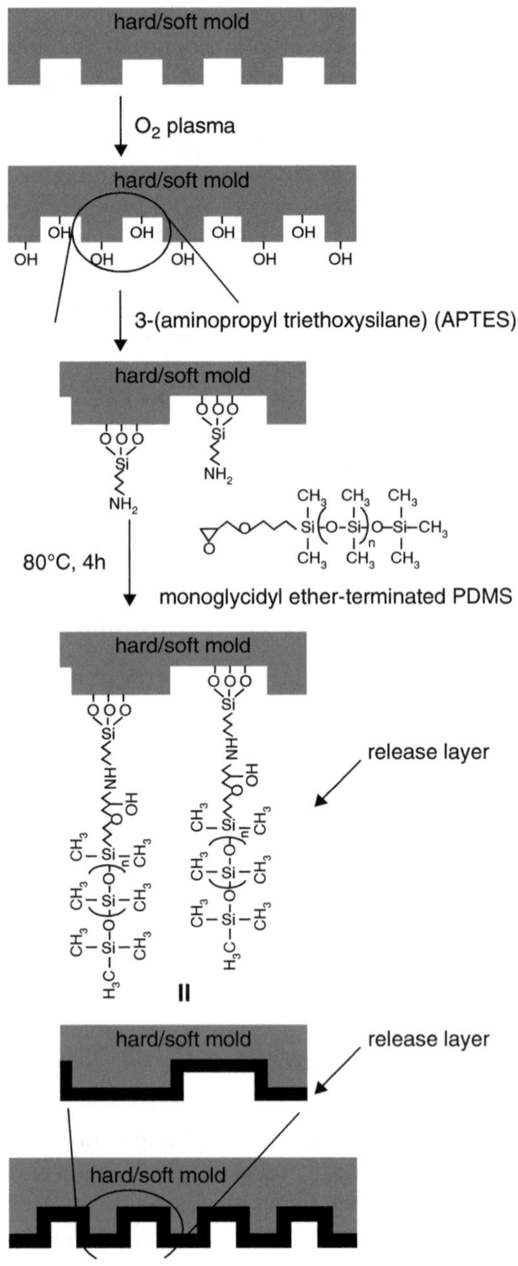

Fig. 4.23 Schematic illustration of the formation of the anti-adhesion PDMS layer on the stamp (Reprinted with permission from [35], Copyright (2006) WILEY-VCH)

Teflon AF is a fluoropolymer (DuPont, USA) that is made by the copolymerization of 2,2-bistrifluoromethyl- 4,5-diflu-oro-1, 3-dioxole (BDD) with other fluorine-containing monomer [33, 34]. It has a good thermal stability and chemical resistance, together with a very low surface energy. It can be used for release agent by spin coating.

PDMS is also used as an anti-adhesion release layer because PDMS has a fairly low surface energy. As we know, siloxane-based PDMS has a surface tension of 21 nN m-1, which is similar to that of the most commonly used anti-adhesion layer, PTFE. The coating mechanism of the PDMS-based anti-adhesion layer is demonstrated in Fig. 4.23 [35]. The aminosilane-grafted stamp is reacted with monoglycidyl ether-terminated PDMS and the epoxy-amine reaction takes place. The thin PDMS antihesion layer can easily be formed on any surface because silanization and epoxy-amine reactions form stable and strong chemical bonds.

References

1. Chou SY, Krauss PR, Renstrom PJ (1995) Imprint of sub-25 nm vias and trenches in polymers. Appl Phys Lett 67:3114–3116
2. Okada M, Kang YJ, Nakamatsu IM, Kanda K, Haruyama Y, Matsui S (2009) Characterization of nanoimprint resin and antisticking layer by scanning probe microscopy. J Photopolymer Sci Technol 22:167–169
3. Brockmann W, Geiß PL, Klingen J (2009) Adhesive bonding: materials, applications and technology. Wiley, Weinheim
4. Alphonsus VP (2002) Adhesion and adhesives technology: an introduction. Hanser, Zurich
5. Matthias W (2009) Hot embossing: theory and technology of microreplication. William Andrew Inc, New York
6. Reinhardt KA (2008) Handbook of silicon wafer cleaning technology. William Andrew Inc, New York
7. Mittal KL (2009) Contact angle, wettability and adhesion. IDC Publishers, Leiden
8. Nalwa HS (2002) Handbook of the thin film materials, vol 5, Nanomaterials and magnetic thin films. Academic, San Diego
9. Yang W, Cui FZ, Qing X (2005) Behavior of phosphatidylcholine adsorption on CNx coated PTFE films. Curr Appl Phys 6:827–832
10. Caro JC, Lappan U, Lunkwitz K (1999) Insertion of sulfur-containing functional groups into polytetrafluoroethylene (PTFE) by low pressure plasma treatment. Surf Coat Tech 116–119:792–795
11. Zelsmann M, Truffier-Boutry D, Francone A, Alleaume C, Kurt I, Beaurain A, Pelissier B, Pépin-Donat B, Lombard C, Boussey J (2009) Double-anchoring fluorinated molecules for antiadhesion mold treatment in UV nanoimprint lithography. J Vac Sci Technol B 27:2873–2876
12. Truffier-Boutry D, Beaurain A, Galand R, Pelissier B, Boussey J, Zelsmann M (2010) XPS study of the degradation mechanism of fluorinated anti-sticking treatments used in UV nanoimprint lithography. Microelectron Eng 87:122–124
13. Beck M, Ling TGI, Graczyk M, Montelius L, Heidari B (2005) Development and characterization of silane antisticking layers on nickel-based stamps designed for nanoimprint lithography. J Vac Sci Technol B 23:575–584
14. Park S, Schift H, Padeste C, Schnyder B (2004) Anti-adhesive layers on nickel stamps for nanoimprint lithography. Microelectron Eng 73–74:196–201
15. Truffier-Boutry D, Galand R, Beaurain A, Francone A, Pelissier B, Zelsmann M, Boussey J (2009) Mold cleaning and fluorinated anti-sticking treatments in nanoimprint lithography. Microelectron Eng 86(4–6):669–672
16. Hilfiker JN, Singh N, Tiwald T, Convey D, Smith SM, Baker JH, Tompkins HG (2008) Survey of methods to characterize thin absorbing films with pectroscopic Ellipsometry. Thin Solid Films 516:7979–7989

17. Veeco Metrology Group, MultiMode™ SPM Instruction Manual (2004) Veeco Instruments Inc.
18. Yasuda Y (1985) Plasma polymerization. Academic, Orlando
19. Gaur S, Vergason G (2000) Plasma polymerization: theory and practice. 43rd annual technical conference proceedings, Denver, 15–20 April 2000
20. Jaszewski RW, Schift H, Schnyder B, Schneuwly A (1999) The deposition of anti-adhesive ultra-thin teflon-like films and their interaction with polymers during hot embossing. Appl Surf Sci 143:301–308
21. Bigelow WC, Pickett DL, Zisman WA (1946) Oleophobic monolayers: I. Films adsorbed from solution in non-polar liquids. J Colloid Interface Sci 1:513–538
22. Lee H, Dellatore S, Miller WM (2007) Mussel-inspired surface chemistry for multifunctional coatings. Science 318:426–430
23. Maboudian R, Ashurst WR, Carraro C (2000) Self-assembled monolayers as anti-stiction coating for MEMS: characteristics and recent developments. Sensor Actuator 82:219–223
24. Beck M, Graczyk M, Maximov I, Sarwe EL, Ling TGI, Keil M (2002) Improving stamps for 10 nm level wafer scale nanoimprint lithography. Microelectron Eng 61–62:441–448
25. Tada Y, Yoshida H, Miyauchi A (2007) Analysis on deterioration mechanism of release layer in nanoimprint process. J Photopolymer Sci Technol 20:545–548
26. Nakai Y, Ommoto S, Kang Y, Okada M, Kanda K, Haruyama Y, Matsui S (2010) Evaluation of heat durability of fluorinated antisticking layers. J Photopolymer Sci Technol 23:55–58
27. Voevodin AA, Donley MS (1996) Preparation of amorphous diamond-like carbon by pulsed laser deposition: a critical review. Surf CoatTechnol 82:199–213
28. Altun AO, Jeong JH, Choi DG, Kim KD, Lee ES (2006) Fabrication of fluorine-doped diamond-like carbon stamps for UV nanoimprint lithography. Nanotechnology 17:4659–4663
29. Schvartzman M, Mathur A, Yang Y, Jahnes C, Hone J, Wind SJ (2008) Fluorinated diamond like carbon templates for nanoimprint lithography. J Vac Sci Technol B 26(6):2394–2398
30. Konijn M (2005) Multilevel nanoengineering for imprint lithography. Dissertation, University of Canterbury
31. Rabilloud G (2000) High performance polymers: chemistry and applications, vol 2. Editions Technip, Paris
32. Faircloth B, Rohrs H, Tiberio R (2000) Bilayer, nanoimprint lithography. J Vac Sci Technol B18:1866–1873
33. Zhao H, Zhang J, Wu NQ, Zhang X, Crowley K, Weber SG (2005) Transport of organic solutes through amorphous teflon AF films. J Am Chem Soc 127:15112–15119
34. Alentiev AY, Shantarovich VP, Merkel TC, Bondar VI, Freeman BD (2005) Gas and vapor sorption, permeation, and diffusion in glassy amorphous Teflon AF1600. Macromolecules 35:9513–9522
35. Lee MJ, Lee NY, Lim JR, Kim JB, Kim M, Baik HK, Kim YS (2006) Antiadhesion surface treatments of molds for high-resolution unconventional lithography. Adv Mater 18:3115–3119

Chapter 5
Nanoimprint Lithography Resists

5.1 Introduction

Resist is a mixture of a polymer or its precursor and other small molecules that the solubility or viscosity is changed by UV light or electron beam or ion beam or X-ray and used in the fabrication of IC field. Resists used during photolithography are called photoresists. It can be functioned as an etching resistant material to protect the substrate. Photoresists can be classified into two groups: the positive and the negative, according to the polarity of pattern structures after photon exposure. A positive resist is a type of photoresist in which the long-chain molecules can be broken by exposure into short chains. The exposed portion becomes soluble to the photoresist developer, and an unexposed portion of the photoresists is insoluble to the photoresist developer. A negative resist is a type of photoresist in which the short-chain molecules can form together to long chains. The exposed portion becomes insoluble to the photoresist developer, and an exposed portion of the photoresists is soluble to the photoresist developer [1]. Figure 5.1 shows the basic microlithography process of photoresists.

An appropriate imprinting resist is one of the most issues to implement NIL for industrial application. The resist must have several requirements. For example, high modulus, low viscosity, low adhesion to the stamp, and high etching selectivity. In nanoimprint lithography process, the resist is mainly used for the pattern transferring layer and removed by plasma etching. Besides, the resist also acts as a component in some designed systems after imprinting, which is generally used in the field of micro-/nanofluid or photonic devices.

The imprint resist (etch barrier) has several types: thermal (thermoplastics and thermosets), UV-curable resist, step-and-flash nanoimprint lithography resist, and roll-to-roll nanoimprint lithography resist. Besides, some novel functional imprint material is also used for imprint resist. The principle of degradable imprint resist is discussed and the development tendency of imprint resist is also given.

W. Zhou, *Nanoimprint Lithography: An Enabling Process for Nanofabrication*,
DOI 10.1007/978-3-642-34428-2_5, © Springer-Verlag Berlin Heidelberg 2013

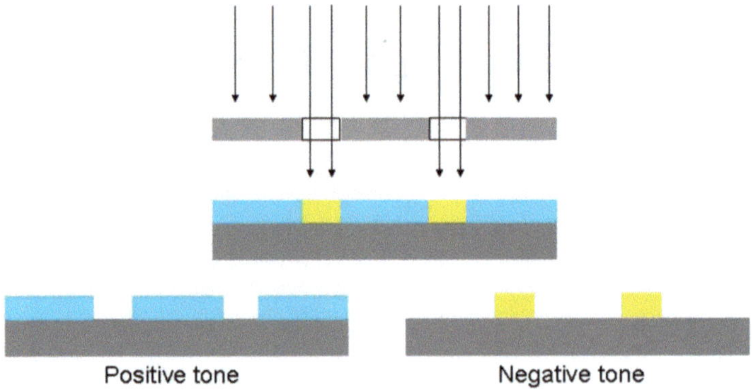

Fig. 5.1 Lithography patterning process for positive and negative tone photoresists

5.2 Characteristics of Nanoimprint Lithography Resist

Nanoimprint lithography is a contact process and differs from conventional photolithography. In this process, the resist is pressed by pressure to generate corresponding structures. The resist was cured by heating or UV light. Other than photoresists, the imprint resist has certain characteristics [2].

5.2.1 Properties of Film Coating

The stamp is pressed on the surface of the resist; therefore, the thickness of the imprint resists is to determinate the depth of the patterns. If the thickness of the resists is lower than the pattern dimension of the stamp, the dimension of the stamp cannot be transferred into the surface of the resist. Therefore, the thickness of the imprint resists must be controlled, usually in a range of 500 and 50 nm. Besides, the film layer homogeneity is ± 5 nm on the 300-mm wafer. Excellent wetting capability on the substrate, such as silicon substrate, is available. The surface roughness of the resists film is within 5 nm. After spin coating of film, the defects of air holes or contaminant must be prohibited.

5.2.2 Material Property

The imprint resist material should have high strength and low viscosity. During the imprint process, the resists are deformable and converted into the desired structures. After curing, the resists patterns should have enough mechanical strength to ensure ease of demolding. The rigidity of the mold is larger than that of the resists [3, 4].

The viscosity of liquid resist determines the fluid ability of the resists. Low-viscosity resists can extend the lifetime of the mold because of the smaller imprinting force [5, 6].

The curing rate has an impact on imprinting efficiency. The thermoplastics have three states: heat state, imprint state, and temperature drop/demolding. The imprinting efficiency depends on the glass transition temperatures. If it is high, the time from high temperature to low temperature is extended. While for thermosets, it depends on the thermal curing rate. If the resist is sensitive to heat, the curing rate is fast and the high imprinting efficiency is obtained. The UV-curable material is cured by UV light, and the photo-curable rate determines the imprinting efficiency.

5.2.3 Etch Resistance

Dry etching is one of the most common pattern transfer method used in the semiconductor field. Although the etching rate of a resist depends on etching gases and process conditions, a resist should have a good resistance to the plasma etching compared to that of a substrate. Etch selectivity is defined by the ratio of etch rate substrate/photoresist. Usually, a minimal 20 nm/min plasma etching rate of resist is necessary, and no residual layer is detected after etching with oxygen plasma. The organic material has low etching resistance because of C–C bonds and –H bonds. The covalent cross-linked network material, including silicate linkages (Si–O bonds) and some organic substituent, has high etch resistance.

5.2.4 Line Width Roughness

Line width roughness (LWR) describes the deviation of both sides of the feature from the desired dimension. LWR has an impact on the performance of the circuit drastically when a small variation in the feature size occurs. The feature roughness depends on the molecular weight of the resist component, molecular weight distribution, molecular structure, etc. The resists with low molecular weight or with low polydispersity is responsible for the reduction of LWR [7].

5.2.5 Shelf Life

Imprint resist has its own shelf life because photoactive compounds may lose their photoactivity with time and the solvent may gradually evaporate. Some resists must be stored at low temperature and be contained in a dark-brown bottle because it is sensitive to heating or ambient light, particularly the UV portion [8].

5.3 Different Types of Imprint Resist

5.3.1 Thermal Imprint Resist

Thermoplastics, also known as a thermosoftening plastic, is a polymer that turns to a polymer when heated and freezes to a very glassy phase when cooled sufficiently. It is elastic and flexible above a glass transition temperature. The most commonly used thermoplastics, such as PMMA, PC, and polystyrene (PS), become viscous fluids when heated above their glass transition temperatures. However, the viscosity of the heated polymers remains high and the imprint pressure requires high. Besides, the thermoplastics are easy to stick to the mold and possibly results in defects during the demolding. Furthermore, amorphous thermoplastics are less resistant. Several thermoplastics, PMMA, PS, PC, PE, PP, PBT, and PET, are now commercially available [9–15].

Although the organic imprint resist can be used in the fabrication of high-resolution patterns, the adhesion between mold and resist is relative high and usually results in the defects of the patterns. In order to decrease the adhesion, the organosilicon or fluoride material is added into the resists.

Apart from the adhesion, the high etch resistances are needed. The polydimethylsiloxane (PDMS)-organic block or graft copolymers are of special interest. Compared to PMMA and organic polymers, siloxane copolymers exhibit significant properties for the high-open, flexible, and mobile Si–O–Si backbone. It renders low surface energy, low glass transition temperature, and high thermal stability. Moreover, the copolymers undergo microphase segregation above their glass transition temperature and results in good adhesion on the substrate and low surface energy, also called "bilayer" structure [16]. Siloxane copolymer resists, including poly(dimethyl siloxane)-polystyrene diblock copolymer(PDMS-b-PS), poly(dimethyl siloxane)-graft-poly(methyl acrylate)-co-poly(isobornyl acrylate)(PDMS-g-PMIA), and poly(dimethyl siloxane)-graft-poly (methyl methacrylate)(PDMS-g-PMMA), are investigated. The copolymer resists have both high etch resistance and high-resolution imprinting performance. The molecular formula of the siloxane copolymer was given in Fig. 5.2. SEM images of the imprinted resist patterns and the stamp are shown in Fig. 5.3.

Thermoset, also known as a thermosetting plastic, is a polymer that irreversibly cures. It is a low viscous fluid and easy to fill the cavity of the stamp. The curing

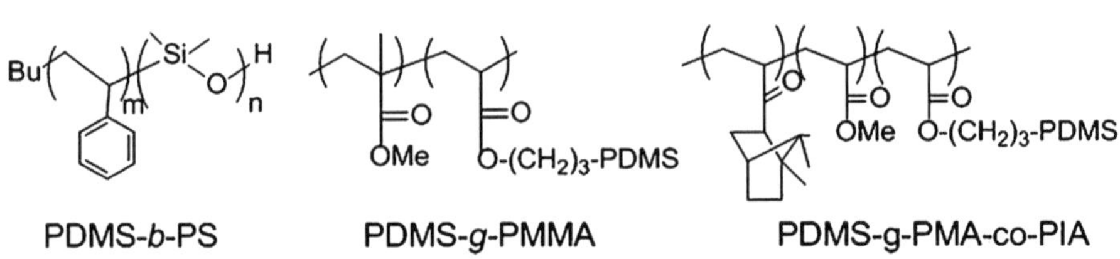

Fig. 5.2 The molecular formula of the siloxane copolymer

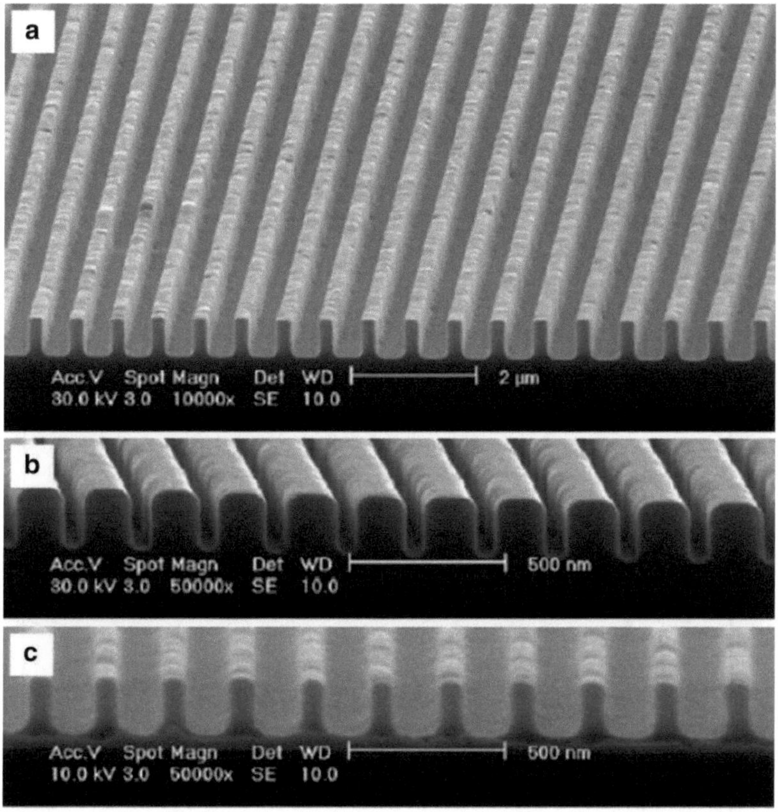

Fig. 5.3 SEM images of the imprinted patterns and the imprint mold. (**a**) 250-nm line width PDMS-b-PS grating. (**b**) Si_3N_4 mold. (**c**) PDMS-g-PMMA of 70-nm pattern using the mold shown in (**b**) (Reprinted from Ref. [3], Copyright (2007), with permission from Wiley-VCH)

process is done by heating and a chemical reaction. It cannot be melted and reshaped after it is cured. After curing, it can be separated from the mold immediately, resulting in high imprint efficiency. Thermoset materials are stronger than thermoplastic materials, therefore, with a relative higher etch resistant. Among these materials, PDMS is a class of thermal curable materials widely used for soft lithography. Table 5.1 lists an overview of some thermal imprint resist.

5.3.2 UV-Curable Resist

The UV imprint resist requires ultraviolet light irradiation to solidify. After being cured, the material does not change its shape. Methyl methacrylate and epoxypropoxy based on precursors are the most common materials used as a resin for UV nanoimprint. UV-curable resists have four sections: photoinitiators, resins (precursors), solvent, and additive.

Table 5.1 Resist materials for thermal nanoimprint [17]

Material	Solvent	Glass transition temp. Tg at molecular weight Mw	Viscosity at temp., Young's modulus	Comments
Poly(methyl methacrylate) (PMMA)	Anisole, Ethyl lactate (safe solvents)	100°C at 25–950 K	105–109 Pas at 170°C, 380–540 MPas	The classic NIL resist
Polystyrene(PS)	Toluene	100°C at 50 K		Integrated optics
Polycarbonate (PC)		145°C at 34 K	2,350 MPas	Integrated optics, $n = 1.6$ High etching resistance
Cyclo olefine copolymer (COC)	Cyclohexanone 1,1,2,2-tetra-chloroethane	60–180°C	2×10^4 Pas at 170°C, 2,600MPas	Highly transparent, chemically resistant, low water absorption Lab-on-chip and optical applications
mr-L 6000	Safe solvent	30°C at 7 K		UV curable, low-Tg NIL resist, used for polymer stamps, and mix-and-match and multilevel patterning
mr-I 7000	Safe solvent	60		Low-T_g NIL resist
mr-I 8000	Safe solvent	115		High-etch-resistance NIL resist with properties as PMMA
mr-I 9000	Safe solvent	65°C at 95 K		Thermocurable NIL resist
Hybrane	Toluene	0–10°C		Room-temperature imprint, SF6 RIE etch resistant
NEB22		80°C at 3 K		Negative EBL resist. High etch resistance in fluoro- and chloro-based plasmas

Table 5.2 General composition and function of a UV-curable resist

Components	Function	Share (%)	Types
Resin	Film formation Basic properties	10–60	Organic, inorganic–organic, etc.
Photoinitiators	Initiation	1–5	Radicals, cations, or anions
Solvent	Viscosity adjustment, x-link density	40–90	Small organic molecules
Additive	Surfactants, stabilizers, etc.	≤5	Fluorine, silicon class

Resins (precursors) are the main ingredient of UV-curable resist and a high-molecular-weight polymer. Under UV exposure, the precursors can react to form a polymer. The property of the polymer depends on the characteristics of precursors. According to the reaction mechanism, the precursors are usually divided into a free radical polymerization and cationic polymerization.

Photoinitiators are molecules that absorb photons upon irradiation with light and form reactive species out of the excited state, which initiate consecutive reactions. The initiating material may be radicals, cations, or anions. The photoinitiators are commercially available in large number from some companies. The most important to the selectivity of the photoinitiators are high absorption at the exposure wavelength and the high yield of formation of the materials [18].

Solvents are classified into the reactive diluents and the nonreactive solvents. Among them the reactive solvent (active diluents) belongs to a small organic molecule containing polymerization of functional groups and is an important component of the curing resists. A high-boiling-point solvent is selected to prevent from its volatility. Besides, the activity of reaction solvent and precursors cannot obviously vary. If the difference is too big, active diluents tend to all together or react with photoinitiators and do not participate in the polymer cross-linking. Nonreactive solvents cannot react with the precursors and only adjust the viscosity of the resists. It will evaporate after coating.

The additive is used to improve some performance of resist, such as smoothness, bubbles, or adhesion between the substrate and resist. An additive containing fluorine or silicon organisms is used. Table 5.2 shows different ingredient and function for UV-curable resist. Typical ingredients of UV-curable imprint resists contain the range of 10–60% oligomeric resins, which are responsible for the film formation and the basic coating characteristics. About 1–5% photoinitiators, together with additives (≤5%), are used to tailor the formulation to resist property requirements. The reactive diluents are introduced to adjust to the viscosity to the requirements of the resists.

A typical organic resist is a product from Toyo Gosei [19]. The structural formula of the polymerizable compound and photoinitiators are shown in Fig. 5.4.

Silsesquioxane resists have high etch resistance and good compatibility with biology. A typical silsesquioxane resist is shown in Fig. 5.5 [20].

Compared to the wafer imprint resist, the step-and-flash imprint resist is a little different. The resist consists of a bulk polymerizable organic monomer, such as

Fig. 5.4 Structural formula of monomer and photoinitiators. (**a**) Radicals. (**b**) Cations

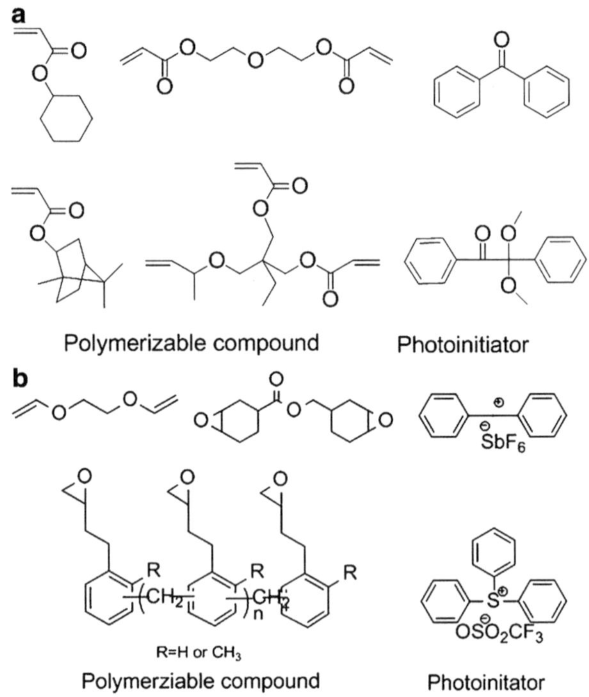

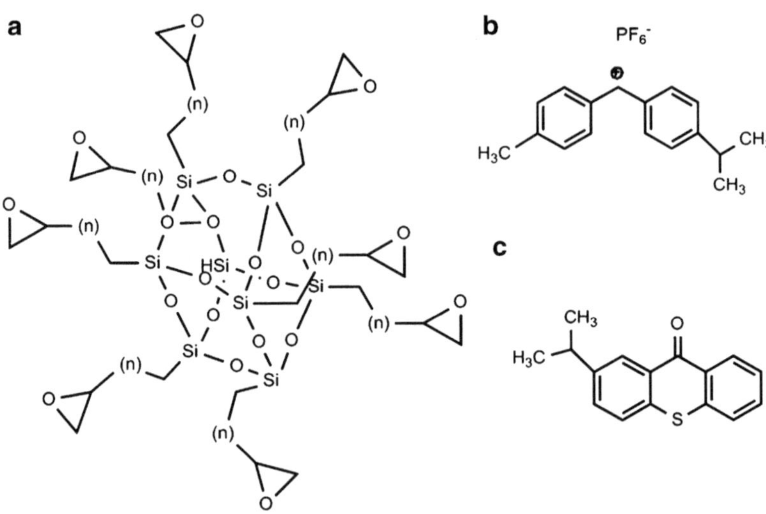

Fig. 5.5 Structural formula of the resist. (**a**) Monomer. (**b**) Irgacure 250. (**c**) Darocur ITX

Table 5.3 Components of SFIL resist

Name	Percent (%)	Components	Purpose
Monomer	25–50	Acrylate, vinyl ether, etc.	Low viscosity
Si monomer	25–50	Si material	Improve etch resistance
Cross-linker	10–30	Functional monomers	High mechanical stability
Photoinitiator	2–6	Irgacure 184, Irgacure 819, Darocur 1173	Cause polymerization

Table 5.4 UV-curable resist composition [22]

Component	Structure	Purpose	Weight percent (%)
Tert-butyl acrylate	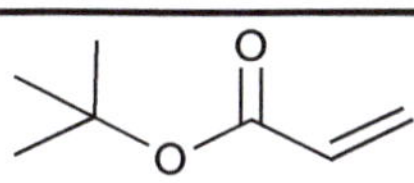	Maintain low viscosity	37
3-[Tris(trimethylsiloxy) silyl]propyl methacrylate	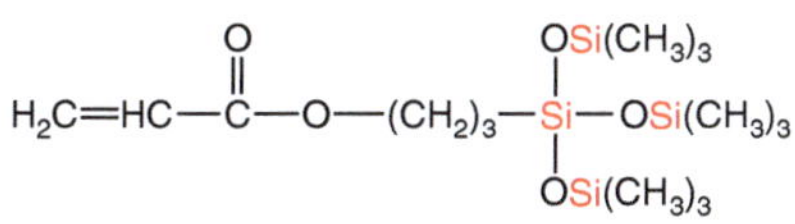	Improve etch resistance Decrease adhesion	44
Diethylene glycol diacrylate	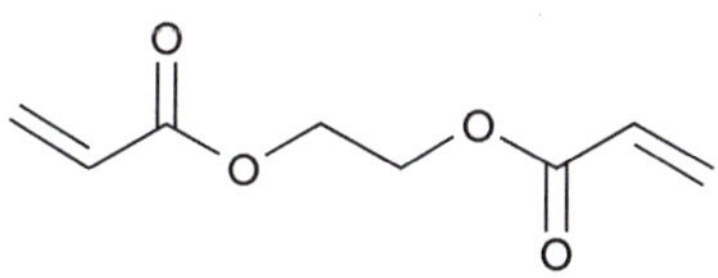	Cross-linker for mechanical stability	15
Darocur 1173	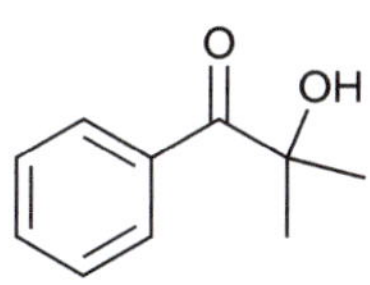	Photoinitiator	1–4

an acrylate or vinyl ether, a silicon-containing or siloxane-containing monomer to provide oxygen etch resistance, a cross-linking agent to provide mechanical strength, a photoinitiator, and fluorinated surfactant to ease mold release [21]. Willson's group proposed components of SFIL resist [22–25], shown in Table 5.3. A resist mixture made of four different components is utilized and given in Table 5.4 [26].

We developed UV resist, a novel fluorine contained material resist ZZJ, which has good mold releasing ability and high dry etching resistance. Especially volatile component of ZZJ is only nontoxic diacetone alcohol and ethyl lactate, which are not unhealthful. Unlike silicone-polymer contained analogue, this silicon-free formula will not generate solid particles during reactive ion etching and can transfer nanostructure to substrate with good fidelity. Figure 5.6 shows SEM images of resist patterns imprinted with using the ZZJ resist. It has been demonstrated that the new resist has a good transferring fidelity.

The XPS spectra of ZZJ are given in Fig. 5.7. The elements' content, C-68.55%, O-23.40%, N-0.21%, and F-7.84%, can be seen from the spectra. Silicon or metal

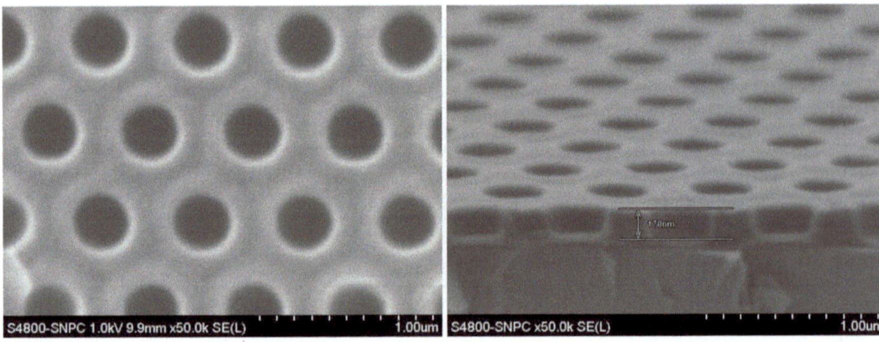

Fig. 5.6 SEM images of patterned resist

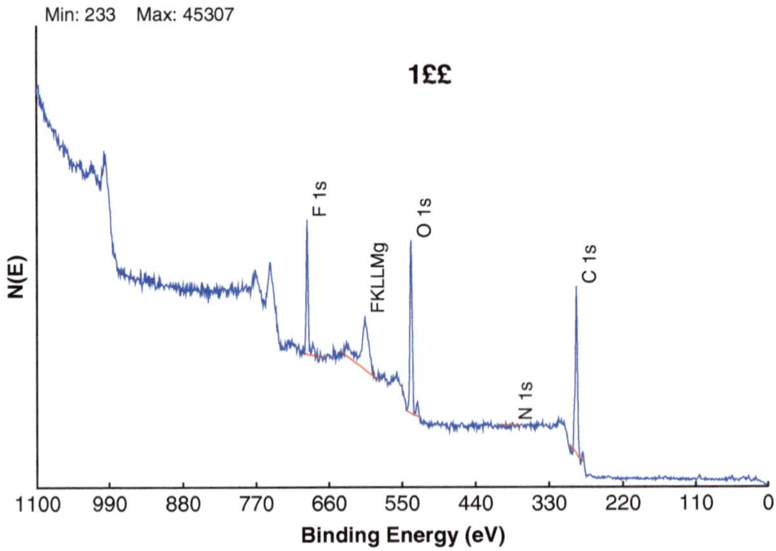

Fig. 5.7 XPS spectra of the novel resist

elements in ZZJ cannot be detected. So solid particles cannot form during the dry etching process, which is good to ensure successful pattern transfer from resist to substrate without affecting the nanoscale structure.

Roller imprint lithography resists differ from that of other imprint lithography. The cured rate is very quick to meet the requirement of continuous imprint processes. The imprint time is short and the pressure is relatively low that needs low-viscosity resist to fill the cavity of the mold. Besides, the low-molecular-weight prepolymer is easy to shrink to create bad fidelity of patterns. Therefore, the roller resists need the considerations of material shrinkage. Roller imprint lithography resist was first used with PMMA [27], but the slow capability of fabricating patterning has an adverse impact on requirement of high efficiency for roller imprint. There are two kinds of resists: thermoset and UV nanoimprint resist [6, 28, 29]. UV resist is suitable for the imprinting requirement. Guo et al. has developed a roller resist that cured within 1 s, and it can meet the rapid roller imprint requirements [29].

5.3.3 Functional Material

Compared to the conventional imprint resist functioning as the etch barriers, the functional imprint resists are patterned as a permanent, functional section of the imprinted devices. Using functional material, photonic crystals can be patterned [30]. Patterned dielectric materials for CMOS interconnect layers have also been created by sol–gel-based dielectric precursors with imprint process. The trench structures for dual damascene process with functional imprint resist material were imprinted [31]. Besides, functional Ag and carbon nanotube materials are directly imprinted [32, 33].

References

1. Levinson HJ (2010) Principles of lithography, 3rd edn. SPIE, Washington
2. Zhao B, Zhang J, Zhou WM, Wang JH, Liu YB, Zhang YP, Shi LY, Zhang JP (2001) Nanoimprint lithography resist. Microelectron Technol 48:606–612
3. Guo LJ (2007) Nanoimprint lithography: methods and material requirements. Adv Mater 19:495–513
4. Hirai Y, Tanaka Y (2002) Application of nano-imprint lithography. J Photopolym Sci Technol 15:475–480
5. Kim EK, Stacey NA, Smith BJ (2004) Vinyl ethers in ultraviolet curable formulations for step and flash imprint lithography. J Vac Sci Technol B 22:131–135
6. Ahn SH, Guo LJ (2008) High-speed roll-to-roll nanoimprint lithography on flexible plastic substrates. Adv Mater 20:2044–2049
7. Manyam J (2010) Novel resist materials for next generation lithography. Dissertation, University of Birmingham
8. Cui Z (2005) Micro-nanofabrication: technologies and applications. Higher Education Press, Beijing
9. McAlpine MC, Friedman RS, Lieber CM (2003) Nanoimprint lithography for hybrid plastic electronics. Nano Lett 3:443–445
10. Babak H (2000) Nanoimprint lithography at the 6 in. wafer scale. J Vac Sci Technol B 18: 3557–3560
11. Schift H, Bellini S, Gobrecht J (2006) Perforated polymer membranes fabricated by nanoimprint. Microelectron Eng 83:873–875
12. Bogdanskia N, Schulza H, Wissena M (2004) 3D-Hot embossing of undercut structures – an approach to micro-zippers. Microelectron Eng 73–74:190–195
13. Schift H (2005) Perforated polymer membranes fabricated by nanoimprint. Microelectron Eng 83:873–875
14. Tallas J, Peyrade D, Lazzarion F (2005) Replication of sub-40 nm gap nanoelectrodes over an 8-in. substrate by nanoimprint lithography. Microelectron Eng 78–79:676–681

15. Kawaguchi Y, Nonaka F, Sanada Y (2007) Fluorinated materials for UV nanoimprint lithography. Microelectron Eng 84:973–976
16. Choi PY (2010) Physicochemical surface interaction of nanoimprint lithography resists and imprinted pore membranes. Dissertation, University of Michigan
17. Bhushan B (ed) (2006) Springer handbook of nanotechnology. Springer, Heidelberg
18. Jürgen Buschow KH, Cahn RW, Flemings MC, Ilschner B (2001) Encyclopedia of materials: science and technology. Elsevier Science, New York, USA
19. Nobuji S (2009) Photo-curable resins for UV-nanoimprint technology. J Photopolym Sci Technol 22:133–145
20. Girolamo JD, Chouiki M, Tortai JH (2008) Epoxy silsesquioxane resists for UV nanoimprint lithography. J Vac Sci Technol B 26:2271–2275
21. Costner EA, Lin MW, Jen WL, Willson CG (2009) Nanoimprint lithography materials development for semiconductor device fabrication. Annu Rev Mater Res 39:155–180
22. Kim EK, Stewart MD, Wu K et al (2005) Vinyl ether formulations for step and flash imprint lithography. J Vac Sci Technol B 23:2967–2972
23. Kim EK, Willson CG (2006) Thermal analysis for step and flash imprint lithography during UV curing process. Microelectron Eng 83:213–217
24. Kim EK, Stacey NA, Smith BJ et al (2004) Vinyl ethers in ultraviolet curable formulations for step and flash imprint lithography. J Vac Sci Technol B 22:131–135
25. Long BK, Keitz BK et al (2007) Materials for step and flash imprint lithography (S-FIL®). J Mater Chem 17:3565–3656
26. Dickey MD, Burns RL, Kim EK, Johnson SC, Stacey NA, Willson CG (2005) Study of the kinetics of step and flash imprint lithography photopolymerization. AIChE J 51:2547–2555
27. Tan H, Gilbertson A, Chou SY (1998) Roller nanoimprint lithography. J Vac Sci Technol B 16:3926–3929
28. Ahn SH, Guo LJ (2009) Large-area roll-to-roll and roll-to-plate nanoimprint lithography: a step toward high-throughput application of continuous nanoimprinting. ACS Nano 3:2304–2310
29. Stuart C, Chen Y (2009) Roll in and roll out: a path to high-throughput nanoimprint lithography. ACS Nano 3:2062–2064
30. Resnick DJ, Gerard S, Mike M, Gary D, Chris J, Dwayne L (2007) Step and flash imprint lithography template fabrication for emerging market applications. Proc SPIE 6607:66070T
31. Lun JW, Palmieri F, Chao B, Lin M, Hao JJ, Owens J, Sotoodeh K, Cheung R, Willson CG (2007) Multilevel step and flash imprint lithography for direct patterning of dielectrics. Proc SPIE 6517(2):65170K.1–65170K.9
32. Choi JH, Lee SW, Jeong JH, Choi DG, Lee ES (2008) Direct imprint of conductive silver patterns using nanosilver particles and UV curable resin. Microelectron Eng 86:622–627
33. Choi JH, Jung SU, Choi DG, Jeong JH, Lee ES (2008) Direct soft UV-NIL with resist incorporating carbon nanotubes. Microelectron Eng 85:195–201

Chapter 6
Nanoimprint Lithography Process

6.1 Introduction

It is well known now that a nanoimprint lithography process generally consists of stamp modification, spin coating of resist, imprinting, and then etching for pattern transfer. The stamp modification has been already demonstrated in Chap. 4. Before imprinting, the substrate surface is coated by a thin film for imprinting; how to create a uniform imprinted film is a crucial issue. The spin coating is generally a common method to control film thickness. There are three main nanoimprint lithography techniques: hot embossing (HE), UV-based nanoimprint lithography (UV-NIL), and soft lithography. Various soft lithography techniques have been proposed such as microcontact printing (μCP), replica molding (REM), microtransfer molding (μTM), micromolding in capillaries (MIMIC), and solvent-assisted micromolding (SAMIM). They can be widely used in various fields. In nanoimprint lithography process, the controlling of pattern defect, alignment, and full area imprinted pattern are among the hot topics. Recently, a great deal of attention has been paid to soft UV nanoimprint because of the full area conformal contact with the substrate.

In the development of the nanoimprint lithography, different types have been proposed and investigated intensively: fast thermal nanoimprint lithography, reverse nanoimprint lithography, combined nanoimprint and photolithography, roller nanoimprint lithography, capillary force lithography, direct imprint functional material, and three-dimensional (3D) nanoimprint lithography.

After imprinting, if necessary, the next step is an etching process. It greatly affects the quality of pattern transfer. In most circumstances, plasma etching is used in the process. Combined with lift-off process, the high aspect ratio nanopattern can be obtained easily.

W. Zhou, *Nanoimprint Lithography: An Enabling Process for Nanofabrication*,
DOI 10.1007/978-3-642-34428-2_6,

6.2 Resist Film for Nanoimprint Lithography

Spin coating is a process that forms uniform thin films on a flat substrate. In brief, the resist is placed on the substrate, which is then rotated at high speed to spread the fluid by centrifugal force. A thin film of the coating material will be deposited on the substrate, which depends on a combination of the spinning speed and the viscosity of the resist.

During the spin-coating process, the four stages are distinctly divided:

- Deposition of the resist material on the substrate.
- Acceleration of the substrate up to the designed speed.
- The substrate is spun at a constant rate. Fluid viscous forces decide the fluid-thinning behavior.
- The substrate is spun at a constant rate. Solvent evaporates until the coating is constant.

The thickness of the imprinted resists depends on the spinning speed and the viscosity of the resist and spinning time. If the other is the same, the maximum thickness is largely a function of the viscosity of the resist. A basic schematic can be presented in Fig. 6.1.

The substrate is held by a vacuum and the resist is deposited in the center of the substrate. The coater is then spun at several thousand rpm, which will cause the resist flow outward and a smooth uniform layer is obtained. A photograph of EVG®101 Advanced Resist Processing System is pictured Fig. 6.2, which ensures low material consumption of precious high-viscosity photoresists or polymers while improving uniformity and resist spreading options.

The resist thickness can be calculated from the equation

$$h = \frac{h_0}{\sqrt{1 + \frac{16\pi^2 f^2}{3\eta} h_0^2 t}},$$

where h_0 is the initial resist thickness (nm), f is the rotational speed (r/s), η the resist viscosity ($m^2\ s^{-1}$), and t is the spinning time (s).

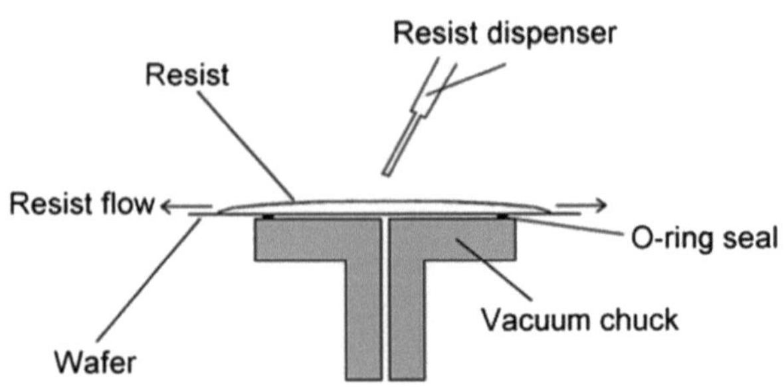

Fig. 6.1 Spin-coating process

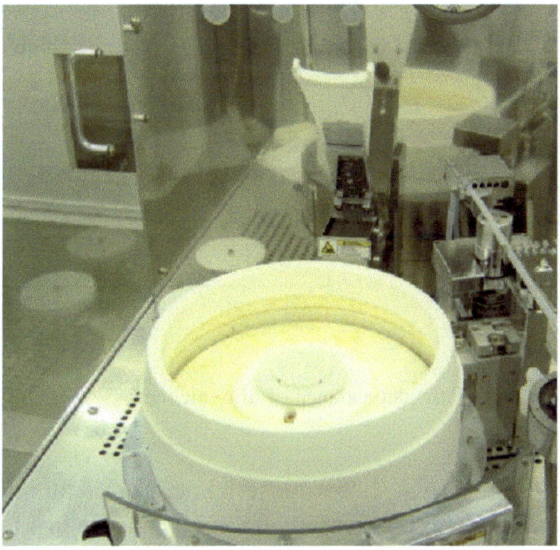

Fig. 6.2 A photograph of spin-coating system

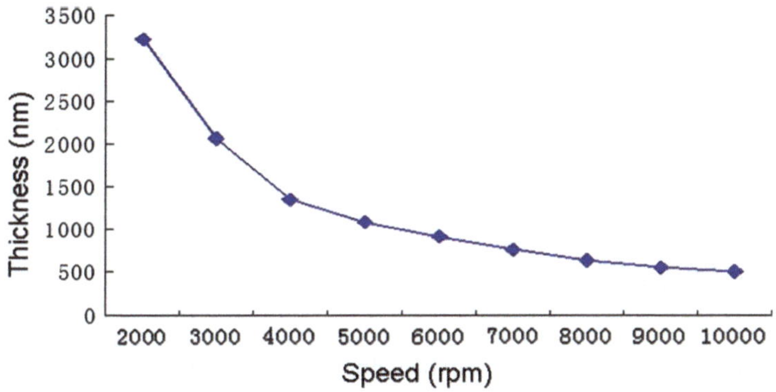

Fig. 6.3 The film thickness as a function of speed

As t increases to about half minute, it finally becomes insignificant. When allows cancellation of the h_0 term, generating equation is as follows

$$h = \frac{\sqrt{3\eta/t}}{4\pi f}.$$

Here, we can see that the resist thickness depends mainly on the spinning speed (f), time (t), and fluid viscosity (η) but only on the condition that t is significant enough.

Film thickness (an imprinting resist) as a function of speed is given in Fig. 6.3. From the figure, it can be seen that the thickness decreases with increased speed.

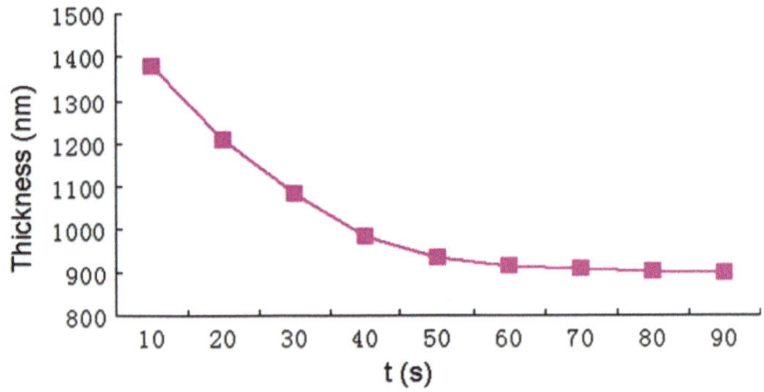

Fig. 6.4 Time dependence of film thickness

Figure 6.4 shows the thickness as a function of time *t*. At early stages, the thickness is decreased gradually and then closed to the constant value of the minimum thickness.

In order to improve the adhesion between the substrate and the resist, a primer layer is pre-spun coated. After spin coating, soft bake is necessary to remove solvent of the resist, which improves the adhesion and remits the internal stress.

Spin coating is the first step for nanoimprint lithography, and the quality of the resist film has an impact on the imprinted patterns. Some defects, such as particle, dust, will destroy the imprinted nanostructures.

6.3 Basic Types of Nanoimprint Lithography

Usually, nanoimprint lithography can be categorized into three major types: thermal NIL, UV-NIL, and soft lithography.

6.3.1 Thermal NIL

Thermal NIL or hot embossing is the earliest nanoimprint lithography developed by Prof. Stephen Chou's group [1]. In the process, a thin film is spin-coated on the substrate. The stamp is brought into contact with the surface of the thin film under certain pressure. Meanwhile heated up above the glass transition temperature of the polymer, the pattern on the mold is pressed into the softened polymer film. After cooling, the stamp and the polymer are separated, and the pattern resist is left on the substrate. A plasma process (reactive-ion etching) is used to transfer the pattern into the substrate. A schematic of hot embossing is shown in Fig. 6.5.

The time and temperature diagram during nanoimprint lithography is shown in Fig. 6.6. The imprint temperature is usually around 70–90°C above glass transition

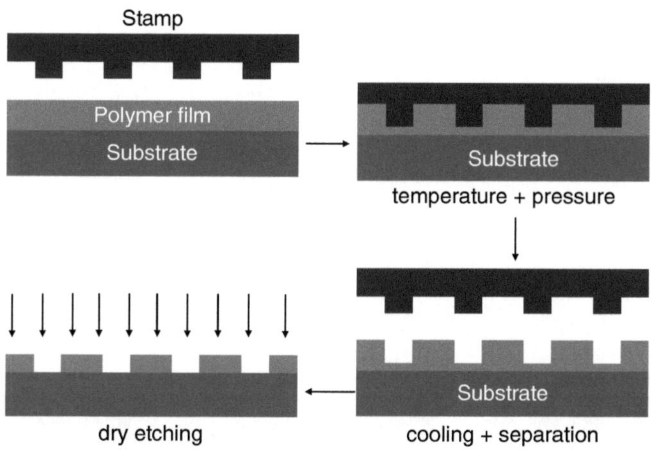

Fig. 6.5 Schematics of the thermal nanoimprint lithography process

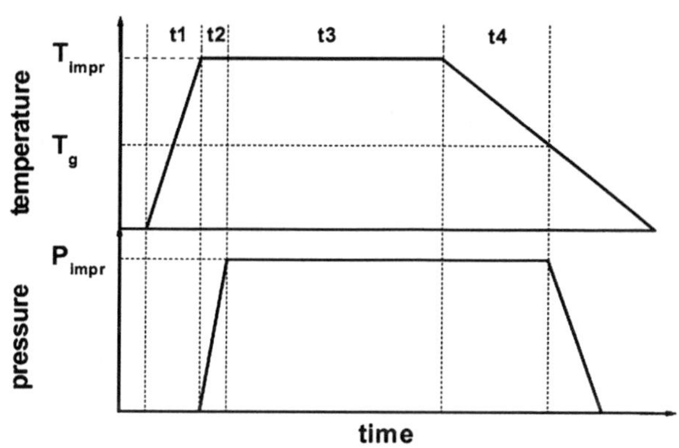

Fig. 6.6 Schematic of temperature and force relation during hot NIL

Table 6.1 Glass transition temperature (T_g) for different polymers

Polymers	Trade remarks	T_g	
PMMA	AR-P 669.04	95–105	Thermoplastic
Polystyrene	168N (BASF)	110	Thermoplastic
PTFE	Teflon AF1601S	160	Thermoplastic
PPM	mr-I8000	107	Thermoplastic
PDAP	mr-I9000	63	Duroplastic

temperature (T_g) so that the polymer reaches a viscous flow state. The glass transition temperature (T_g) of some materials for hot embossing is presented in Table 6.1.

Ideally, imprinted resist has good flowability and incompressibility. A high viscosity of polymer has low flowability. By heating polymer and decreasing the polymer viscosity, it easily fill the cavities of the stamp. Three different types of amorphous polymer are schematically given in Fig. 6.7 [2].

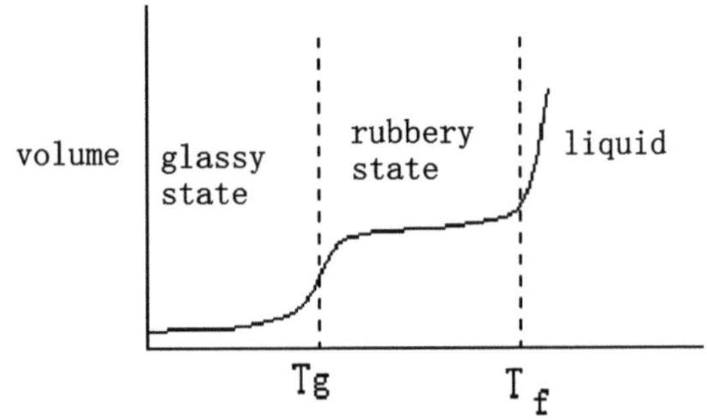

Fig. 6.7 Volume vs. temperature plot for an amorphous polymer

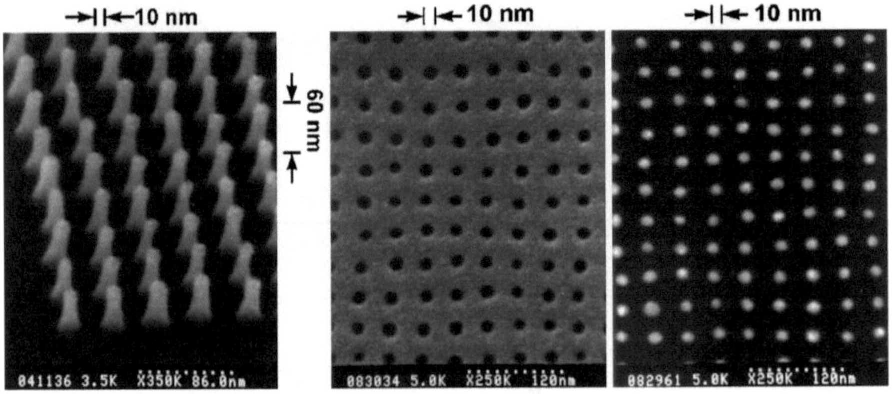

Fig. 6.8 SEM images. (**a**) SiO_2 stamp of pillars of 10 nm diameter, 60 nm height, and 40-nm period. (**b**) Imprinted pattern on the PMMA surface. (**c**) 10-nm metal dots after lift-off process (Reprinted with permission from [3]. Copyright (1997) American Vacuum Society)

1. Glass Transition: As the temperature of a polymer is below T_g, it behaves in an increasingly brittle manner. When the temperature increases above the T_g, the polymer becomes more rubberlike.
2. Rubbery state: The temperature between glass transition and freezing point, T_f.
3. Liquid state: As the temperature increases over T_f, the motion of entire chains takes place, and the polymer flows by chain sliding. The region is of particular interest for nanoimprint lithography. The imprinted polymer melt at sufficiently high temperature will have a viscous flow.

To fill fully into the cavity of the stamp, the pressure is as high as 50–100 bar. Therefore, the stamp material is with high mechanical strength, such as Si, SiO_2, Si_3N_4, diamond, or metal. High thermal stability, low thermal expansion coefficients, low Poisson's ratio, low roughness, and high Young's modulus are necessary for hot embossing stamp material.

Figure 6.8 shows Chou's reported the experimental results [3]. Figure 6.8a is a SEM image of SiO_2 stamp with a 40-nm-period array of pillars with 60 nm. The imprinted temperature is 175°, and pressure is 4.4 Mpa. Figure 6.8b is a SEM image of imprinted 10-nm-diameter holes in a thin polymer film (PMMA). Combined with

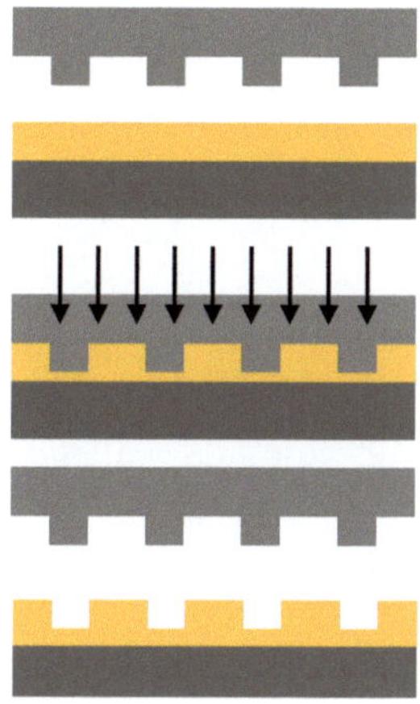

Fig. 6.9 Schematic of UV-NIL: imprint resist is coated on the substrate and align; the stamp is imprinted into the resist, and UV curing is applied; separation of the stamp and the substrate

lift-off process, 10-nm metal dots using the thin polymer layer as a mask have been obtained. It has been shown that hot embossing is a high-throughput and low-cost process.

6.3.2 *UV-NIL*

In 1996, a scientist named Jan Haisma from Philips Research Laboratories published an interesting result of patterning with dimensions 100 nm or less on a thin UV curable resist layer. In his experiment, a transparent stamp was used for imprinting a pattern on the UV curable film and then cured by UV exposure. This technique is the basis of UV curable nanoimprint lithography (UV-NIL) technology [4].

Compared to thermal NIL, the viscosity of the resist for UV-NIL is much smaller. Meanwhile, the stamp is transparent to UV light. UV-NIL has advantages of room temperature operation, a lower applied pressure and alignment for IC devices. The schematic process of UV-NIL is given Fig. 6.9.

Stamp of UV-NIL is used with hard material (quartz) or soft polymer material (PDMS). Especially, when used with soft stamp, named soft UV-NIL, it can bring into conformal contact with a rough surface and satisfy with a large area imprint surface (8 in. full pattern) [5].

6.3.3 *Soft Lithography*

Soft lithography technique was proposed by Prof. Whiteside from Harvard University for the first time [6]. It can replicate micro/nanostructures by the template/master made out of an elastic, deformable material (soft material). Some good advantages of various soft materials, replication on smooth or rough surface, are preferable to the fabrication of structures.

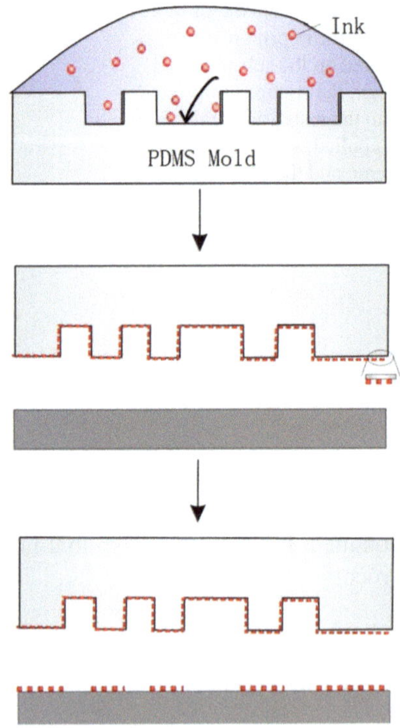

Fig. 6.10 The schematic illustration of μCP

Basically, three steps are involved in soft lithography: (1) mother template fabrication, fabricated by conventional lithography or unconventional lithography. (2) A copy of a deformable material stamp (daughter stamp) is formed from a mother stamp. Usually, the soft material is PDMS polymer. (3) Imprint patterns with the daughter stamp. A variety of soft lithography techniques have been proposed and realized, such as microcontact printing (μCP), replica molding (REM), microtransfer molding (μTM), micromolding in capillaries (MIMIC), solvent-assisted micromolding (SAMIM), and nanotransfer lithography (nTP). All the techniques have their specific advantages or drawbacks and are useful for specific applications.

6.3.3.1 Microcontact Printing (μCP)

Microcontact printing (μCP), first reported by Whiteside [7], is a method in which the template is used to transfer molecules (self-assembled monolayers of ink) on a target substrate by conformal contact. The molecules are used as masking material in a successive etching step. The ink is a thiol solution. The PDMS stamp is immersed in or coated with the ink. The thiols reside both on the surface and on the bulk of the stamp material because of the highly hydrophobic PDMS material. Let the stamp dry, and the stamp with an ink reservoir is created. The schematic illustration of μCP is shown in Fig. 6.10.

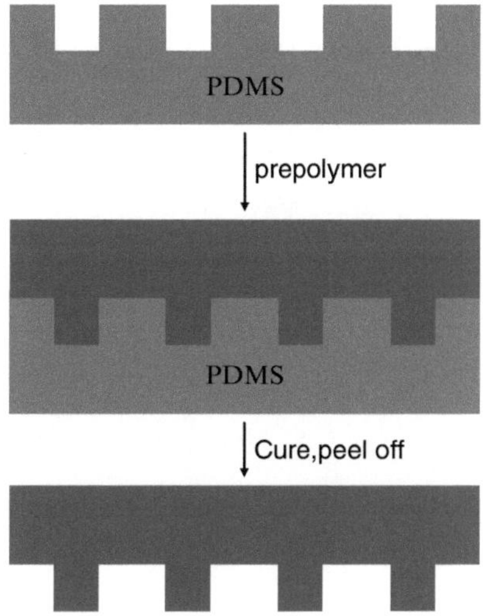

Fig. 6.11 The schematic illustration of REM

The technique mainly depends on the rapid formation of alkanethiols on gold and on the properties of SAMs. After μCP, the Au region without SAMs coating will be washed with a dilute solution containing the second molecule. Or alkanethiols grated some molecule is self-assembled and used for sensor devices.

The technique is attractive for simplicity, low cost, and flexibility. It can contact with other surfaces (metal such as Au, Ag, Cu, and Pt) either smooth or nonplanar, which conventional lithography will fail. But the applications concentrate on chemical and biological purpose.

6.3.3.2 Replica Molding (REM)

Replica molding (REM) is a method in which the stamp itself is used as a mother master for a casting or embossing process. Previously, it is used with an elastomeric mold rather than a rigid stamp [8]. The schematic illustration of REM is given in Fig. 6.11.

Later, the mold is also used with rigid material. Willson has developed step and flash lithography methods that a rigid and transparent (usually quartz) master is used in replica molding. In fact, the process is nanoimprint lithography. The modified quartz stamp is applied on the surface of photocured polymer and then UV cured [9].

The replica molding with elastomeric molds can be obtained with a resolution on the nanometer scale. Once the mother stamp is created by conventional lithography, the structures could be duplicated into multiple copies by replica molding with organic polymers. The technique has also been used for the fabrication of complex, functional surface that is difficult to make with other techniques.

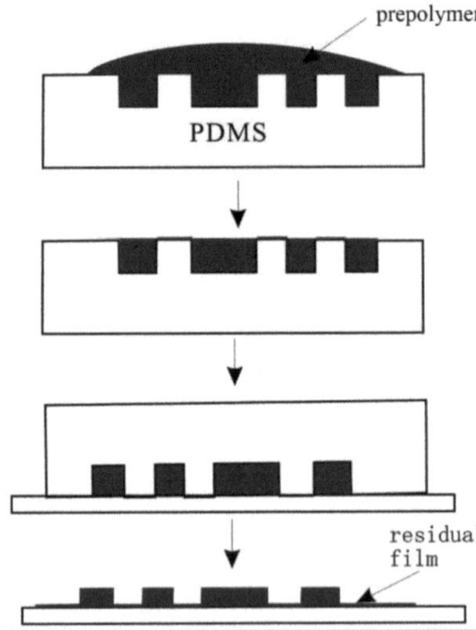

Fig. 6.12 The schematic illustration of μTM

6.3.3.3 Microtransfer Molding (μTM)

Microtransfer molding (μTM) is a method in where the cavities of the stamp are filled with polymer material, which is later on molded onto a substrate. This is to say, the liquid is applied to the patterned surface of PDMS stamp, and the excess liquid is removed by blowing off with N_2 gas or scarping with a flat PDMS sheet. Then, the filled mold is brought into contact with a substrate surface and cured by UV light or heated. The mold is separated from the substrate leaving a patterning on the surface. The schematic illustration of μTM is given in Fig. 6.12. The residual polymer film could be removed by plasma etching.

It has advantages of fabrication on nonplanar surface and 3D structures layer by layer. Besides, the technique can create microstructures over large areas (~3 cm^2) within a short period of time (~10 min) [10].

6.3.3.4 Micromolding in Capillaries (MIMIC)

Micromolding in capillaries (MIMIC) is a method that a stamp is placed on a substrate, and a material is put at the open ends of the channels. The resulting channels are filled with a material by capillary action. After removal of PDMS mold, the cured material is left on the surface of the substrate [11]. The schematic illustration of MIMIC is given in Fig. 6.13.

The principle of MIMIC depends on capillary filling. If the channel is short (~1 cm), it can be filled shortly. Otherwise, the channel is not full because of the adhesion of materials. To solve the problem, the material with solvents is proposed. The solvents were evaporated after the solutions have filled the channels.

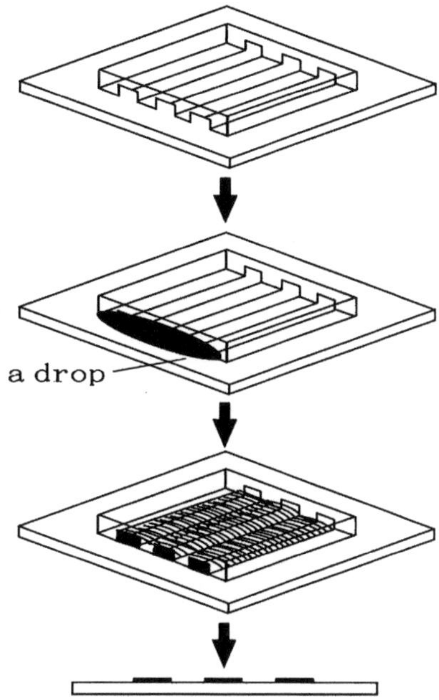

Fig. 6.13 The schematic illustration of MIMIC

Various materials can be filled, such as precursors for carbon, ceramics, polymer beads, colloids, inorganic salts, sol–gel materials, and biological macromolecules.

6.3.3.5 Solvent-Assisted Micromolding (SAMIM)

Solvent-assisted micromolding (SAMIM) is a method in which a solvent wets the stamp. The substrate carries a polymer sensitive to this solvent. After bringing into contact, the solvent dissolves a thin layer of the polymer and fills the cavities of the mold, as shown schematically in Fig. 6.14 [12].

In fact, the technique is similar to hot embossing. The cured approach is not heated but dissolves or swells the polymers and then fills the cavities of PDMS mold. The matching solvent is a key issue for SAMIM. The solvents can dissolve the polymer of the substrate surface and wet the mold surface and not swell the PDMS mold. The selective solvent should have a relatively high vapor pressure and a moderately high surface tension (e.g., Methanol, ethanol, and acetone) to ensure rapid evaporation of the excess solvent and minimal swelling of the PDMS mold [6]. The corresponding solvent for PMMA is acetone, and for photoresists is ethanol.

6.3.3.6 Nanotransfer Lithography (nTP)

Nanotransfer printing (nTP) is similar to microcontact printing, which prints thiol "inks" onto Au films. nTP prints Au films onto thiol-terminated monolayers. First,

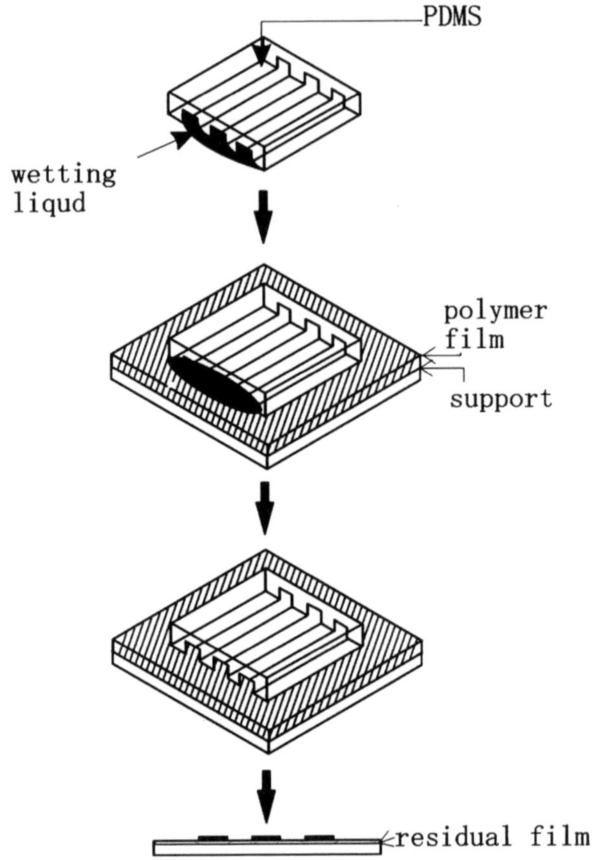

Fig. 6.14 The schematic illustration of SAMIM

Au layer is deposited by physical vapor deposition onto the surface of the stamp. The substrate attached a monolayer can bond to the Au coating when the PDMS mold contacts the substrate. The methods can create single and multilayer 3D nanostructures [13]. Other soft lithography, such as phase-shifting edge lithography [14], decal transfer lithography [15], nanoskiving [16], please refer to these related references.

6.4 Variations of Nanoimprint Lithography Process

Since the technique was proposed, it has many benefits. To enhance the capability, some novel approaches have been developed intensively. The key issue concentrates on the efficiency of NIL. As an example, large area nanoimprint lithography and direct functional material nanoimprint lithography are introduced as follows:

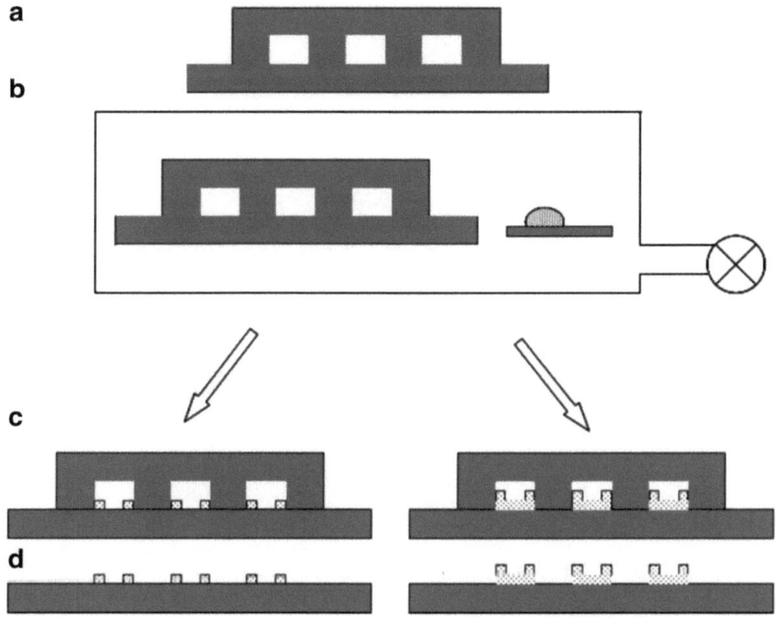

Fig. 6.15 Procedure of direct gas-phase patterning of organosilane thin films using a PDMS stamp template. (**a**) Reversible bonding of PDMS stamp to substrate. (**b**) Deposition of organosilane in desiccator. (**c**) Removal of substrate and stamp from desiccator after exposure. (**d**) Removal of stamp from substrate

6.4.1 Fast Thermal Nanoimprint Lithography

In conventional thermal nanoimprint lithography, stamp and substrate are both heated and can cause mismatch of the substrate and the resist, which will affect the fidelity of the patterns. In 2002, Prof. Chou proposed a novel technique named laser-assisted direct imprint (LADI) [17]. First, a quartz mold is brought into contact with the silicon substrate, and the pressure is applied. Then, an excimer laser pulse melts a thin surface layer of Si, and the molten is imprinted until it rapidly solidifies. Finally, the Si substrate and the stamp are separated, and a negative pattern is achieved. The process saves the following etching steps and renders low cost.

Research on heating and cooling process is two key points for thermal nanoimprint lithography. A novel method has been developed by George [18]. First, the PDMS mold reversibly contacts on the substrate. The gas transports and deposits on the cavities of the mold, and the height of deposited materials mask for the cavities and the inner walls are different. The deposited mask is used for etching. The detailed schematic of the process is shown in Fig. 6.15.

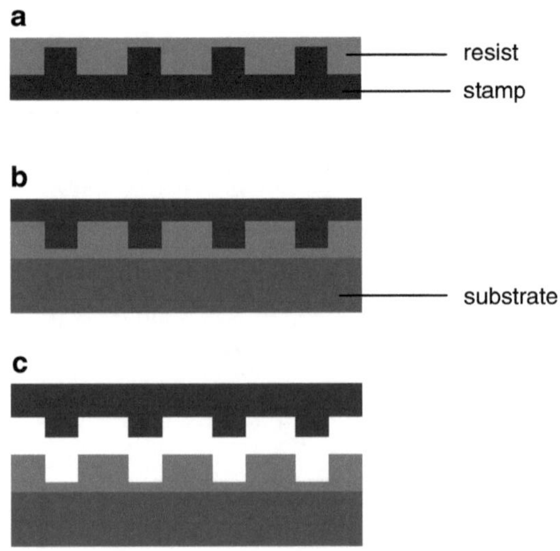

Fig. 6.16 Reversal nanoimprint lithography process. (**a**) Spin-coating stamp with imprinted resist. (**b**) Reversal imprinting by thermal press or UV curing. (**c**) Demolding

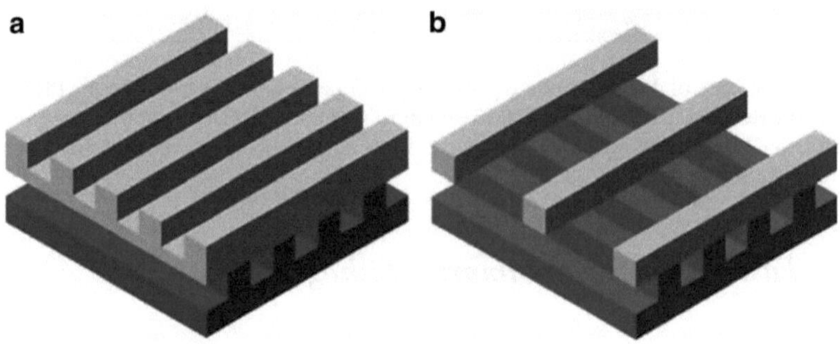

Fig. 6.17 Reverse nanoimprint lithography. (**a**) High pattern density. (**b**) Low pattern density

6.4.2 Reverse Nanoimprint Lithography

Usually, nanoimprint lithography is a process that a stamp is pressed into a resist film on the substrate surface. The process is also, in other word, reverse nanoimprint lithography. The process is schematically illustrated in Fig. 6.16. The polymer is spin coated on the surface of the stamp and fills up all the stamp cavities. The filled polymer stamp is brought into contact with the substrate and then cured. After separation, the pattern is left on the substrate surface [19]. Different types of reverse nanoimprint lithography are shown in Fig. 6.17.

The success of reverse nanoimprint lithography is to make sure that surface energy of substrate is greater than that of stamp, and the patterned polymer can separate from the stamp but firmly stick to the substrate surface. The stamp and the substrate both need to be modified.

It can be used on the substrate either nonplanar or difficult to spin coating. Its unique benefit is the fabrication of multiple levels of polymer structures by repeating reverse nanoimprint lithography. The residual layer always appears. But by modifying the pattern density on the stamp and the properties of polymer, it cannot be created. For dense pattern on a stamp (pattern protrusion area >25% of the whole stamp area), the whole polymer sheet and the residual layer is likely to be transferred onto the substrate. If is a sparsely stamp (protrusion area <25% of the whole stamp area), only the residual layer on the top of the stamp protrusions can be transferred onto the surface [20].

6.4.3 Combined Nanoimprint and Photolithography

The process is combined with nanoimprint lithography and photolithography and termed as CNP [21]. The schematic of combined nanoimprint and photolithography is presented in Fig. 6.18. The hybrid stamp has a pattering for UV light transparent and patterning for blocking UV light. When the hybrid is pressed into the polymer resist, UV light shines through the transparent quartz stamp and part of UV light is blocked by the opaque chrome patterns on the stamp. After separation, the substrate is immersed into the developer solution. The area without UV irradiation, the resist (negative photoresist) will be removed. Finally, the pattern is imprinted parts and the developed features.

The technique is to enable patterning both small and large features in one step. As we know, the small cavities can easily be filled quickly, and large cavities are difficult to fill on the same stamp and often generate defects on the imprinted patterns. CNP can solve the problem.

Based on the theory of CNP, the opaque layer can be created on the protrusion features of a stamp. A zero residual layer is easily generated by CNP process, as shown schematically in Fig. 6.19 [22].

When imprinted, the UV light is blocked at the protrusion patterns, and the area is not cured. The residual layer can be removed by wet etching in developer solution without RIE. The technique can generate high aspect ratio imprinted patterns. Certainly, the developer cannot damage the imprinted pattern structures.

6.4.4 Roller Nanoimprint Lithography

Roller presses are used in the cement industry and to imprint into a flexible thin film that can conform to the shape of the roller. The roller press is composed of a roller, a movable platform, and a hinge. Compared to flat nanoimprint lithography, roller nanoimprint lithography (RNIL) has some advantages of better uniformity, lower pressure, and the ability to repeat imprint on a large scale.

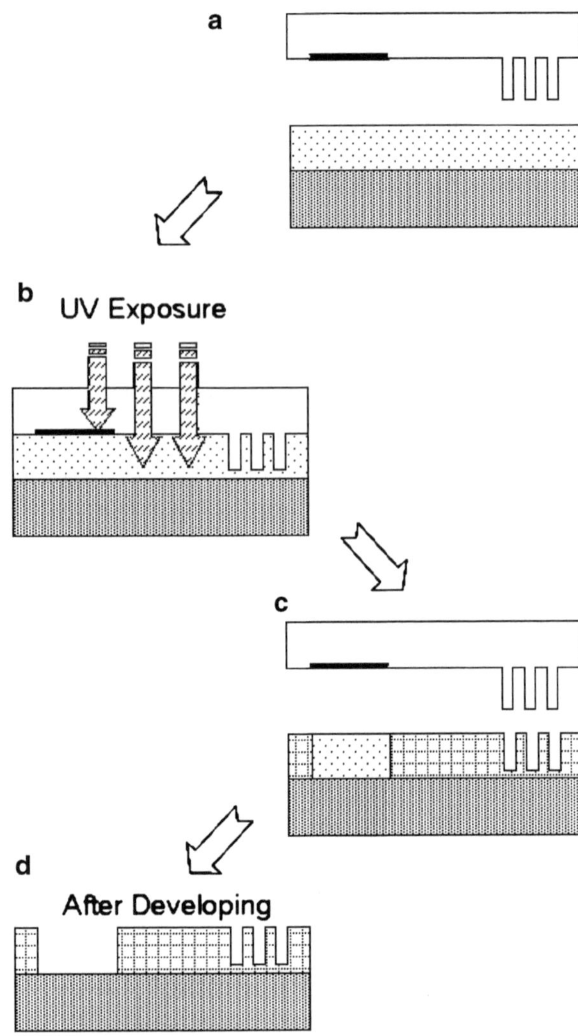

Fig. 6.18 Schematic of combined nanoimprint and photolithography. (**a**) Hybrid NIL stamp. (**b**) UV exposure. (**c**) Demold. (**d**) Developed resist

In 1998, Tan et al. first presented the novel RNIL. Using the roller process, sub-100-nm resolution pattern has been created. The RNIL machine design is schematically shown in Fig. 6.20. The system consists of a roller, a movable platform, a hinge, and heaters [23]. Two different types of RNIL are shown in Fig. 6.21.

Based on the aforementioned theory, two methods have been developed: (1) the cylinder mold with nanopatterns is imprinted into the PMMA resist. (2) A flat mold is imprinted into PMMA resist, and the smooth cylinder roller is rotated over the mold. In the two methods, the roller temperature is set above the glass transition temperature (T_g) of the resist; the temperature of the platform is set below T_g.

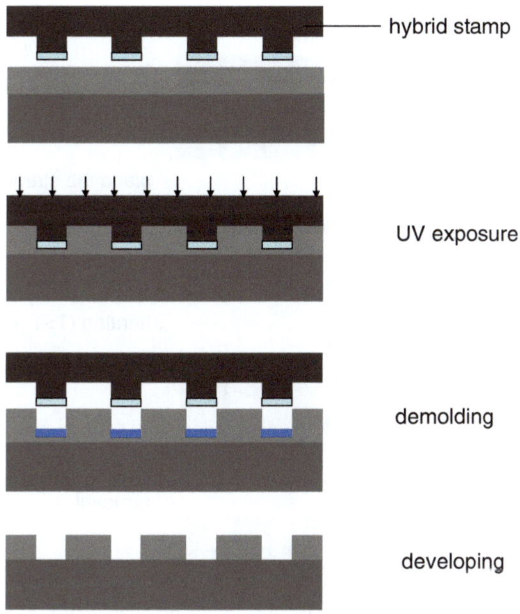

Fig. 6.19 Zero residual layer by CNP process

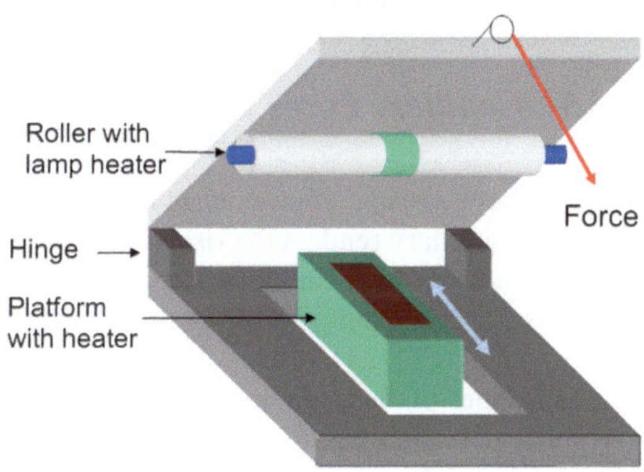

Fig. 6.20 Schematic of a roller nanoimprint system (Reprinted with permission from [23]. Copyright (1998) American Vacuum Society)

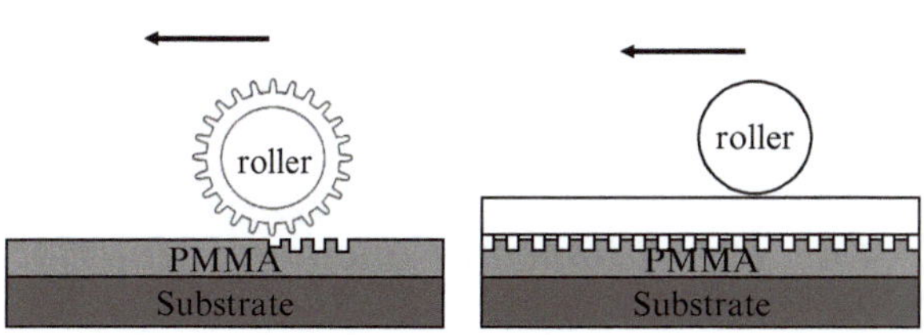

Fig. 6.21 Two different types of RNIL

Because of the advantage of RNIL, many institutes are dedicated to the process and application. The other RNIL process will be introduced in the section of large area imprint lithography.

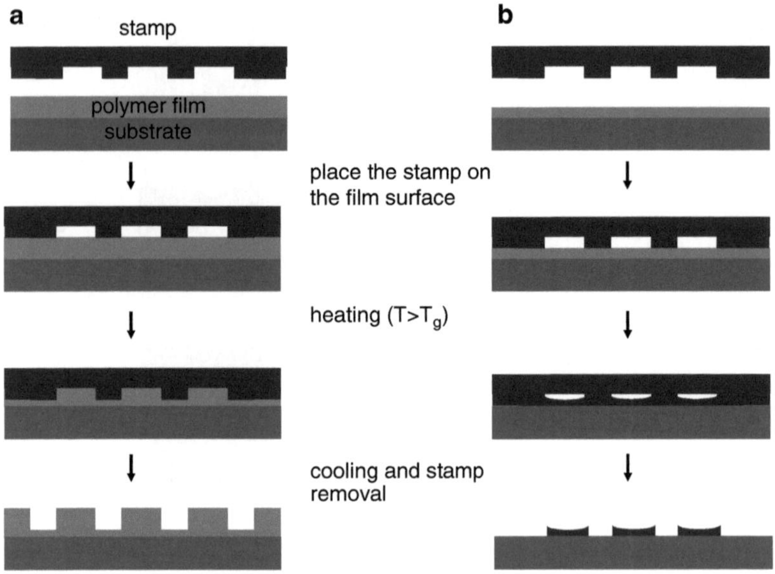

Fig. 6.22 Schematic diagram of capillary force lithography (CFL) (**a**) when the film thickness is relatively thick with respect to the mold step height and (**b**) when it is thin

6.4.5 Capillary Force Lithography

In the solvent-assisted microcontact molding process, the polymer is wetted by solvents and capillary renders the rise of the soften liquid. Capillary force lithography is a method in which capillary forces drive the polymer (at a temperature above the glass transition temperature of the polymer) to conform to the mold pattern, as shown in Fig. 6.22 [24]. If the polymer is relatively thick, the sufficient polymer can fill the cavities of the mold, resulting in a negative copy of the mold, with a residual layer that remains on the substrate. If the film is thin, insufficient polymer cannot completely fill the cavities of the mold, leaving a zero residual layer on the surface. Moreover, the critical dimension of the pattern can be adjusted by the thickness of the polymer.

6.4.6 Direct Imprint Functional Material

Functional materials, including organic and inorganic composite material, sol–gel material, and metal material [25–29], have been directly imprinted, and the as-imprinted patterns can be used for a part of some devices. The patterns do not need the conventional etching process. It is well known that semiconductor nanocrystals or quantum dots (QDs) may have superior optical properties than bulk crystals

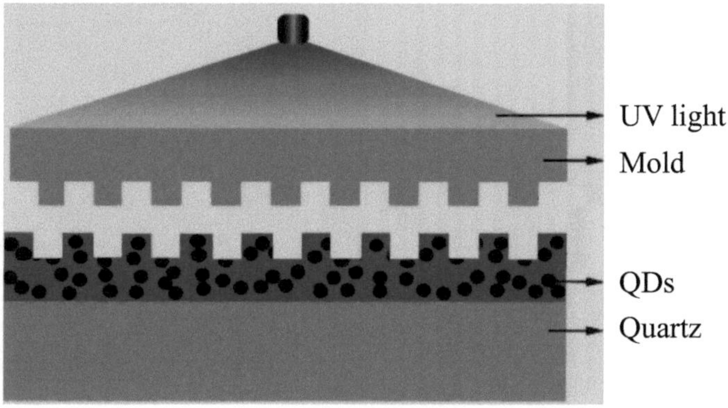

Fig. 6.23 Schematic of functional materials (QDs) patterns imprinted

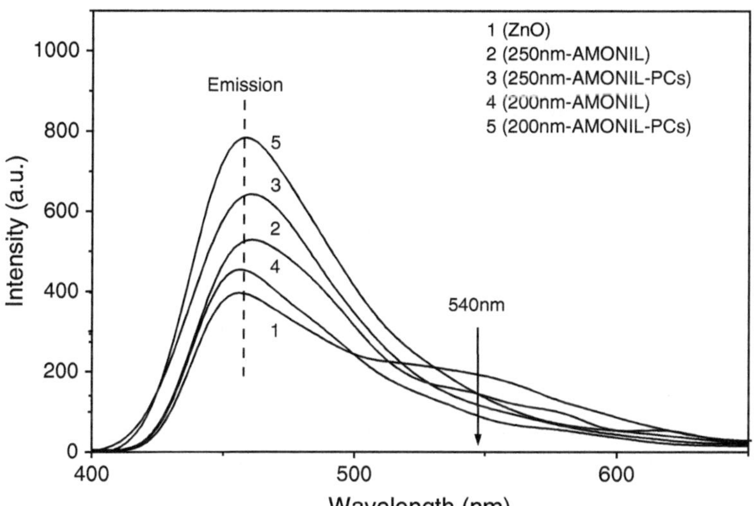

Fig. 6.24 PL spectra of different ZnO QD

owing to quantum confinement effect. We report on the fabrication process in which 2D-PC patterns of the colloidal ZnO QD films were introduced for the first time. The schematic of functional imprinted material is shown in Fig. 6.23.

PL intensity measurement was carried out on the bare colloidal ZnO QD film, ZnO film combining AMONIL resist without imprinted pattern, and the incorporation 2D-PC pattern samples with the AMONIL film thickness of 250 and 200 nm, respectively (Fig. 6.24). It is shown that there is an obvious emission intensity enhancement in the visible range as 2D-PC patterns integrated on. Moreover, the maximum emission output is nearly double as that of the ZnO QD film without PC structure. The result demonstrates a new way to enhance the QDs emission

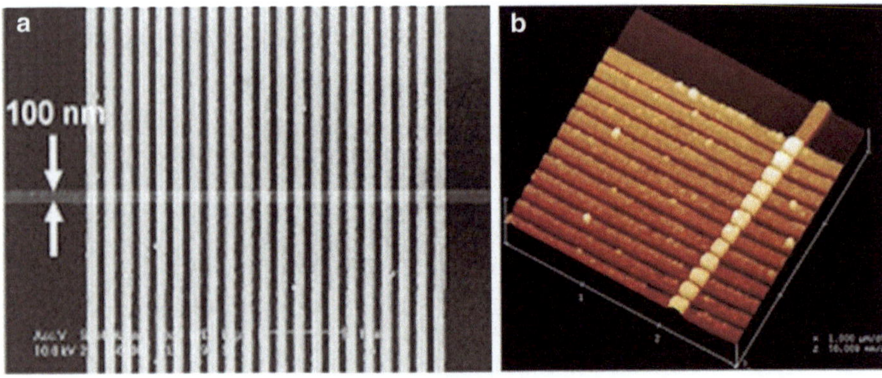

Fig. 6.25 1 × 21 crossbar array. (**a**) SEM image. (**b**) AFM image (Reprinted with permission from [30]. Copyright (2010) American Chemical Society)

with nanostructure, which is no doubt to push the QDs for wide applications, for example, the development of biological tagging methods or the proposed devices incorporating QDs such as LEDs.

6.4.7 Three-Dimensional Nanoimprint Lithography

Three-dimensional (3D) structure in the microsystem field has a very important prospect. Small structure, light weight, high sensitivity, low cost, and superior performance of the device can be fabricated by 3D fabrication technology. It has greatly promoted the development of the product miniaturization and portability, which can increase system density, information density, and interconnection density, and more importantly, it can greatly save materials and energy.

A unique benefit of nanoimprint lithography is the ability to pattern 3D structures. Its unique physical transfer process cannot only reduce costs, with high resolution, but significantly reduce the process of 3D structures. The 3D pattern fabricated by conventional lithography still suffers from these problems: (1) efficiency and precision, (2) limited to some certain products, (3) complexity of process and expensive tools.

3D NIL is a key point in NIL process and application. The fabrication process for 3D structure is still challenging. Application of 3D NIL will bring about a breakthrough of nanofabrication. The crossbar array has been developed by HP Company [30]. 3D structures are imprinted only one step, which has significantly reduced the processes. Figure 6.25 show SEM and AFM images of crossbar array of nanodevices. The area of each device is 100 × 100 nm^2 at cross points.

6.5 Large Area Nanoimprint Lithography

Since NIL was proposed, it has attracted many researchers' interest for its ability to fabricate high resolution patterns with low cost and large area. Many great progresses were made in those years. However, there are still many challenges to overcome for high volume production. One of these key issues is a repeating imprinted pattern on a large scale.

6.5.1 Full Wafer Nanoimprint Lithography

In a full wafer nanoimprint lithography process, all the patterns are defined in a single imprint area and are transferred in a single imprint step.

Nanoimprint lithography can image sandwich structures which the middle layer is the imprinted resist, and top layer and bottom layer are the stamp and the substrate, respectively. Compared to conventional optical lithography, nanoimprint lithography is a mechanical contact of stamp with substrate. A full conformal contact of stamp and substrate is necessary. Usually, the solid plate imprint lithography technique has disadvantages, as shown in Fig. 6.26. (1) Solid two-parallel-plate press has a serious problem of relative shift between the stamp and substrate, nonuniform pressure distribution for the roughness of the two planes. (2) The top plate or the under plate is uneven or an existing particle, stress concentration will damage the stamp or the substrate. (3) Non-parallelism between two plates causes a relative shift between mold and substrate, resulting in a damage to the stamp and the imprinted patterning. (4) Particles on the stamp or the nonuniform thickness of the resist render local contact between the stamp and the resist.

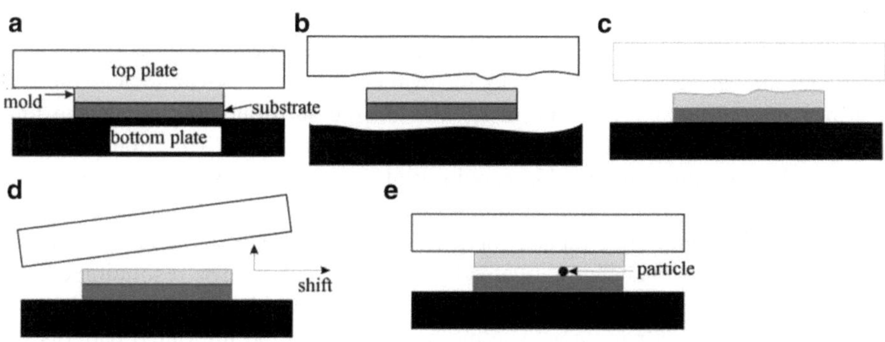

Fig. 6.26 Schematic of the solid parallel-plate imprinting. (**a**) Ideal imprinting. (**b**) Imperfect plate surface. (**c**) Uneven stamp/substrate backside. (**d**) Non-parallelism between solid plates. (**e**) Particle

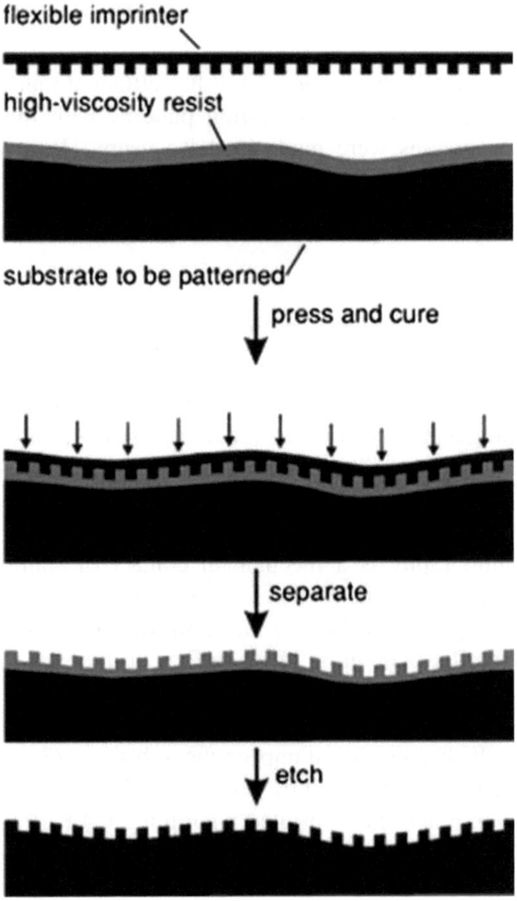

Fig. 6.27 A flexible stamp can achieve the full imprinting and the uniform residual layer thickness

Soft stamp can enable the full area imprinting pattern because of its elastomeric character. It can adapt its shape to the waviness of the substrate and thereby conformably contacts the whole resist film on the substrate. Moreover, the uniform residual layer is achieved on the full wafer and ensures the uniformity of imprinted patterning. The flexible nanoimprint lithography is schematically shown in Fig. 6.27.

The flexible stamp material is a polymer, such as PMDS, PUA, ETFE, and PVC. The flexible stamp has been fabricated by mold replication. The resolution of 2.4-nm features has been demonstrated by the h-PDMS mold based on the replication of single wall carbon nanotube. The mother stamp and the imprinted patterns are shown in Fig. 6.28 [31].

Applying air pressure to press the cushion for contact and imprint to realize ultra-uniformity over the whole imprint field, Philips Research and SUSS MicroTec have developed a novel soft nanoimprint lithography named substrate conformal imprint lithography (SCIL) [32], as shown in Fig. 6.29. SCIL can imprint large area

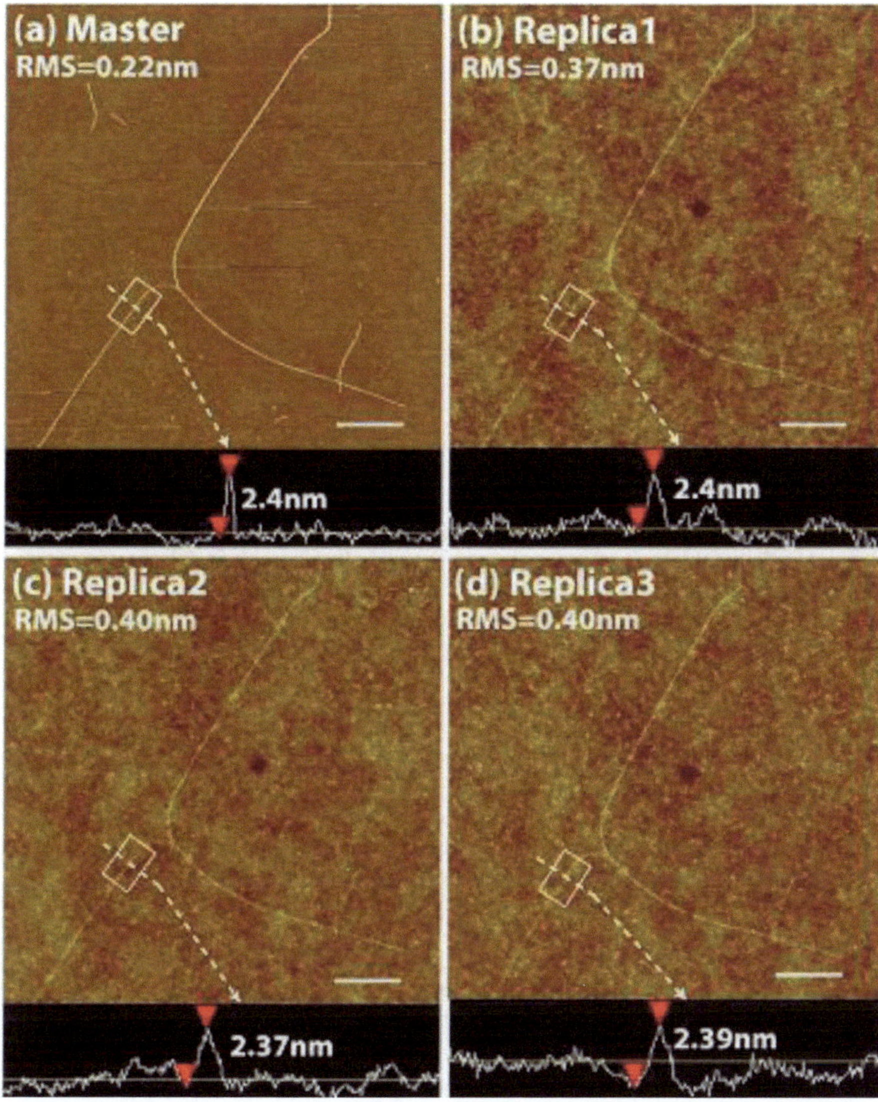

Fig. 6.28 AFM images. (**a**) A master. (**b–d**) Three separate polymer nanostructures imprinted using PMDS mold based on single-walled carbon nanotube (SWNT) replication. The *white scale bars* are 1 μm (Reprinted with permission from [31]. Copyright (2004) American Chemical Society)

patterning up to 6 in. wafer. The SCIL stamp is a composite bilayer (h-PDMS and PDMS), which is attached on a thin glass carrier. The PDMS layer compensates the nonplanarity of substrates and renders a good contact. The thin glass carrier voids lateral deformation. SCIL enables large area full contact little by little by gas pressure. After full contacting, the patterns are cured by UV light. The demolding is

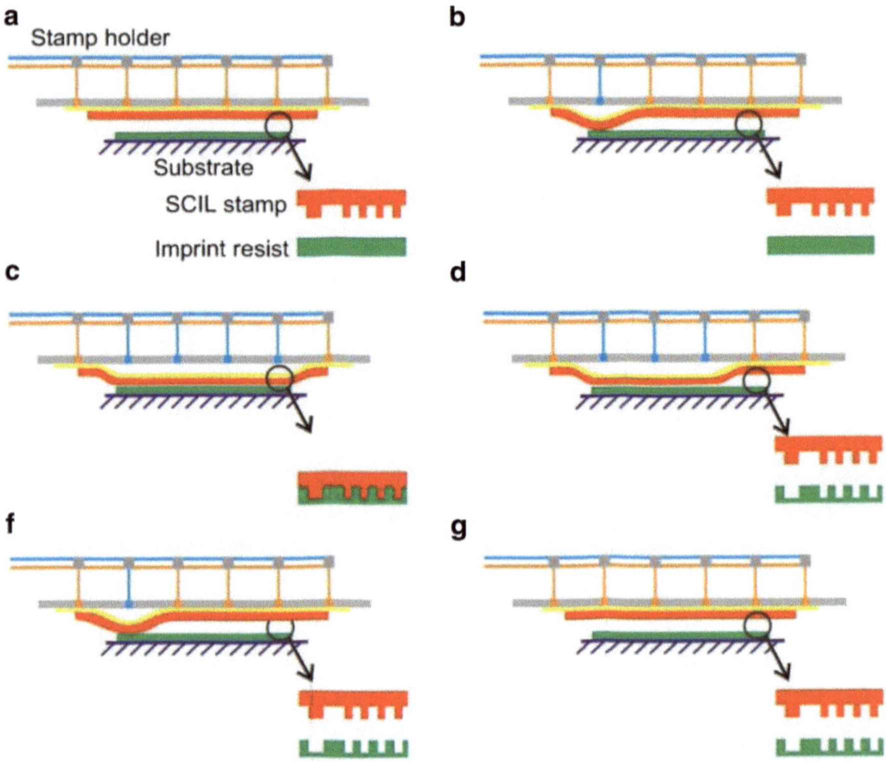

Fig. 6.29 Schematic of SCIL process. (**a**) The SCIL stamp is fixed on the stamp holder by vacuum. (**b**) Imprint process starts from one side of the stamp, then to the other side. (**c**) Imprint on full substrate. The resist is cured. (**d, e**) The demolding process starts from one side of the stamp to the other side. (**f**) Completion of the demolding process (Reprinted with permission from [32]. Copyright (2010) Elsevier)

also from one side to the other side of the stamp by gradually evaluating the vacuum grooves, resulting in a low peeling force that avoids the damage of the pattern of the resist and the stamp.

Dusty particle causes damage to IC industry and regards as "killer." SCIL enables imprint pattern on the substrate containing particles because soft PDMS stamp is not sensitive to particles. Figure 6.30 is a SEM image of SCIL imprint over a particle.

6.5.2 Step and Repeat Nanoimprint Lithography

In1998, the idea proposed by Prof. Willson from University of Texas at Austin originated from step and flash nanoimprint lithography technique. The step and repeat nanoimprint lithography (SRNL) process is shown schematically in Fig. 6.31.

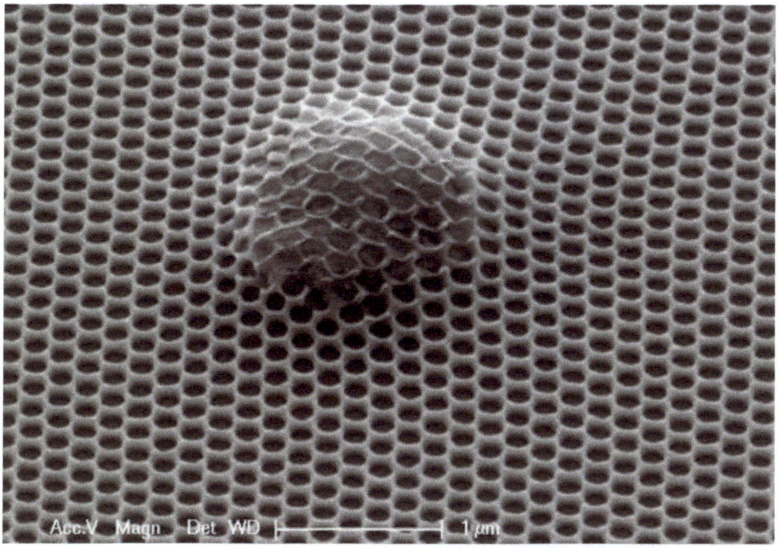

Fig. 6.30 SCIL imprint over a particle (Reprinted with permission from [32]. Copyright (2010) Elsevier)

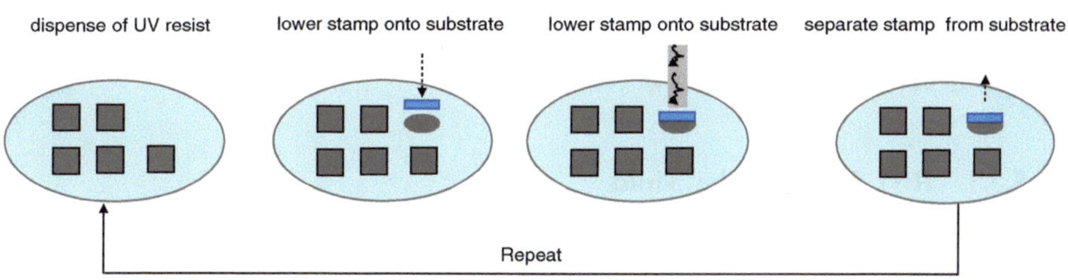

Fig. 6.31 Schematic of step and repeat nanoimprint lithography

A stamp is pressed into a resist to create the first imprint by UV cured .Then, the stamp is lifted and moved to the next site to generate the next pattern and so forth. Finally, the wafer is full of the many times imprinted patterning [33]. The substrate is spin-coated with a resist. The UV exposure is only at every imprinted area. The repeated operations enable a large area patterning with a small area imprint stamp. SRNIL is in principle a relative flexible process which can be sued for multilayer imprinting, thermal imprinting.

6.5.3 Roller Nanoimprint Lithography

In Sect. 6.4.4, roller nanoimprint lithography (RNL) has been introduced. It is a continuous process for imprinting large area patterning: in this section, detailed roller nanoimprint lithography has been demonstrated.

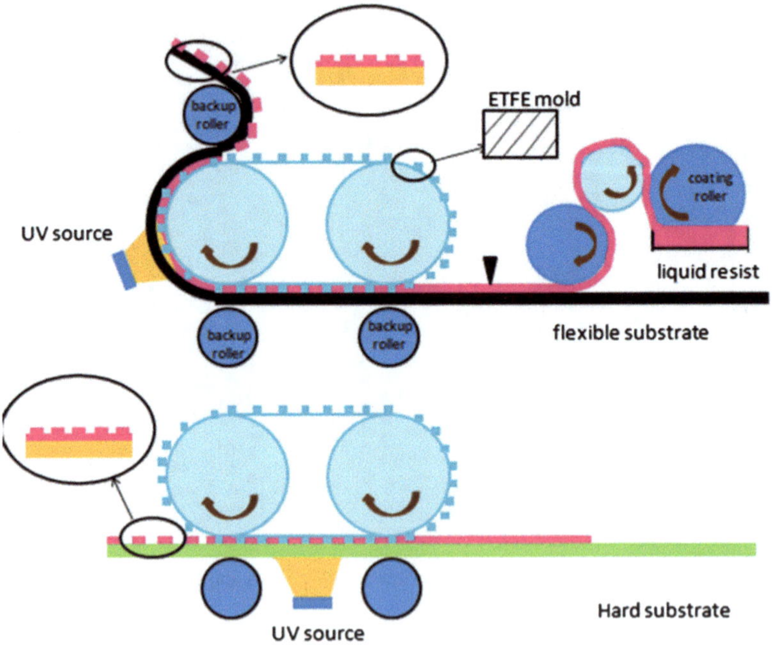

Fig. 6.32 Schematics (**a**) of R2RNIL and (**b**) of R2PNIL processes (Reprinted with permission from [34]. Copyright (2009) American Chemical Society)

6.5.3.1 RNL Using Roller Mold

RNL is widely used in thermal nanoimprint lithography or UV nanoimprint lithography. In the thermal nanoimprint lithography, roller stamp is pressed into the polymer at the temperature over the glass transition criterion. After cooling, the patterns of the roller stamp were transferred into the polymer. In UV nanoimprint lithography, UV light is illuminated at the field where the roller stamp contacts the resist surface. The resist films are cured by UV light.

The fabrication of mold is very important for thermal and UV RNIL. It is a barrier to the development of RNIL.

A continuous roll-to-roll nanoimprint lithography technique has been intensively developed [34]. Large area (4 in. wide) continuous imprinting of nanogratings was fabricated by using a newly developed apparatus capable of roll-to-roll imprinting (R2RNIL) on flexible web and roll-to-plate imprinting (R2PNIL) on rigid substrate. The 300-nm line width grating patterns are continuously transferred on either glass substrate or flexible substrate by these methods.

The schematic of RNIL is shown in Fig. 6.32. Two steps are involved: the coating process and the resist curing process. The resist is coated on the flexible or hard substrate. Stamp material is a flexible fluoropolymer, ethylene tetrafluoroethylene (ETFE).

6.5.3.2 RNIL Using Flat Mold and Smooth Roller

Using flat mold in the RNIL process cannot realize a continuous replication. The gap between the roller and the flat mold must be controlled to achieve good quality of the imprinted patterns. AIST presented a process for large area imprinting, as shown in Fig. 6.33 [35].

The micro-/nanostructures can be replicated into the polymer by heating and roller. An automatic mold releasing mechanism is installed to avoid transferring pattern deformation error by the mismatch of thermal expansion ratio between a polymer substrate and a metal mold.

A continuous rigiflex imprinting (CRIM) to realize pattern replication with nanoscale has been demonstrated by Seoul National University. The mold both has rigid and flexible enough to realize continuous imprinting. The as-imprinted patterns were cured by UV light after the continuous imprinting [36].

6.6 Alignment

Nanoimprint lithography is widely used in the fabrication of nanodevices, making it a competitive candidate for next generation lithography. Overlay is a critical technique for nanoimprint lithography, and alignment must be very accurate.

Alignment for nanoimprint lithography differs from that of conventional photolithography. In optical lithography, the mask is a noncontact process. However, nanoimprint lithography is a contact process. Usually, in hot nanoimprint lithography, the marks on the stamp backside are first fabricated. The marks on the substrate is recorded and stored in a computer. Then, the stamp is moved to overlap. The submicrostructures can be aligned. Or Alignment camera is installed between stamp and substrate. The two methods are pre-alignment technique. However, misalignment often takes place. One main reason is a mechanical shift between the stamp and the substrate during the imprinting process. Non-parallelism between two plates causes a relative shift between mold and substrate, and other particles can cause stress concentration and local deformation. Another reason is the difference in thermal expansion coefficient between stamp and substrate, especially the solid plate of different materials. In all, alignment remains a key issue for nanoimprint lithography.

There are several possible alignment methods: direct imaging, amplitude-sensitive, and phase-sensitive methods.

Figure 6.34 shows the schematic of direct imaging alignment. The image of the stamp alignment marks is stored, and then the stamp is lowered to close to the substrate (wafer) until the marks are clear in microscope field. The alignment marks on the substrate are aligned to the stored stamp alignment marks. The method is used for micro-patterning imprinting. Alignment accuracy is $\sim$1 um, and usually a few microns. With the submicro alignment, multilevel nanoimprint lithography could be achieved over 4 in. Si wafer [37].

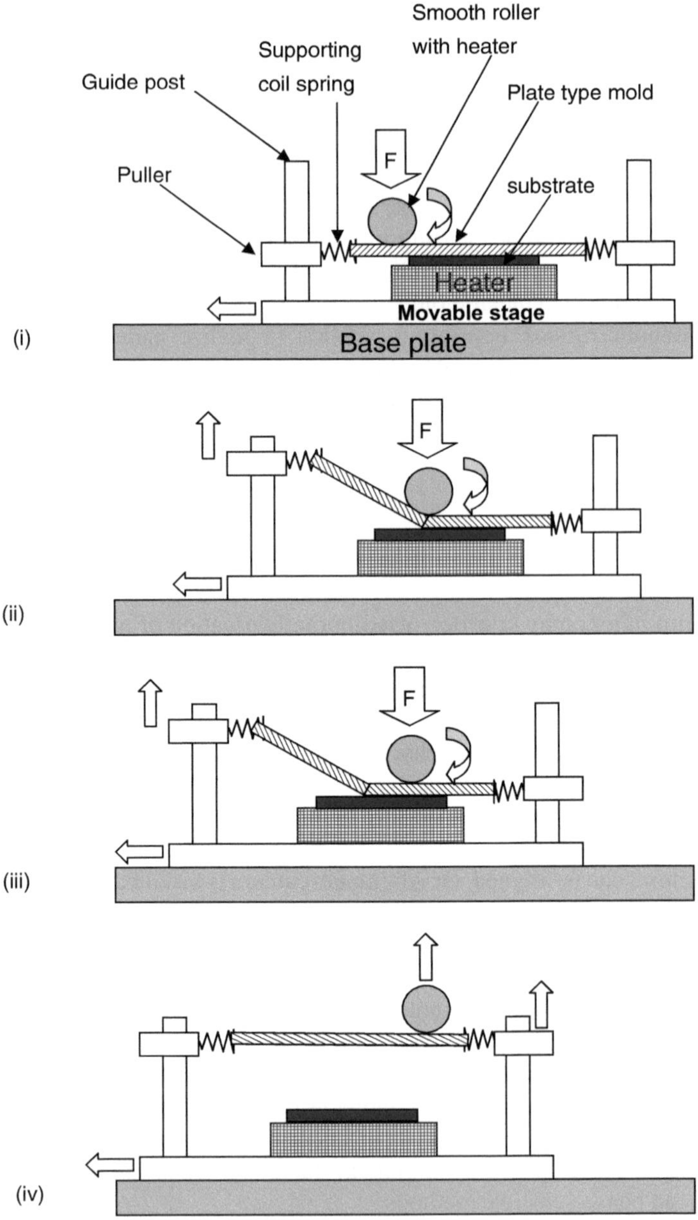

Fig. 6.33 The schematic of flat mold and smooth roller

Interferometric alignment technique is also used for nanoimprint lithography. The technique has been previously developed for X-ray lithography. The theoretical overlay accuracy is down to the sub-20 nm [38]. The interferometric alignment method is based on phase shift when different arms of a light beam propagate

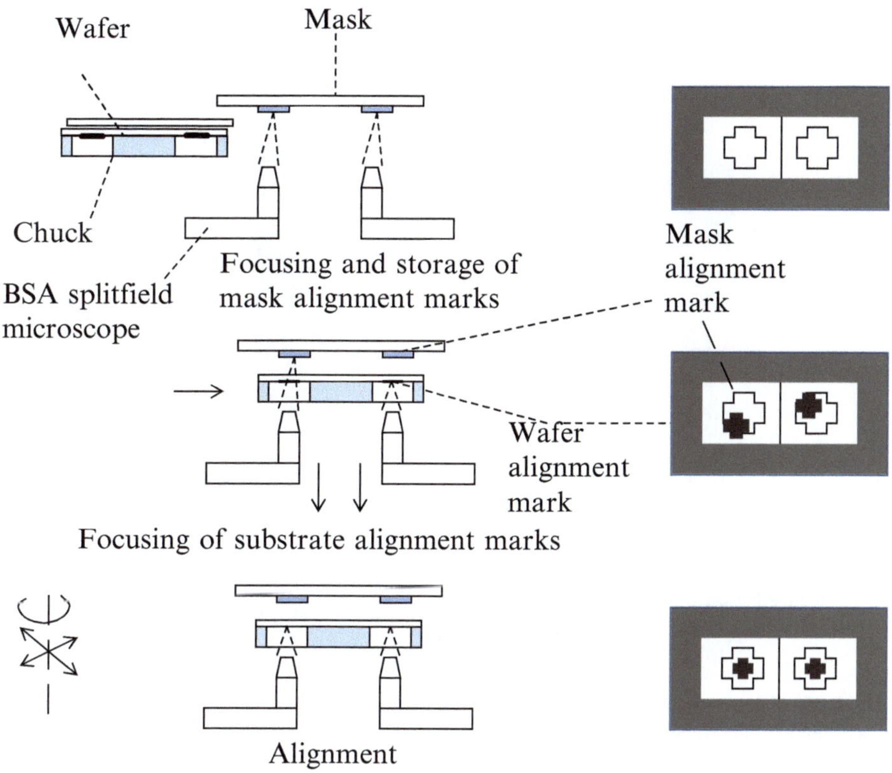

Fig. 6.34 The schematic of submicron alignment

through the mediums of different diffraction coefficients. If the detectors can show equal intensities, which means the gratings are oriented highly symmetrically, their alignment is perfect.

Moiré patterns are used in the high-resolution alignment. It is an interference pattern when two grids are overlaid at an angle or have slightly different periods. The period length of *moiré* pattern is given as [39]

$$P_{\mathrm{m}} = \frac{P_1 P_2}{P_1 - P_2},$$

where P_1 and P_2 are the grating period on substrate and on stamp, respectively. The interference pattern has much larger period length than that of the alignment grating. A misalignment Δ between stamp and substrate will cause a distance of $d = \Delta(P_{\mathrm{M}}/P_2)$. A little shift will create larger patterns, which is easy to see from an optical microscope. Sub-20-nm alignment accuracy has been completed by this fine alignment technique. Figure 6.35 is a video image taken during alignment of substrate and stamp [39].

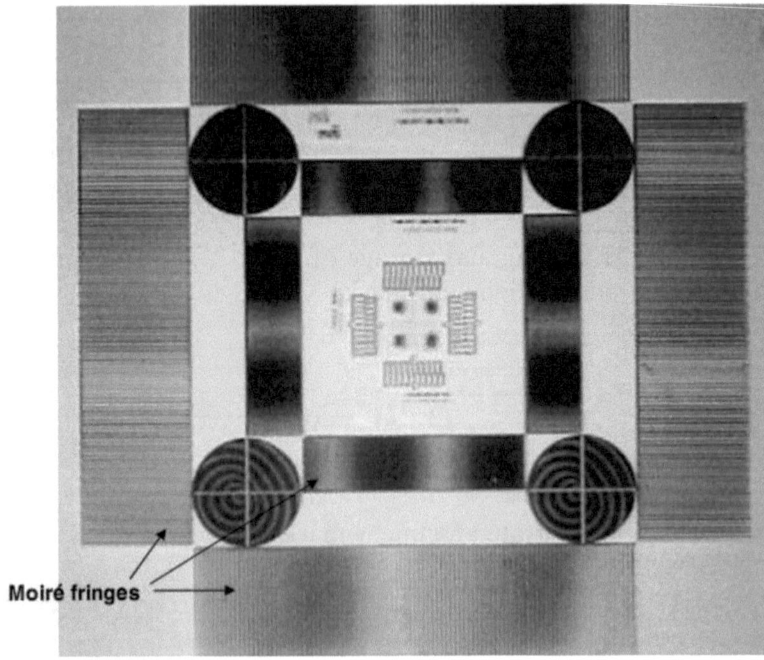

Fig. 6.35 Video image taken during alignment, showing interference dark fringes of moiré pattern (Reprinted with permission from [39]. Copyright (2007) Elsevier)

6.7 Pattern Transfer

The pattern transfer process consists of two steps: lithographic resist patterning and the subsequent etching. Etching process can be done by chemical method (means wet etching), physical damage (means dry etching), or some combination of the two. Wet chemical etching is usually characterized by low anisotropy and high selectivity. However, chemical etching is difficult to control the etch rate, which is prone to high defect levels due to solution particulate contamination. The dry etching is characterized by a high degree of anisotropy, but the etch rate is nearly independent of the substrate material.

Three steps must take place during etching process, namely, transport of etchants to surface, surface reaction, and removal of product species. The basic reactions in etching are as follows:

1. Wet etching

$$\text{Solid} + \text{liquid etchant} \rightarrow \text{soluble products}$$

$$Si\,(s) + 2OH^- + 2H_2O \rightarrow Si(OH)_2(O^-)_2\,(aq) + 2H_2\,(g)$$

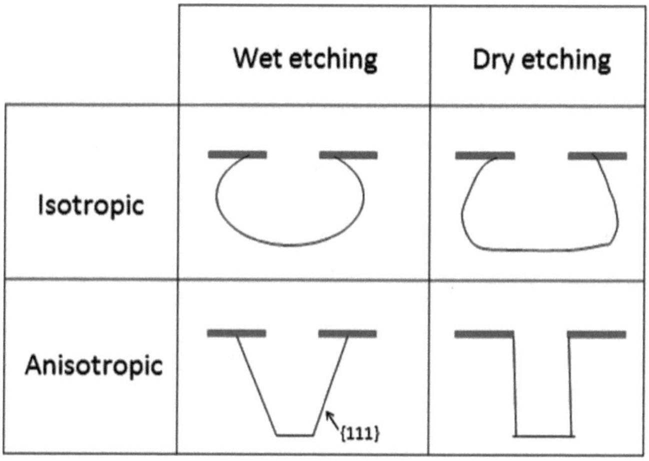

Fig. 6.36 Schematic illustration of cross-sectional trench profiles resulting from four different types of etch methods

2. Dry etching

$$\text{Solid} + \text{gaseous etchant} \rightarrow \text{volatile products}$$

$$SiO_2\ (s) + CF_4\ (g) \rightarrow SiF_4\ (g) + CO_2\ (g)$$

Isotropic etchants etch uniformly in all directions, which results in rounded cross-sectional features. By contrast, anisotropic etchants etch in some directions preferentially over others, leading to trenches or cavities delineated by flat and well-defined surfaces, which need not be perpendicular to the surface of the wafer, as shown in Fig. 6.36.

6.7.1 *Wet Etching*

Wet etching is used for cleaning, removing surface damage, polishing, shaping 3D structures, and characterizing structural and compositional features. The most important parameters in chemical etching include bias (undercut), tolerance, etch rate, anisotropy, over-etch, selectivity, feature size control, and loading effects.

6.7.1.1 Isotropic Wet Chemical Etching

The etching rate and the resulting etched feature depend on many things: the type of substrate, the specific chemistry of the etchant, the choice of masking layer, the temperature, and whether or not the solution is well stirred (which affects the rate of arrival of fresh reactants at the surface). Depending on the temperature and mixing conditions, the etching reaction can be either reaction-rate controlled, dominated by temperature, or mass-transfer limited, determined by the supply of reactants or the rate of removal of reaction products.

Most wet etching is isotropic, in that the etching rate does not depend on the orientation of the substrate. The most common group of silicon isotropic wet etchants is HF/nitric/acetic acid (HNA). It is a mixture of hydrofluoric (HF), nitric (HNO_3), and acetic (CH_3COOH) acids, although water may replace the acetic acid. In the chemical reaction, the nitric acid oxidizes silicon, which is then removed by the hydrofluoric acid. The etch rate of silicon can vary from 0.1 to over 100 μm/min, which depends on the proportion of the acids in the mixture. Etch uniformity is normally difficult to control but is improved by stirring [40].

6.7.1.2 Anisotropic Wet Chemical Etching

When etching a single crystal, certain etchants exhibit orientation-dependent etch (ODE) rates. Specifically, strong bases, such as potassium hydroxide (KOH), tetramethyl ammonium hydroxide (TmAH), and ethylenediamine pyrocatechol (EDP), exhibit highly orientation-dependent etch characteristics in silicon [41]. Hydrofluoric acid exhibits similar orientation-dependent effects when etching single-crystal quartz [42].

KOH is by far the most common ODE. Etch rates typically happen in the [100] direction, corresponding to the etch front being the (100) plane. The etching rate of {110} planes are in KOH about twice as rapidly as {100} planes, while {111} planes are etched at a rate about 100 times slower than for {100} planes [43]. The latter feature is routinely used to form V-shaped grooves and trenches in (100) silicon wafers, which are precisely delineated by {111} crystallographic planes.

Because anisotropic etches require long exposure to the etchant, the etch rate ratio of the fast-etching planes to the {111} plane is extremely important. The effect of the masking layer on the etch rate is also important. Possible masking materials for KOH are silicon nitride, which hardly etches at all, SiO_2, which etches at about 1.4 nm/min, and p + silicon, which offers between a 10:1 and 100:1 reduction in etch rate over lightly doped silicon, depending on the etchant and the etch temperature. As an example, to etch all the way through a 500 μm thick wafer, which requires about 6 h in 20% KOH at 85°C, would require at least a 504 nm of silicon dioxide as a masking layer [44].

6.7.2 *Dry Etching*

Currently, almost all dry etching method is performed based on plasma, that is to say, dry etching is also called plasma etching. Plasma etching is a key process

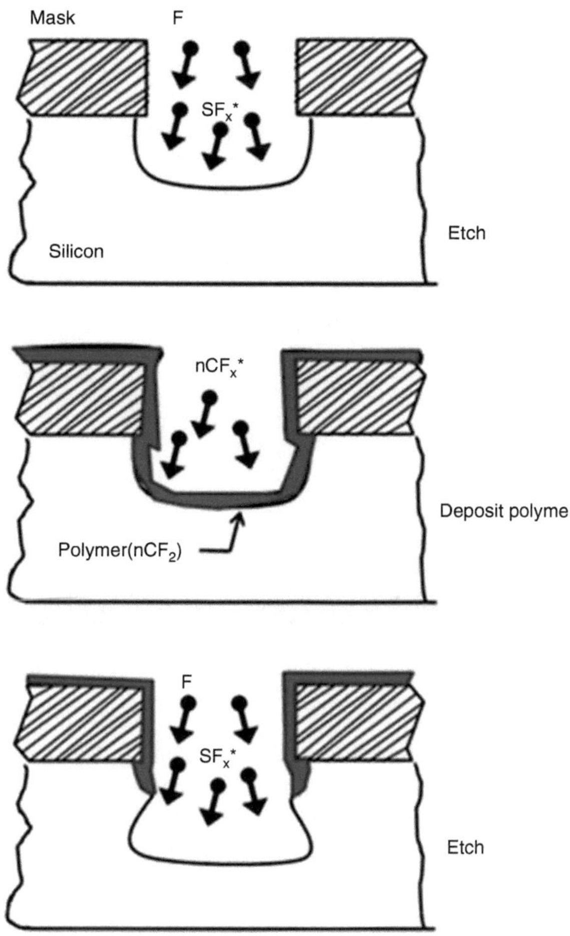

Fig. 6.37 DRIE process with the Bosch process. The process cycles between an etch step using SF6 gas and a polymer deposition step using C_4F_8. The polymer protects the sidewalls from etching by the reactive fluorine radicals. The scalloping effect of the etch is exaggerated

in the semiconductor industry. Plasma etching has several significant advantages compared to wet etching. Plasmas are much easier to start and stop than simple immersion wet etching. Furthermore, plasma etch processes are much less sensitive to small changes in the temperature of the wafer. These two factors make plasma etching more repeatable than wet etching. Most important for small features, plasma may have high anisotropic etching. Plasma environments may also have far fewer particles than liquid media. Finally, plasma etch process produces less chemical waste than wet etching.

Plasma etching generates chemically reactive neutrals (e.g., F, Cl) and ions (e.g., SF_x^+) that are accelerated under the effect of an electric field toward a target substrate. The collision of molecules happens in a reactant gas (e.g., SF_6, CF_4, Cl_2, $CClF_3$, NF_3) with a cloud of energetic electrons excited by an RF electric field, which results in the reactive species (neutrals and ions). When the etch process is purely chemical, powered by the spontaneous reaction of neutrals with silicon, it is

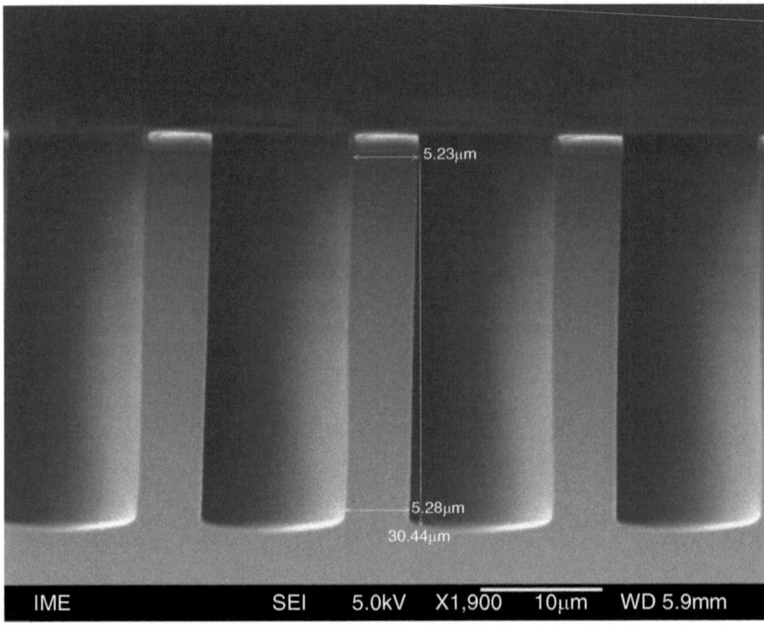

Fig. 6.38 SEM image of features etched in silicon with DRIE

colloquially referred to as plasma etching. But if ion bombardment of the silicon surface plays a synergistic role in the chemical etch reaction, the process is then referred to as reactive-ion etching (RIE) [40].

A dramatic and commercially highly important variant of reactive-ion etching is a relatively new process called deep reactive-ion etching (DRIE). This process has the advantage of a side effect of a glow discharge, the tendency to create polymeric species by chemical cross-linking. In fact, most plasma processes are a critical race between deposition of polymeric material from the plasma and the removal of material from the surface.

Figure 6.37 shows the DRIE process. A photoresist mask can be used. For very deep etches, a combination of photoresist and oxide may be necessary. The etch proceeds in alternating steps of reactive-ion etching in a SF_6 plasma and polymer deposition from a C_4F_8 plasma. During the etch process, the polymer is rapidly removed from the bottom of the feature but lingers on the sidewall, protecting it from the SF_6 etchant. As a result, the silicon beneath the first cut is removed during the second etch cycle, but the top of the feature does not become wider. Eventually the polymer protecting on the sidewalls is eroded. At that point, another polymer deposition step is used, and the cycle is repeated. Finally, a deep trench feature is obtained [40].

Figure 6.38 shows a photograph of deep features etched into silicon with the Bosch process [45]. DRIE is a powerful tool to form deep trenches with near-vertical sidewalls. However, process development is required to develop and optimize good uniformity, high verticality, high speed, small scalloping, and small footing. In general, all factors cannot be considered simultaneously. Sequentially running different processes or slowly tuning the process parameters may be done for the best result.

References

1. Chou SY, Krauss PR, Renstrom PJ (1995) Imprint of sub-25 nm vias and trenches in polymers. Appl Phys Lett 67:3114–3116
2. Perez J (1998) Physics and mechanics of amorphous polymer. A.A Balkema, Rotterdam
3. Chou S, Krauss P, Zhang W, Guo L, Zhuang L (1997) Sub-10 nm imprint lithography and applications. J Vac Sci Technol 15:2897–2904
4. Haisma J, Verheijen M, van den Heuvel K, van den Berg J (1996) Mold-assisted nanolithography: a process for reliable pattern replication. J Vac Sci Technol B 14:4124–4128
5. Landis S, Chaix N, Gourgon C, Perret C, Leveder T (2006) Stamp design effect on 100 nm feature size for 8 inch nanoimprint lithography. Nanotechnology 17:2701–2709
6. Xia YN, Whiteside GM (1998) Soft lithography. Angew Chem Int Ed 37:550–575
7. Whiteside GM (1994) Microfabrication by microcontact printing of self-assembled monolayers. Adv Mater 6:600–604
8. Xia Y, Ki E, Zhao XM, Rogers JA (1996) Complex optical surfaces formed by replica molding against elastomeric masters. Science 273:347–349
9. Colburn M, Johnson S, Stewart M, Damle S, Bailey T, Choi B (1999) Step and flash imprint lithography: a new approach to high-resolution patterning. Proc SPIE 3776:379–389
10. Zhao XM, Smith SP, Waldman SJ, Whiteside GM (1997) Demonstration of waveguide couplers fabricated using microtransfer molding. Appl Phys Lett 71:1017–1019
11. Kim E, Xia Y, Whitesides GM (1995) polymer microstructures formed by moulding in capillaries. Nature 376:581–584
12. King E, Xia Y, Zhao XM, Whitesides GM (2004) Solvent-assisted microcontact molding: a convenient method for fabricating three-dimensional structures on surfaces of polymers. Adv Mater 9:651–654
13. Jeon S (2004) Three-dimensional nanofabrication with rubber stamps and conformable photomasks. Adv Mater 16:1369–1371
14. Rogers JA, Paul KE, Jackman RJ, Whitesides GM (1997) Using an elastomeric phase mask for sub-100 nm photolithography in the optical near field. Appl Phys Lett 70:2658–2660
15. Childs WR, Nuzzo RG (2002) Decal transfer microlithography: a new soft-lithographic patterning method. J Am Chem Soc 124:13583–13596
16. Xu Q, Rioux RM, Dickey MD, Whitesides GM (2008) Nanoskiving: a new method to produce arrays of nanostructures. Acc Chem Res 41:1566–1577
17. Chou SY, Keimel C, Gu J (2002) Ultrafast and direct imprint of nanostructures in silicon. Nature 417:835–883
18. Geogge A, Blank DH, Ten Elshof JE (2009) Nanopatterning from the gas phase: high resolution soft lithographic patterning of organosilane thin films. Langmuir 25(23):13298–13301
19. Huang XD, Bao LR (2002) Reversal imprinting by transferring polymer from mold to substrate. J Vac Sci Technol B 20:2872–2876
20. Cui Z (2008) Nanofabrication. Springer, Heidelberg
21. Cheng X, Guo LJ (2004) One-step lithography for various size patterns with a hybrid mask mold. Microelectron Eng 71:288–293
22. Guo LJ (2004) Recent progress in nanoimprint technology and its applications. J Phys D: Appl Phys 37:R123–R141

23. Tan H, Gilbertson A, Chou SY (1998) Roller nanoimprint lithography. J Vac Sci Technol B 16:3926–3928
24. Suh KY, Kim YS, Lee HH (2001) Capillary force lithography. Adv Mater 13:1386–1389
25. Inkyu P, Heng P, Seung H, Albert P (2008) Micro/nanoscale structure fabrication by direct nanoimprinting of metallic and semiconducting nanoparticles. In: Proceedings of the ASME International Mechanical Engineering Congress and Exposition, Seattle, 2007, pp 307–314
26. Choi JH, Lee SW, Choi DG, Kim KD (2008) Direct UV-imprint lithography using conductive nanofiller-dispersed UV-curable resin. J Vac Sci Technol B 26:1390–1394
27. Choi JH, Lee SW, Kim KD, Chio DG (2009) Nanosilver particles-based conductive patterns developed by direct soft imprint lithography. Current Applied Physics 9:S138–S140
28. Vieira SMC, Teo KB, Milne WI, Gröning O, Gangloff L, Minoux E (2006) Investigation of field emission properties of carbon nanotube arrays defined using nanoimprint lithography. Appl Phys Lett 89:12301–12303
29. Wang PT, Guo JB, Wang HH, Zhang Y, Wei J (2009) Functionalized multi-walled carbon nanotubes filled ultraviolet curable resin nanocomposites and their applications for nanoimprint lithography. J Phys Chem C 113:8118–8123
30. Xia QF, Yang JJ, Wu W (2010) Self-aligned memristor cross-point arrays fabricated with one nanoimprint lithography step. Nano Lett 10(8):2909–2914
31. Hua F, Sun Y, Gaur A, Meitl MA, Bilhaut L, Rotkina L, Wang JF, Geil P, Shim M, Rogers JA (2004) Polymer imprint lithography with molecular-scale resolution. Nano Lett 4(12):2467–2471
32. Ran J, Michael H, Verschuurenb MA, Van de Laar R, Van Eekelen J, Ulrich P, Michael M, Christian M (2010) UV enhanced substrate conformal imprint lithography (UV-SCIL) technique for photonic crystals patterning in LED. Microelectron Eng 87:963–967
33. McMackin I, Choi J, Schumaker P, Nguyen P (2004) Step and repeat UV nanoimprint lithography tools and processes. Proc SPIE 5374:222–231
34. Ahn SH, Guo LJ (2009) Large-area roll to-roll and roll-to-plate nanoimprint lithography: a step toward high-throughput application of continuous nanoimprinting. ACS Nano 3:2304–2310
35. Ratchev S (2010) Precision assembly technologies and systems: 5th IFIP WG 5.5 international precision assembly seminar. Springer, Heidelberg
36. Choi SJ, Yoo PJ, Beak SJ, Kim TW, Lee HH (2004) An ultraviolet-curable mold for sub-100-nm lithography. J Am Chem Soc 126:7744–7746
37. Zhang W, Chou SY (2001) Multilevel nanoimprint lithography with submicron alignment over 4 in. Si wafers. Appl Phys Lett 79:845–847
38. Fuchs A, Vratzov B, Wahlbrink B, Georgiev Y, Kurz H (2004) Interferometric in situ alignment for UV-based nanoimprint. J Vac Sci Technol B 22:3242–3245
39. Muhlberger M (2007) A Moiré method for high accuracy alignment in nanoimprint lithography. Microelectron Eng 84:925–992
40. Maluf N, Williams K (2004) An introduction to microelectromechanical systems engineering. Artech House Inc., London
41. Kovacs GTA, Maluf NI, Petersen KE (1998) Bulk micromachining of silicon. Proc IEEE 86:1536–1551
42. Hedlund C, Lindberg U, Bucht U, Soderkvist J (1993) Anisotropic etching of Z-cut quartz. J Micromech Microeng 3:65–73
43. Seidel H (1990) Anisotropic etching of crystalline silicon in alkaline solutions. J Electrochem Soc 137:3612–3632
44. Senturia SD (2002) Microsystem design. Kluwer Academic Publishers, New York
45. Larmer F, Schilp P (1994) Method of anisotropically etching silicon. German Patent DE 4241045

Chapter 7
Modeling and Simulation of NIL

7.1 Introduction

Nanoimprint lithography is a method for fabricating nanoscale patterns by pressing stamp into viscous materials. To enhance a better understanding of nanoimprint lithography process, NIL simulation provides an efficient and accurate use of nanoimprint lithography. By simulation, we can know how to design the stamp, control the process for nanoimprint lithography, and decrease the defect of imprinted patterns.

As we know, in thermal nanoimprint lithography process, the temperature and holding time are two parameters to define the quality of imprinted patterns. The optimized force and time must be selected. Similarly, the UV intensity and holding time must also be chosen for UV nanoimprint lithography. The filling process, residual layer thickness, and demolding process can also be evaluated. Therefore, not only can simulation be used to guide the selection of process parameter, it can be used to help optimize the design itself and controlling of processes.

There are a lot of experimental works in the published papers that describe nanoimprint lithography process optimized parameters [1–5]. Much of this work made an attempt to elucidate the results through numerical simulation. Finite element approaches, coarse grain approaches, contact mechanics-based approaches, and molecular dynamics approaches are widely used for understanding of the related processes.

Nowadays, some software have been developed for nanoimprint lithography. Simprint Core, one of those softwares, enables the users to obtain a fast, full-chip simulation of residual layer thickness and stamp cavity filling in thermal NIL. It is at least 1,000 times faster than conventional (e.g., finite element-based) simulation approach and balances simulation speed with spatial resolution according to the specific requirements. For detailed information, please refer to the website [6].

Understanding the nanoimprint lithography will facilitate high-throughput streaming of product manufacturing at high quality and low cost. The simulation results are to provide process guidelines and design rules for industrial application.

W. Zhou, *Nanoimprint Lithography: An Enabling Process for Nanofabrication*,
DOI 10.1007/978-3-642-34428-2_7, © Springer-Verlag Berlin Heidelberg 2013

In this chapter, some modeling and simulation cases for nanoimprint lithography have been demonstrated, which may provide the readers an insight into the principles of nanoimprint lithography.

7.2 Polymer Flow

7.2.1 Flow Mechanism

An imprint pattern with good fidelity is that the pattern should maintain its mechanical integrity after stamp-resist separation. First of all, the resist should fully fill the cavity of the stamp. Understanding of polymer flow is necessary to elucidate the mechanism.

As we know, liquid can flow and solid can deform. Hooke's law states the elasticity behavior of solid. All fluids are compressible to some extent, resulting in changes of density. If the changes are so sufficiently small that its density is negligible, the fluid can be modeled as an incompressible flow. It is usually expressed by Newton's law of viscosity:

$$\tau = \mu \frac{\mathrm{d}\theta}{\mathrm{d}t} \tag{7.1}$$

where τ is shear stress, μ is liquid viscosity, and $\frac{\mathrm{d}\theta}{\mathrm{d}t}$ is strain rate.

In fact, the polymer flow is a complex process. The elasticity and viscous flow take place once the force is used. Once the force is removed, only elasticity section can be a recovery.

A thermoplastic material is used as the hot imprint resist. When the material is heated above its glass transition temperature (T_g), it becomes a viscous liquid and can flow. The Young's modulus and the viscosity decrease by several orders of magnitude compared to their values at room temperature. At a temperature below T_g, the major contribution to the deformation comes from the elongation of the atomic distance, and the deformation is ideally elastic. At a temperature above T_g, local motion of chain segments takes place. When the temperature is further increased, the viscous liquid flow state is reached. The polymer flows by chain sliding, and the deformation is irreversible. T_g is the onset temperature for molecular motion of polymers. The viscosity of a polymer material depends on the temperature and the polymer's molecular weight (M_w). The polymer molecular weight is critical because it determines its physical properties (T_g, stiffness, strength, viscoelasticity, and viscosity).

The relation of the number-average molecular weight and the glass transition temperature is followed by the Flory-Fox equation, as shown below:

$$T_g = T_g(\infty) - \frac{K}{M} \tag{7.2}$$

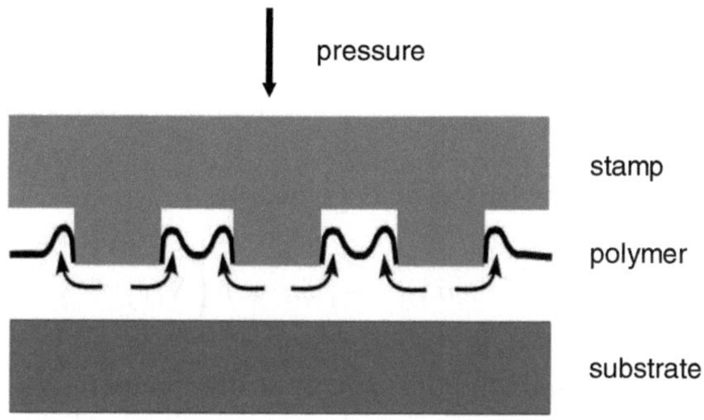

Fig. 7.1 Polymer flow during nanoimprint process

where T_g is the glass transition temperature of the molecular weight M, $T_g(\infty)$ is the glass transition temperature of an infinite molecular weight polymer, and K is constant.

Liquid viscosity tends to decrease as its temperature increase. The temperature dependence of liquid viscosity is usually expressed by the following model:

$$\mu(T) = \mu_0 \exp\left(\frac{E}{RT}\right) \tag{7.3}$$

where T is temperature, μ_0 is a coefficient, E is the activation energy, and R is the universal gas constant.

When the stamp is contacted on the polymer film by force, the polymer flows along the stamp as shown in Fig. 7.1. The film tends to be compressed and extruded from the bottom of the structure. Shear deformation takes place near the stamp surface and corner, and the polymer flows along the sidewall of the stamp due to the friction force and fills the cavity of the stamp. After curing, the polymer is separated from the stamp. In the process, the parameters such as polymer thickness, process time, temperature, and cavity size are responsible for the deformation of the polymer.

7.2.2 Navier–Stokes Equation

Fluid mechanics is described by Navier–Stokes. Newton's second law was applied to a volume element of fluid and presented by [7].

$$m\dot{v} = \sum_i F_i. \tag{7.4}$$

For a unit volume, it is rewritten as

$$\rho\dot{v} = \frac{1}{V}\sum_i F_i = \sum_i f_i \tag{7.5}$$

where f_i is an effect density of force from the force F_i acting on the volume V.

The time derivative of the velocity vector is given by

$$\dot{v} = \frac{\partial v}{\partial t} + (v \cdot \nabla)v. \tag{7.6}$$

The stress tensor, σ, is given by

$$\sigma_{ij} = -p\delta_{ij} + \eta\left(\nabla_i v_j + \nabla_j v_i - \frac{2}{3}\nabla \cdot v\delta_{ij}\right) \tag{7.7}$$

where $-p\delta_{ij}$ is the pressure contribution along the coordinate axis, $\nabla_i v_j + \nabla_j v_i$ are the viscous contribution due to velocity gradient, and $-\frac{2}{3}\nabla \cdot v\delta_{ij}$ is tensile stress from compressible fluid.

The force density is given by

$$\sum_i f_i = f_{\text{body}} + \sum_i \nabla_j \sigma_{ij}. \tag{7.8}$$

Combining the above-mentioned equations, we get

$$\rho\left(\frac{\partial v}{\partial t} + (v \cdot \nabla)\right) = -\nabla p + \eta\nabla^2 v + \frac{1}{3}\eta\nabla(\nabla \cdot v) + f_{\text{body}}. \tag{7.9}$$

Equation 7.9 is called Navier–Stokes equation.

For an incompressible fluid, $\nabla \cdot v = 0$, Navier–Stokes equation is rewritten as

$$\rho\left(\frac{\partial v}{\partial t} + (v \cdot \nabla)\right) = -\nabla p + \eta\nabla^2 v + f_{\text{body}}. \tag{7.10}$$

If a thin film is considered, all body forces can be neglected. The Navier–Stokes equation is given by

$$\nabla p = \eta\nabla^2 v. \tag{7.11}$$

7.3 Simulation Method

7.3.1 Finite Element Method

At present, the commonly used numerical simulation methods in the engineering field are finite element method, boundary element method, discrete element method, and finite difference method. But considering the practicality, the most widely used method is finite element method (FEM).

The basic idea of the FEM is to find the solution to a complicated problem by replacing it with a simpler one. Since the actual problem is replaced by a simpler one to find the solution, we will be able to find only an approximate solution rather than the exact solution. When mathematical tools cannot be sufficient to find solutions of practical problems, it will be preferred to find an approximate solution. In the FEM, the solution region is considered as built up of many small, interconnected subregions called finite elements [8].

From a physical view, in FEM, a nonindividual body can be replaced by a connected combination unit, which connects with each other only on the notes of the elements. Namely, the ready-for-analysis nonindividual body can be divided into several connected units. From the feature analysis of elements, characters of the nonindividual body can be obtained. From a mathematical view, it can change a continuous unlimited mobility problem into a discrete limited mobility problem, and the problem will be simplified maximum. It is obvious that with the increase of the number of units (the reduction of the element dimension), the approximate solution degree will be improved. So the precision of result in FEM has much to do with the computer technique.

Because of the absorption of the latest development of computer technology and calculation method, Finite Element Analysis Software has combined the FEM, computer graphics, and optimization technique, and it has become an essential powerful tool to solve modern engineering problems.

The basic steps of FEM can be described as follows:

- Problems and solving field definition
  In this step, according to the practical problem, determine the physical properties of approximate solving domain and geometric area.
- Solving domain discretization
  Solving domain is divided into discretization domain, which includes limited elements. This step is usually named finite element network division, and it influences the precision of the FEM directly.
- State variable and control method determination
  A specific physical problem usually can be described by a set of differential equations, which includes boundary conditions of the problem state variables. In order to be suitable for finite element analysis, the differential equation usually should be translated into functional form.
- Finite element derivation
- A suitable approximate solution of the element, that is a derivation of finite element formulation, is constructed. The approximate solution includes reasonable unit coordinate system, unit trial function establishment, and each unit state variables discretization. The aim of this step is to form the unit matrix.
- Simultaneous equation solution
  FEM eventually leads to simultaneous equations. There are three methods, namely, direct method, iterative method, and random method, to obtain the results. The result is an approximate state variable in every unit node.
- Results analysis
  Quality of the calculation results will be evaluated though the comparison with permissible value from design criteria and decided whether the repeat calculation need be carried out.

In a word, the finite element analysis can be divided into three stages: preprocessing, processing, and post-processing. In preprocessing, the finite element model is built and finite mesh elements are divided. In the post-processing, the analysis results are collected then users can simply extract information.

Many simulation techniques have been used in nanoimprint lithography. In the preceding years, molecular dynamics method is used as the common method in this area. But with the development of theory and technology in nano and computer fields, FEM has become more and more valued.

Up to date, numerous FEM techniques have been applied to nanoimprint experiment, generally as tools to help understand the nanoimprint itself, to develop improvements in the analytical methods commonly used to extract mechanical properties from the data, and to develop a flexible set of tools which use FEM for routine analysis of materials.

For example, Liu et al. performed FEM simulations to investigate the nanoimprinting of aluminum polyimide bilayer substrates at temperatures ranging from 25 to 200°C [9]. When under low-temperature conditions (<100), the imprint pressure reduces as the aluminum-to-polyimide thickness ratio decreases. However, if temperature is higher than 150°C and a polyimide-to-substrate thickness ratio of more than 40%, the imprint pressure reversely enlarged. The results provide a convenient method of processing technique for an actual application.

7.3.2 *Molecular Simulation*

MD method is one of the microscopic simulation methods with computer and has developed rapidly in the past couple of decades. The advantages of MD method is that not only quantitative information such as energy, temperature, pressure, and velocity distributions can be obtained but also a complete microscopic description of the dynamics processes involved in laser ablation can be provided by it. Moreover, direct comparisons of the results from the simulations with experimental data help to deepen the comprehension of problems being researched and to elucidate the new phenomenon that is found to be imperceptible in experiments.

7.3.2.1 The Equations of Motion

Molecular dynamics (MD) is a kind of computer simulation technique that can predict the time evolution of a system by solving a set of classical equations of motion for all particles in the system. The motion of each particle is governed by Newton's second law. Newton's equations of motion for a system consisting of N particles are as follows:

$$\dot{r}_i = \frac{P_i}{m_i}$$

$$m_i \ddot{r}_i = F_i = -\nabla_{r_i} V \tag{7.12}$$

where P_i, F_i, r_i, and m_i stand for momentum, composite force, position, and mass of atom *i*, respectively. *V* is the total potential energy of system.

7.3.2.2 Interparticle Interaction Potentials

It is important to select an available potential function since composite force of each particle is defined by the first derivative of it. Accuracy, computational speed, and transferability need to be satisfied to some extent in the choice of interaction potentials.

Generally, the potentials can be separated loosely into two classes: pair potentials and many-body potentials. Pair potentials were used mainly in the early period of MD methods application. Pair potentials are simple since the interaction between any pair of atoms is not affected by the presence of other atoms. Hard spheres, Lennard-Jones, and Morse and Coulomb potentials are usually used in pair potentials, which are expressed by:

1. Hard spheres potential

$$E = \infty,\ r_{ij} \leq r_0$$
$$E = 0,\ r_{ij} > r_0 \tag{7.13}$$

2. Lennard-Jones potential

$$U(r_{ij}) = 4\varepsilon\left[\left(\frac{\sigma}{r_{ij}}\right) - \left(\frac{\sigma}{r_{ij}}\right)\right]^{12} \tag{7.14}$$

3. Morse potential [10]

$$U(r_{ij}) = D[\exp\{-2a(r_{ij} - r_0\}\} - 2\exp\{-a(r_{ij} - r_0\}] \tag{7.15}$$

4. Coulomb potential

$$U(r_{ij}) = \frac{q_i q_j}{r_{ij}} \tag{7.16}$$

Hard spheres potential is the simplest potential without any cohesive interaction and is useful in theoretical investigations of some idealized problems. Lennard-Jones potential established to describe van der Waals interaction in inert gases and molecular systems is sometimes adopted for body-centered cubic metals such as chrome, molybdenum, and tungsten. Morse potential is more suitable for face-centered cubic metals. Coulomb is often added to other functional forms for the purpose of taking charge-charge interaction or polarization into account.

Many-body potentials that can avoid Cauchy relation of pair potentials have to be carried out when the interaction of a pair of atoms is modified by the presence of other atoms. EAM potential is a typical kind of many-body potential that can be used to calculate the interparticle force efficiently and eliminate the limitation of Cauchy relation [11]. The total potential energy U_{tot} is given in terms of the EAM function by

$$U_{tot} = -\sum_{i} {\rho_i}^{\frac{1}{2}} + \frac{1}{2}\sum_{i}\sum_{j(j\neq i)} V_{ij} \tag{7.17}$$

where ρ_i is local electron density and V_{ij} is pair-potential term.

7.3.2.3 Algorithms

Gear predictor-corrector algorithm [12] and Verlet algorithm [13] are commonly employed in molecular dynamics calculations. There are three steps in Gear predictor-corrector algorithm. Firstly, new position $r^{p}(t+\delta t)$, new velocity $v^{p}(t+\delta t)$, and new acceleration $a^{p}(t+\delta t)$ are predicted according to Taylor expansion; secondly, acceleration $a(t+\delta t)$ is calculated based on forces at predicted position; Finally, position and velocity terms are corrected by the discrepancy of $a^{p}(t+\delta t)$ and $a(t+\delta t)$ to obtain the velocity and acceleration values of current step. Gear predictor-corrector algorithm has an advantage of high accuracy, but at the same time it has the disadvantages of poor stability and requiring more storage for large time steps.

The simpler Verlet algorithm has been utilized widely with increasing of calculations and MD system volume. Verlet algorithm is such a method that can calculate positions $r(t)$ at time t, accelerations $a(t)$ at time t, and positions $r(t-\delta t)$ at time $t-\delta t$ and has many strong points such as simple algorithm, high calculation speed, and energy conservation and less storage requirements for short time. There are also some Verlet-like algorithms, for example, velocity-Verlet algorithm [14] and leapfrog algorithm [15]. They have the nearly same accuracy, whereas the only difference is the number of variables.

7.3.2.4 Boundary Conditions

The goal of simulating atomic or molecular systems by MD methods is to obtain information about the properties of a macroscopic sample. Generally, MD systems cannot be too big as far as the present calculated capacity. Periodical boundary conditions make a real small part of a large system simulation possible.

A sample containing N particles can be treated as the primitive cell of an infinite periodical lattice of identical cells. The movement of particles in the sample can be computed only according to boundary conditions around the primitive cell so as to reduce the calculated quantity greatly.

The program is constructed as follows [16, 17]:

- We read the parameters that specify the conditions of the run (e.g., initial temperature, number of particles, density, time step).
- We initialize the system (i.e., we select initial positions and velocities).

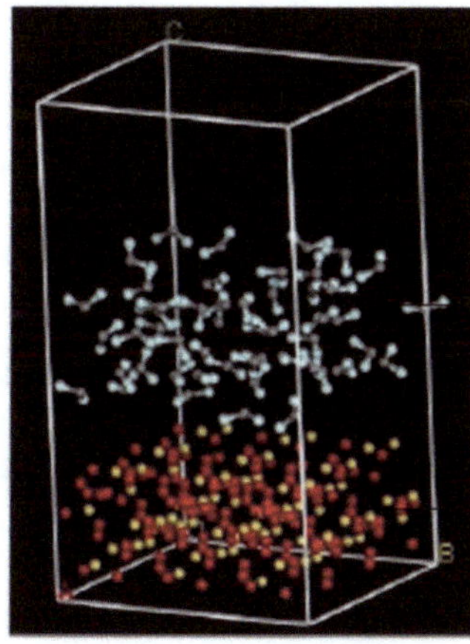

Fig. 7.2 Contact schematics of simulation mode

- We compute the force on all particles.
- We integrate Newton's equations of motion. This step and the previous one make up the core of the simulation. They are repeated until we have computed the time evolution of the system for the desired length of time.
- After completion of the central loop, we compute and print the averages of measured quantities, and stop.

As we know, the adhesion between stamp and resist has an impact on the quality of imprinted patterns. Usually, the stamp surface is self-assembled with fluoride to decrease the stamp surface energy. The contact schematics of simulation mode for the quartz stamp and polymer is shown in Fig. 7.2.

By MD simulation, their energy is as follows:

$$E_{\text{total}} = 1,57,127.43\,\text{kcal/mol},\ E_{\text{polymer}} = 1,013.66\,\text{kcal/mol},\ E_{\text{SiO}_2} = 1{,}56{,}128.2\ \text{kcal/mol}.$$

The interaction energy ($E_{\text{interaction}}$) is calculated by

$$E_{\text{interaction}} = E_{\text{total}} - (E_{\text{SiO}_2} + E_{\text{polymer}}) = -14.43\ \text{kcal/mol} = -6.06 \times 10^4\ \text{J/mol} < 0.$$

This means that the energy relaxes from the surface and decreases. The CF_2 is an alternative material for stamp adhesion.

7.4 Simulation of Nanoimprint Lithography Process

7.4.1 Polymer Flow During Imprinting

Understanding of polymer flow during filling of the mold cavity can possibly predict the physical driving mechanism governing flow and provide the guidelines for

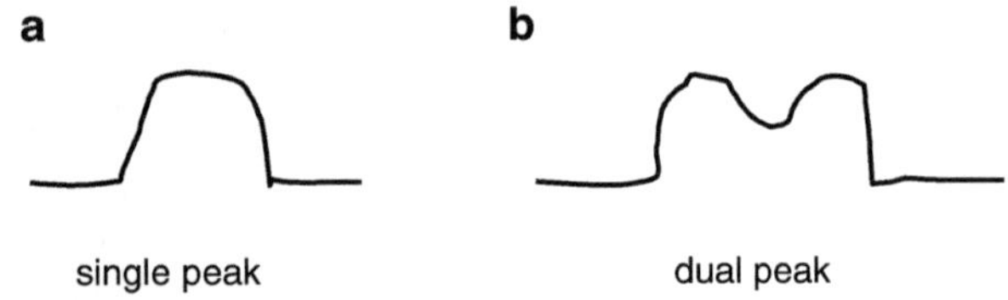

Fig. 7.3 Nanoimprint lithography polymer deformation modes. (**a**) Single peak deformation. (**b**) Dual peak deformation

imprinting. During imprinting, polymer deformation profiles are shown for four times: the start of imprinting, half the time to onset of filling, onset of filling, and the time when polymer peak first touches the mold cavity floor [18].

The simulation showed that the types of single and dual peak deformation are decided by the ratio of cavity half width (W) and the initial film thickness (h). When $W/h < 1$, single peak occurs; if $W/h > 1$, dual peak deformation occurs. Figure 7.3 shows single and dual peak viscous flows, which are reported in polymer deformation of nanoimprint lithography. The single peak means there is one peak formed in the cavity during the filling. The dual peak mode indicates the polymer flows into the cavity and forms two peaks near the cavity sidewall. Differences in deformation mode are attributed to viscous flow and stress, surface tension, and so on.

7.4.2 Process Simulation

Kim et al. used MD simulation to investigate the deformation behavior of the polymer film and the effect of adhesion and friction with the change of a pattern aspect ratio. Figure 7.4 shows the analysis model, which consists of a quartz stamp, a polymer film, and a nickel substrate. The quartz mold is a line pattern with 4.5 nm width and a height of 5 nm (low aspect ratio pattern) or 15 nm (high aspect ratio pattern). The imprinted polymer is an amorphous PMMA, and the molecular weight of each molecule is 10.016. The substrate is composed of four layers of nickel atoms with FCC structure. Atoms in the quartz molds are in rigid body motion during the simulation, while atoms in the nickel substrate stay in their position.

Figure 7.5 shows snapshots of several stages in the NIL simulation. The pattern aspect ratio is 2.5, and the imprint depth is 8.0 nm. The speed is 10 m/s for imprint and demolding. Figure 7.5a shows different colored atoms in the polymer film to further understand the deformation behavior of the polymer. As the mold is approaching into the resist (polymer), the resist deforms according to the mold shape, as shown in Fig. 7.5b. Then, the top part of polymer film begins to be compressed and extruded from the bottom of the pattern. The resist fills the cavity of the mold, as shown in Fig. 7.5c, d. During the demolding process, large amount of springback is clearly observed where the stamp imprints. Furthermore, the bulge of the resist film is elongated due to the friction force. After the mold is totally separated, the pattern is transferred to the polymer film and shown in Fig. 7.5f.

Fig. 7.4 Schematics of simulation model (Reprinted with permission from [19])

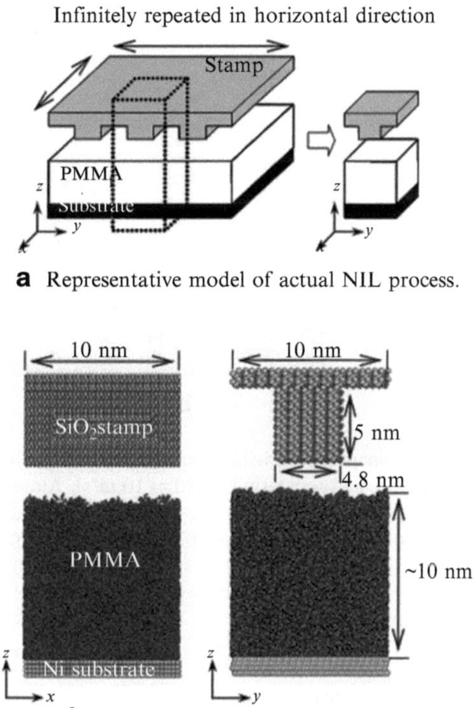

b Generated atomistic model.

Besides, MD simulation reveals that the adhesion force is nearly constant regardless of the pattern aspect ratio and the friction force increases as the pattern aspect ratio grows. It means that the effect of friction becomes dominant over the adhesion force as the aspect ratio of the pattern grows higher.

The effects of mold shape and sidewall roughness on the formation of atomic scale structure in a metal direct imprinting process have been demonstrated. With increasing taper angle, the imprint force required decreases. The geometries of the imprint patterns are affected by sidewall roughness [20].

7.4.3 Residual Layer Thickness

In the nanoimprint lithography, a uniform thickness of the residual layer over the entire imprinted area is essential. A schematic of the cross section of the stamp is shown in Fig. 7.6 [21].

To simplify the problem, some assumptions are as follows: (1) contact between the surface of the substrate and the stamp is parallel. (2) Their surfaces are without waviness. (3) Surface tensions and capillary forces are neglected.

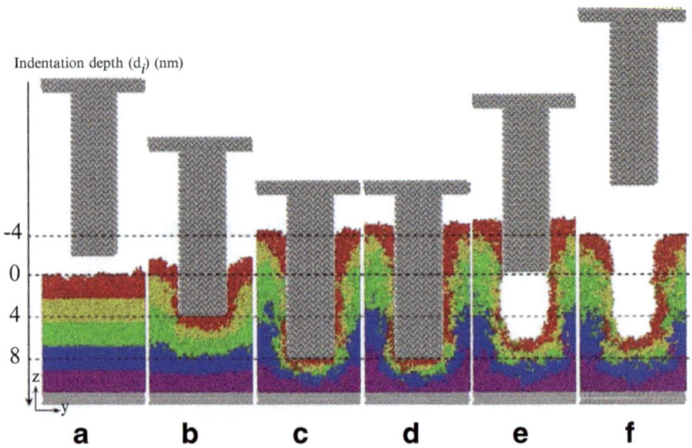

Fig. 7.5 Snapshots of MD simulation of NIL (pattern aspect ratio = 2.5, imprint depth = 8.0 nm, indentation, and separation speed = 10 m/s). Atoms are differently colored as layers according to their initial positions. The time sequence represents how the polymer film deforms. From *left to right*: (**a**) start of indentation, 0 ns, di = −1.5 nm; (**b**) during indentation, 0.55 ns, di = 4.0 nm; (**c**) end of indentation, 0.95 ns, di = 8.0 nm; (**d**) start of separation, 1.7 ns, di = 8.0 nm; (**e**) during separation, 2.5 ns, di = 0.0 nm; (**f**) end of separation, 3.5 ns, di = −8.0 nm (Reprinted with permission from [19])

Fig. 7.6 Schematic of the cross section of the stamp

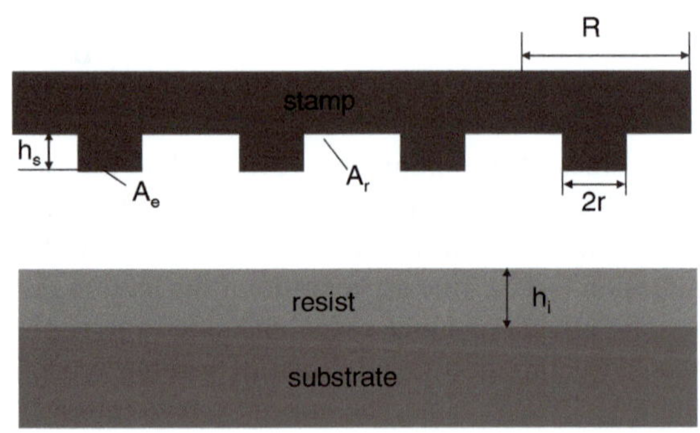

The geometry of the round stamp is shown in Fig. 7.6. Radius R and radius r are circular pattern on the stamp and its height is h_s. There are N protrusion, each protrusion area is A_e and recessed area is A_r.

According to Navier–Stokes equation, the force from a single stamp protrusion, F, is given by

$$F = \frac{3\pi \eta u R^4}{2{h_i}^3} \tag{7.18}$$

where η is viscosity and u is a rate.

An equation for the residual resist thickness (RLT) h_r is presented by

$$\begin{cases} h_r = \sqrt[3]{n\frac{3\pi\eta u r^4}{4F}} & F \le F_1 \\ h_r = h_i - \frac{h_s}{1+A_e/A_r} & F_1 \le F \le F_2 \\ h_r = \sqrt[3]{n\frac{3\pi\eta u r^4}{4F}} & F \ge F_2 \end{cases} \quad (7.19)$$

where

$$F_1 = n^* \frac{3\pi\eta u r^4}{4\left(h_i - \frac{1}{1+A_e/A_r} h_s\right)^3}$$

$$F_2 = \frac{3\pi\eta u R^4}{4\left(h_i - \frac{1}{1+A_e/A_r} h_s\right)^3}.$$

If $F < F_1$, the recessed areas on the stamp have not yet been filled. h_r depends on the applied force F and the size of the pattern on the stamp.

If $F_1 < F < F_2$, h_r depends only on the initial film thickness h_i and geometric of the stamp. The cavities are filled.

If $F > F_2$, h_r depends on the radius and the applied force F.

References

1. Li H, Amemiya K, Talke FE (2010) Finite element simulation of static contact of a slider with patterned media. J Adv Mech Des Syst Manuf 4:42–48
2. Son JW, Song NH, Rhim SH, Oh SL (2007) Prediction of defects in nano-imprint lithography using FEM simulation. Key Eng Mat 345–346:665–668
3. Marín JM, Rasmussen HK, Hassager O (2010) 3D Simulation of nano-imprint lithography. Nanoscale Res Lett 5:274–278
4. Zhang YH, Van DG, Nicola L (2010) Dislocation dynamic simulations of metal nanoimprinting. Mat Res Soc Symp Proc 1224:199–204
5. Kim SM, Woo YS, Lee DE, Lee W (2007) Numerical simulation of NIL process based on continuum hypothesis. NSTI Nanotechnology Conference and Trade Show – NSTI Nanotech 2007. Tech Proc 4:442–445
6. Simprint Core v2.1: NIL process simulation software. http://simprintnanotech.com/pg/products. Accessed 12 Dec 2011
7. Doering CR, Gibbon JD (1995) Applied analysis of the Navier–Stokes equations. Cambridge University Press, London
8. Pepper DW, Heinrich JC (1992) The finite element method: basic concepts and applications. Hemisphere Publishing Corporation, Washington, D.C
9. Liu DS, Tsai CY, Lu YT, Sung CK, Chung CL (2010) Finite element method investigation into nanoimprinting of aluminum/polyimide bi-layer substrates. Microelectron Eng 87:2361–2367
10. Girifalco LA, Weizer VG (1959) Application of the Morse potential function to cubic metals. Phys Rev 114:687–690

11. Daw MS, Weizer VG (1984) Embedded atom method derivation and application to impurities surfaces and defects in metals. Phys Rev B 29:8486–8495
12. Gear CW (1971) Numerical initial value problems in ordinary differential equation. Prentice-Hall, Inc., Englewood Cliffs
13. Verlet L (1967) Computer 'experiments' on classical fluids. I. Thermodynamical properties of lennare-Jones molecules. Phys Rev 159:98–103
14. Swope WC, Anderson HC, Berens PH, Wilson KR (1982) A computer method for the calculation of equilibrium constants for the formation of physical clusters of molecules: application to small water clusters. J Chem Phys 76:637–649
15. Honcy RW (1970) The potential calculation and some applications. Methods Comput Phys 9:136–211
16. Frenkel D, Smit B (1996) Understanding molecular simulation from algorithms to application. Academic, San Diego
17. Rapaport DC (1995) The art of molecular dynamics simulation. Cambridge University Press, Cambridge
18. Rowland HD, King WP, Sun AC, Schunk PR (2006) Impact of polymer film thickness and cavity size on polymer flow during embossing: towards process design rules for nanoimprint lithography. Sandia report, SAND
19. Kang JH, Kim KS, Kim KW (2007) Molecular dynamics study of pattern transfer in nanoimprint lithography. Tribol Lett 25:93–102
20. Yao CH, Chang CH, Hsieh CW, Sung CK (2010) Effect of mold shape and sidewall roughness on nanoimprint by molecular dynamics simulation. Microelectron Eng 87:864–868
21. Otto M, Bender M, Hadam B, Spangenberg B, Kurz H (2001) Characterization and application of a UV-based imprint technique. Microelectron Eng 57–58:361–366

Chapter 8
Application of NIL in Light-Emitting Diode

8.1 Introduction

Light-emitting diode (LED) is a very popular semiconductor diode available today. In the background of the world energy crisis, it has been proved to be highly efficient and energy saving and have a long lifetime. Compared to some conventional lamps, LED has given birth to the new light technology (solid-state lighting). Among them, GaN-LED is widely considered as one of the most promising next-generation light sources due to its reliability, durability, and efficiency. GaN-LED has been widely used in displays, traffic signals, and backlights [1–5]. However, the external quantum efficiency of the GaN-LED is generally much lower than the theoretical expectations due to total internal reflection effect on its surface. Consequently, one of the current essential research interests in this area is to find out methods of enhancing efficiency of GaN-LED. How to enhance efficiency of GaN-LED is a key subject in solid-state light.

Currently, all rigorous methods have been applied to enhance light extraction efficiency of LED, such as surface-texturing micro-pyramid, but the results are still unfavorable. The external quantum efficiency of LED is less than 30%. Photonic crystal (PC) structure has been demonstrated as an effective approach to enhance the light extraction efficiency of GaN-LED. Usually, the PC pattern is fabricated by electron beam lithography (EBL) or focused ion beam lithography (FIB) followed by the etching process. However, these techniques have disadvantages of inherent characteristics with low efficiency and high cost. In 2005, a scientist from Korea fabricated PC structures on GaN surface by holographic lithography [6]. Compared to EBL technique, this technique can fabricate PC patterns on GaN surface with large area.

Nanoimprint lithography, with low cost, large area, and high output efficiency is somewhat superior to the fabrication technique of PC patterns. Therefore, the technique can meet the requirements of large-area PC nanostructures in GaN-LED field.

W. Zhou, *Nanoimprint Lithography: An Enabling Process for Nanofabrication*,
DOI 10.1007/978-3-642-34428-2_8,

In this chapter, our research achievements on light extraction efficiency of PC-LED with nanoimprint lithography technique have been demonstrated. To realize PC nanostructure, imprinted mold is first designed by way of simulation. The fabrication process route is also designed to enable better compatibility with conventional LED process. This result is exciting and lays a good basis for the technique used for PC-LED.

8.2 Basic Physical Concepts of Photonic Crystals

8.2.1 Properties of Photonic Crystals

Photonic crystals (also known as photonic bandgap materials) are periodic optical nanostructures that are designed to affect the motion of photons. It is similar to the manner in which the periodicity of a semiconductor crystal affects the motion of electrons. Photons that propagate through these nanostructures are decided by their wavelength. The bandgap forbids propagation of a certain frequency range of light. By this property, it enables to control light that is used for conventional optics.

Photonic crystals have been reported by Eli Yablonovitch and Sajeev John in 1987, which opens a new era [7, 8]. After that, it has been intensively researched from theory understanding to some applications. Photonic crystals can be divided into three categories by the dimension of its period—1D photonic crystals, 2D photonic crystals, and 3D photonic crystals—as shown in Fig. 8.1.

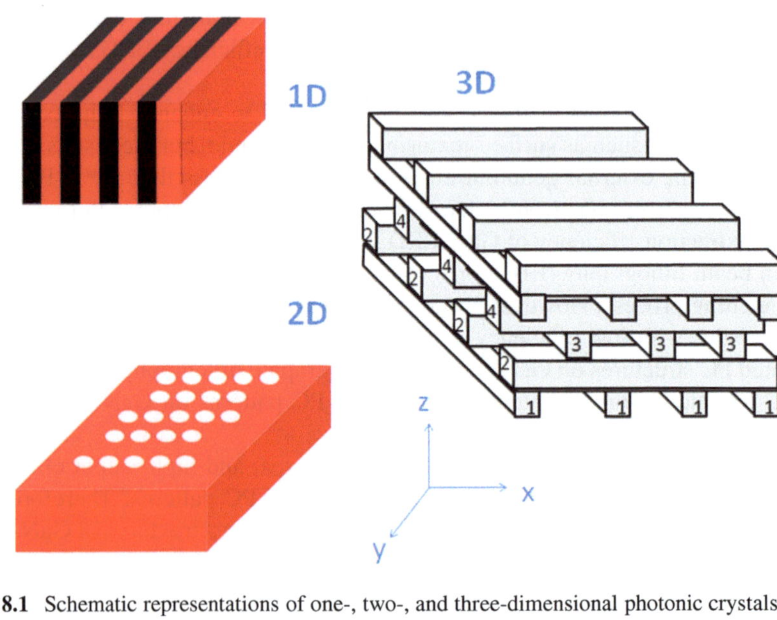

Fig. 8.1 Schematic representations of one-, two-, and three-dimensional photonic crystals

In 1D photonic crystals, layers of different dielectric constant may be deposited together to form a bandgap in a single direction. In 2D photonic crystals, the air holes can be fabricated on a substrate which has triangular or square lattices of holes. In 3D photonic crystals, several structures have been fabricated, such as woodpile structure, inverse opals, or inverse colloidal crystal spheres.

8.2.2 *Computational Methods of Photonic Band Structure*

Understanding of the photonic bandgap (PBG) is necessary to construct the desired patterns. The most common computational methods and modeling tools applicable to photonic crystal analysis and design are plane wave expansion method (PWEM), transfer matrix method (TMM), finite element method (FEM), finite-difference time-domain method (FDTD), etc.

Plane wave expansion method (PWM) refers to a computational technique in electromagnetics to solve the Maxwell's equations by formulating an eigenvalue problem out of the equation. It was used in some of the earliest studies of photonic crystals. However, the accuracy is a growing function of the decoupling between adjacent supercells, and it results to be computationally very expensive for the theoretical analysis of complex photonic crystal [9–11].

Transfer matrix method (TMM) is based on representing Maxwell's equations on a discrete lattice of real space points. The equations are recast into the form of a transfer matrix that connects magnetic and electric fields in one layer of lattice points to those in the next layer [12, 13].

Finite-difference time-domain (FDTD) is one of the primary available computational electrodynamic modeling techniques and a time-domain method. It reproduces numerically the propagation in real space of time-varying electromagnetic waves. In 1966, Yee presented Yee cell used for FDTD [14].

Maxwell's equation is as follows:

$$\nabla \times \vec{E}(\vec{r},t) = -\mu \frac{\partial \vec{H}(\vec{r},t)}{\partial t} \quad \nabla \times \vec{H}(\vec{r},t) = -\varepsilon(r) \frac{\partial \vec{E}(\vec{r},t)}{\partial t}. \tag{8.1}$$

As we know, electromagnetic field in the 2D simulation model can be decomposed into two polarization directions that are TE mode and TM mode. According to FDTD, the space differential can be replaced by a central difference quotient and the time differential can be described by Taylor expansion. Maxwell's equations can be discredited by Yee-cell technique. For TE mode, the Maxwell's differential equations can be given as follows:

$$E_x^{n+1}(i,j) = E_x^n(i,j) + \frac{H_z^{n+\frac{1}{2}}(i,j+\frac{1}{2}) - H_z^{n+\frac{1}{2}}(i,j-\frac{1}{2})}{\Delta y} \cdot \frac{\Delta t}{\varepsilon(i,j)}, \tag{8.2}$$

$$E_y^{n+1}(i,j) = E_y^n(i,j) + \frac{H_z^{n+\frac{1}{2}}(i+\frac{1}{2},j) - H_z^{n+\frac{1}{2}}(i-\frac{1}{2},j)}{\Delta x} \cdot \frac{\Delta t}{\varepsilon(i,j)}, \tag{8.3}$$

$$\begin{aligned} H_z^{n+\frac{1}{2}}(i,j) = H_z^{n-\frac{1}{2}}(i,j) &+ \frac{E_z^n(i,j+\frac{1}{2}) - E_x^n(i,j-\frac{1}{2})}{\Delta y} \cdot \frac{\Delta t}{\mu} \\ &- \frac{E_x^n(i+\frac{1}{2},j) - E_x^n(i-\frac{1}{2},j)}{\Delta x} \cdot \frac{\Delta t}{\mu}, \end{aligned} \tag{8.4}$$

where n is the number of discrete time step; i, j are the location of discrete lattice on the x–y plane; and Δx, Δy are the space cell size in the x, y direction.

The time interval restrictions (numerical stability condition) meet the following equation:

$$\Delta t \leq \frac{1}{c\sqrt{(\Delta x)^{-2} + (\Delta y)^{-2} + (\Delta z)^{-2}}}. \tag{8.5}$$

But FDTD simulation of calculating field and computational domain is finite. By inserting artificial boundaries into simulation space, the errors can be minimized. Most modern FDTD simulations use a special absorbing "material" called a perfectly matched layer (PML) to implement absorbing boundaries.

8.3 The Light Extraction Efficiency of GaN-LEDs

8.3.1 Analysis of the Light Extraction Efficiency

The schematic of LED is shown in Fig. 8.2. The active layer is capable of emitting region and multiple quantum wells (MQWs) structure of less than 1 μm in thickness in most cases.

The active layer can be considered as a thin film containing a large body of point light sources radiating photons in all directions. But these photons cannot fully escape from LED system because there is total internal reflection (TIR) at GaN/air interface. The critical angle of total internal reflection is θ_c, which is given by

$$\theta_c = \sin^{-1}(n_{air}/n_{GaN}) = 24°, \tag{8.6}$$

where refractive index of GaN and air is $n_{GaN} = 2.4$ and $n_{air} = 1$, respectively. Only the light within the cone with an angle $2\theta_c$ (light cone) can escape from LED system. As the angle of incidence is greater than the critical angle, no light can be transmitted on the interface and all light is reflected. The schematic of the light cone is given in Fig. 8.3.

Because of the existence of the total internal reflection, the quantum efficiency (QE) of GaN-LEDs is low, while its internal quantum efficiency is close to 100%.

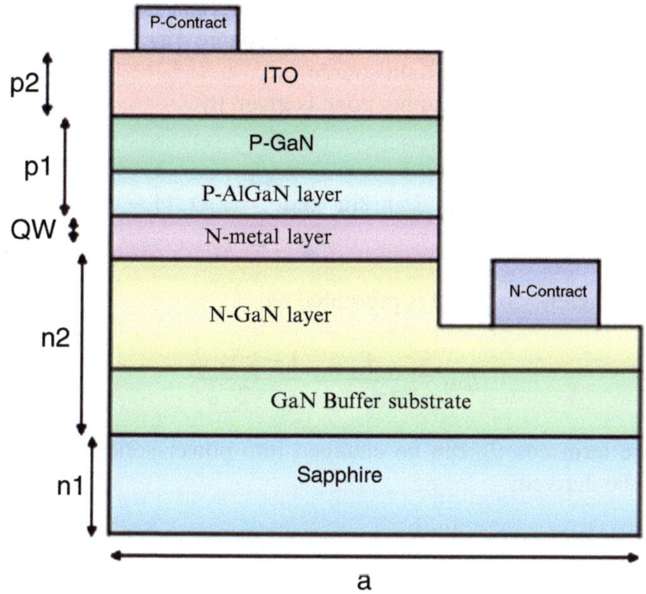

Fig. 8.2 Schematic diagram of LED structure

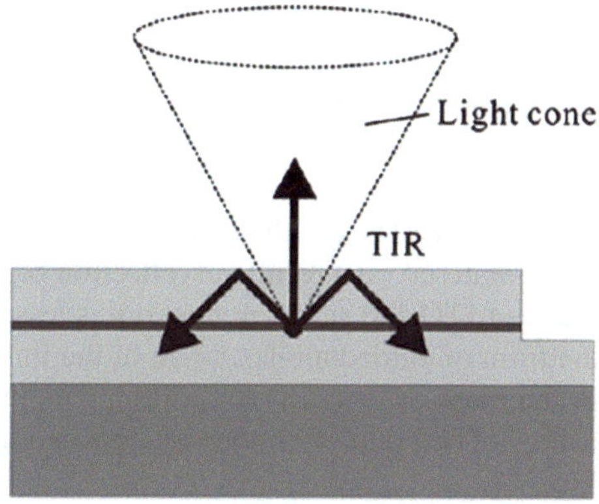

Fig. 8.3 Schematic of light cone. Light within $2\theta_c$ angle can escape

The external quantum efficiency of LEDs, η_{ex}, depends on internal quantum efficiency, η_{in}, and extraction efficiency, η_{ext}:

$$\eta_{ex} = \eta_{in} . \eta_{ext}. \tag{8.7}$$

The external quantum efficiency is determined by semiconductor material, match of crystals, and internal structures. As mentioned just before, internal quantum efficiency is very high, and the property of LED efficiency depends on external quantum efficiency.

In 3D space, the point light source emits photons isotropically in spherical directions. The radiation area is the area of the sphere $4\pi r^2$ (r denotes the distance from the wavefront to the point light source).

The radiation area of the escaping cone is given by

$$A = \int \mathrm{d}A = \int_{\theta=0}^{\theta_c} 2\pi r \sin\theta r \mathrm{d}\theta = 2\pi r^2(1 - \cos\theta_c). \tag{8.8}$$

Extraction efficiency of LED is presented by

$$\eta_{\text{ext}} = \frac{A}{4\pi r^2} = \frac{1 - \cos\theta_c}{2}. \tag{8.9}$$

The cosine term $\cos\theta_c$ can be changed into power series, and the following equation can be derived:

$$\cos\theta_c \approx 1 - \frac{{\theta_c}^2}{2!} + \frac{{\theta_c}^4}{4!} - \cdots. \tag{8.10}$$

Therefore, extraction efficiency η_{ext} is

$$\eta_{\text{ext}} \approx \frac{1}{4}{\theta_c}^2 = 4\%. \tag{8.11}$$

As seen from Eq. 8.11, light extraction efficiency (calculated above) from the active region surface to air is only 4%, and large amount of light is lost in the internal system.

In GaN-LEDs, the difference between high index material (GaN, ITO) and low index material (air) results in reflection of most of fractional lights. Meanwhile, the index of ITO and sapphire material is less than that of GaN, the light transforms in the form of guided modes. Most of the light coupled into guided modes propagates in the plate of the LED and disappears in the internal of LEDs, resulting in heat energy. The coupled modes of light from LED chip are schematically shown in Fig. 8.4:

1. External modes: propagate into air
2. ITO modes: the total internal reflection in ITO layer
3. ITO/GaN/substrate waveguide mode

Therefore, light extraction efficiency of LEDs can be improved by introducing photonic crystal structures in different interfaces. Selecting an optional parameter is to decrease the total internal reflection, increase external modes, and decrease waveguide mode.

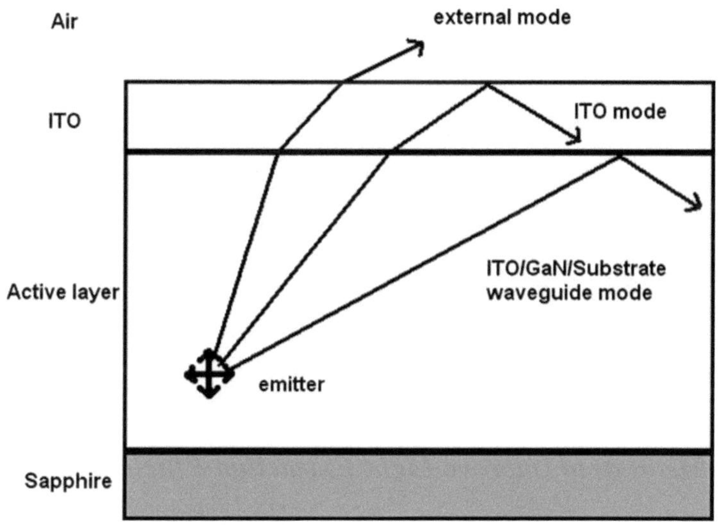

Fig. 8.4 Three propagation modes of the emitted light from LED chips

There are two possible methods by which photonic crystals can improve LED efficiency [15]. The first is by the Purcell spontaneous emission rate enhancement. Purcell enhancement factor is given by

$$f = \frac{3Qg(\lambda/2n)^3}{2\pi V_{\mathrm{CAV}}}, \tag{8.12}$$

where Q is the quality factor, g is the mode degeneracy, n is the refractive index, λ is the wavelength, and V_{CAV} is the cavity volume.

The second possibility is due to the fact that photonic crystals can improve LED efficiency by Bragg scattering. Light extraction in the case is classified into two possible ways: one is the factor of 2D PBD that forbids the lateral guided modes in the bandgap region and forces the emitted light to radiate out of the plane. The other is photonic crystals acted as grating effect. The enhancement factor is given by

$$f = \frac{1}{2} \int_{\theta_c - \theta_g}^{\theta_c + \theta_g} \eta_{\mathrm{dif}} \sin\theta \, d\theta, \tag{8.13}$$

where η_{dif} and θ_g are diffraction efficiency of photonic crystals, and angle changes due to diffraction effect. If the diffraction efficiency is high, light extraction efficiency of photonic crystals is also high. Diffraction efficiency of photonic crystals depends on parameters of photonic crystals, ratio of air material, and depth of air holes.

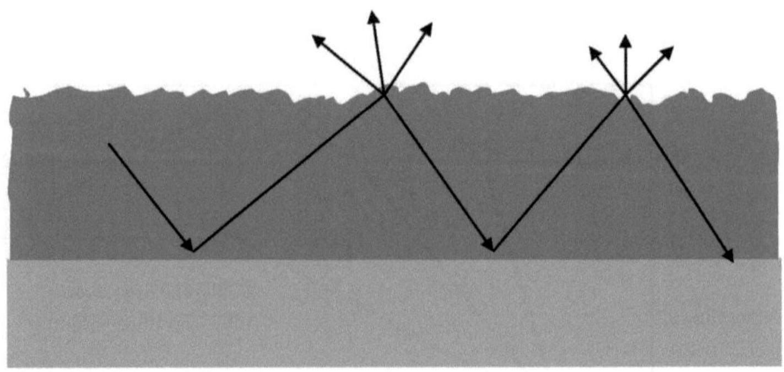

Fig. 8.5 The schematic of surface texturing

8.3.2 Methods to Improve Light Extraction Efficiency

Numerous methods have been demonstrated to enhance light extraction efficiency including flip-chip mounting, surface texturing, photonic crystals, and reflectors with high reflectivity [16–19].

Surface texturing is often used in light extraction efficiency of LED, which can enhance the scattering effect. Figure 8.5 shows the light is extracted by surface texturing technique. However, the technique degrades the electrical characteristics of LEDs, resulting in the rise of the forward voltage.

By introducing a textured surface, the light output power of the textured LED was increased by 29.4%, and the wall-plug efficiency was increased by 33% at an injection current of 20 mA, compared to that of the non-etched LED [20].

Flip chip is another approach in which light is extracted through the transparent sapphire substrate. In flip-chip LED, the light emitted toward metal contacts is reflected up, increasing light extraction.

Photonic crystals have shown to enhance light extraction efficiency of LED and can alter the fundamental process of light generation in the material itself [21]. The behavior of photonic crystal layer is more like scattering. It scatters light into two main directions, out of the slab and into the slab, which depends on the design parameter of the photonic crystals. Hence, a desired design can render photonic crystal function as a good extractor or reflector [22].

Patterned sapphire substrate (PSS) technique is used to reduce the dislocation density of GaN and can effectively improve the light output power. Besides, the PPS is similar to the roughened surface to enhance the light extraction efficiency for the multiple reflections at the GaN and sapphire interface, as shown in Fig. 8.6. It can improve both the internal quantum efficiency and the extraction efficiency of LEDs. PSS technique is widely used for high output LED, and the patterns are on the order of micrometers. A 25% enhancement in output power was achieved in the PSS LED as compared with that of the conventional LED sample [23]. The nanoscale patterned sapphire substrate (NPSS) is believed to significantly enhance extraction efficiency.

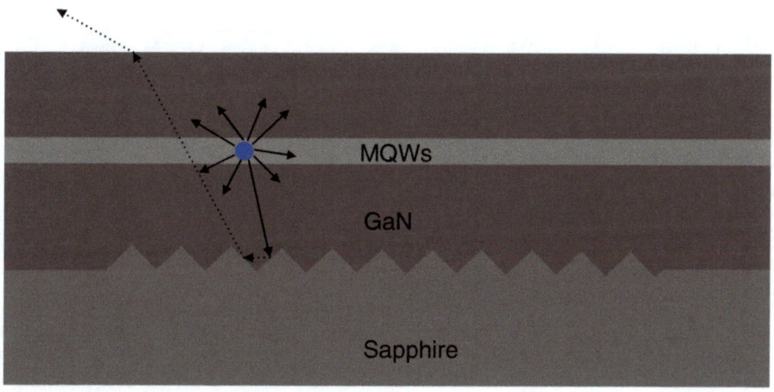

Fig. 8.6 Possible light traveling paths in the PSS LED

8.4 Structure Design to Improve Light Extraction Efficiency

In LED field, photonic crystals are arranged in the form of square lattice and triangular lattice. The structures can be the dielectric rods or air rods. Round period holes are further studied because of lower contact resistance and ease of fabrication. The effect of film thickness of photonic crystals on light efficiency depends on bandgap density, property of photonic crystal, and so on. In the research, two arranged photonic crystals were used, as shown in Fig. 8.7.

8.4.1 Construction of FDTD Simulation

The theory research was carried using FDTD software. First, the FDTD solutions model of LED multilayer was constructed. The parameters include dielectric constant, refractive index, and permeability for different materials. The basic model is a prototype waveguide module. Setting x–z plane shape and position, the horizontal setting of the periodic structure can be completed by matrix output. Besides, the location and orders of cross-sectional shape of waveguide and different modules are also set. Because the software provides only top view and cannot observe the real-time information of editing the geometry, the vertical direction of the construction must have a specific design data. Besides, the complex graphics setting should be avoided in the x–z or the z–y plane and placed on the x–z plane. The periodic photonic crystal structure is set to the horizontal distribution, as shown in Fig. 8.8.

Modeling of photonic crystal is done mainly using its physical parameters and geometric parameters. Physical parameters include dielectric constant, refractive index, and permeability. Geometric parameters include pitch (a), height (h), and duty ratio (R_p, $2r/a$ for round rods, $1/R_p$ for square). Considering blue GaN-LED, the refractive index $n = 2.4 + 0.002i$ and wavelength is 460 nm.

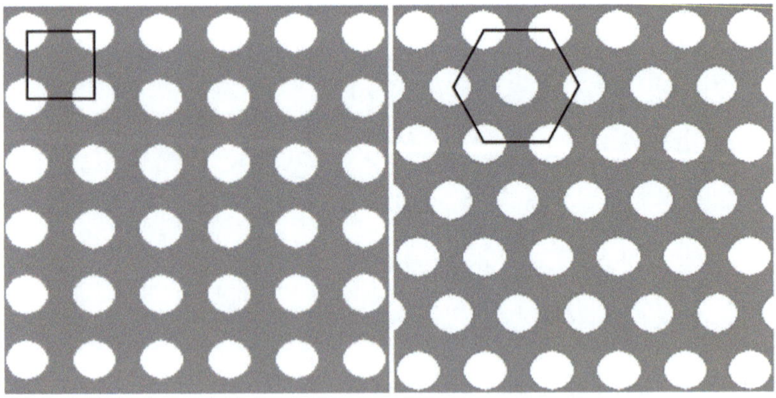

Fig. 8.7 Plane schematic of circular photonic crystals: *square* and *hexagon* arrangement

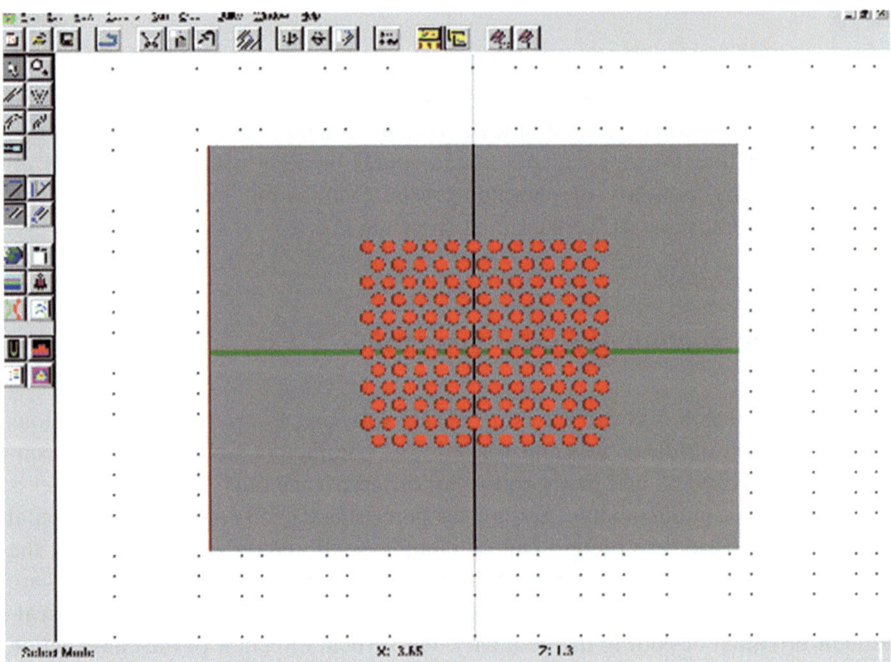

Fig. 8.8 The modeling interface of CAD tool

A 200*50*4 μm^3 area region is enough to accurately calculate the effect of photonic crystals on extraction efficiency. But the area is too large for FDTD with small grid size (<0.5 μm). In the simulation, the space (1.9*1.9*2.6 μm^3) is selected. In order to accurately restore, periodic boundary is set around x–z plane. Meanwhile, the PML boundary condition is also set on the y-axis.

8.4.2 Square Arrangement

The resulting band diagrams with different R_p are shown in Fig. 8.9. There are no photonic bandgap (PBG) for $R_p = 0.1$ or 0.2. When R_p is 0.3, there are three PBG regions in the considered frequency range. The normalized frequency is a range of 0.445–0.491 and 0.948–0.977 (TE). The PBG width is 0.046 and 0.029 (a/λ), respectively. The two midgap frequencies are 0.468 and 0.963, respectively.

When R_p is 0.4, there is one PBG region in the considered frequency range. The normalized frequency is a range of 0.392–0.469 (TE). The PBG width is 0.077 (a/λ). The midgap frequency is 0.43. When R_p is 0.5, 0.6, 0.7, 0.8, and 0.9, there are two, three, four, three, and zero PBG regions, respectively.

PBG can enhance light extraction efficiency of LED. In terms of square-arrayed circle photonic crystals, when R_p is 0.7, there are four PBG regions in the considered frequency range. The resulting midgap frequencies are 0.329, 0.549, 0.722, and 0.897, respectively. From the normalized parameters, we can retrieve the dimensions of the real photonic crystals. For this, we require the midgap frequency corresponds to the wavelength $\lambda_0 = 460$ nm and obtain consequently for the lattice constant:

$$a = f_0 \times \lambda. \tag{8.14}$$

So, the lattice constants are 151.34, 252.54, 332.12, and 412.62 nm, respectively. In consideration of fabrication, the value of 412 is selected, and the radius is 144.2 nm. When a is 412, R_p is 0.7, and the height is 250 nm, the maximum extraction efficiency is obtained.

8.4.3 Hexagon Arrangement

The resulting band diagrams with different R_p are shown in Fig. 8.10. There are no photonic bandgap (PBG) for $R_p = 0.1$ or 0.2 or 0.3 or 0.4. When R_p is 0.5, there is one PBG region, the normalized frequency is a range of 0.284–0.321. The PBG width and midgap are 0.037 and 0.303, respectively. If R_p is 0.6, the frequency range is 0.299–0.358. The PBG width and midgap are 0.059 and 0.329, respectively. When R_p increases to 0.7 and 0.8, the PBG width and midgap are also increased. When R_p is 0.9, there are two PBG regions. Only one PBG region is responsible for the value that R_p is 1.

The maximum extraction efficiency reaches in the condition of hexagonal arrays when R_p is 0.7. It corresponds to the air holes of hexagonal arrays, and R_p is 0.8. We require that the midgap frequency corresponds to the wavelength $\lambda_0 = 460$ nm and obtain consequently for the lattice constant (430 nm), and radius is 172 nm. The relative extraction efficiency (E_r) as a function of etch depth is shown in Fig. 8.11. The maximum efficiency is 5.3 times when the depth is 110 nm.

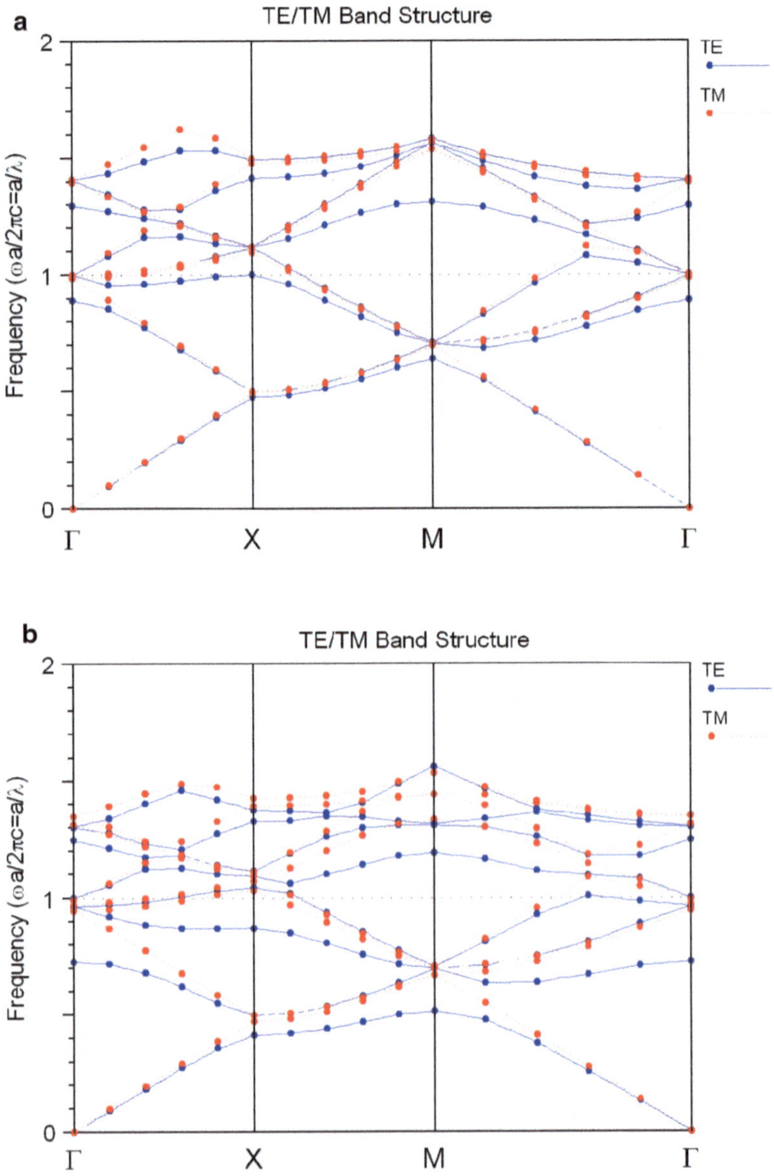

Fig. 8.9 Band diagram for square-type photonic crystals made of air holes in GaN. (**a**) $R_p = 0.1$. (**b**) $R_p = 0.2$. (**c**) $R_p = 0.3$. (**d**) $R_p = 0.4$. (**e**) $R_p = 0.5$. (**f**) $R_p = 0.6$. (**g**) $R_p = 0.7$. (**h**) $R_p = 0.8$. (**i**) $R_p = 0.9$

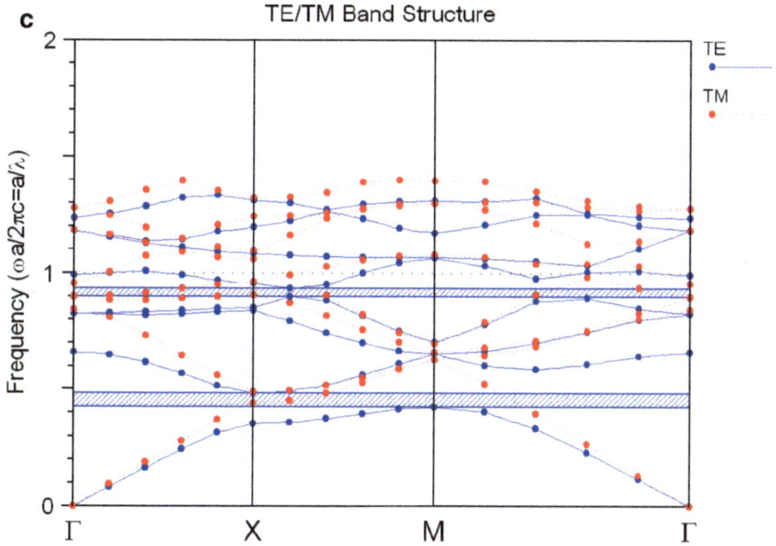

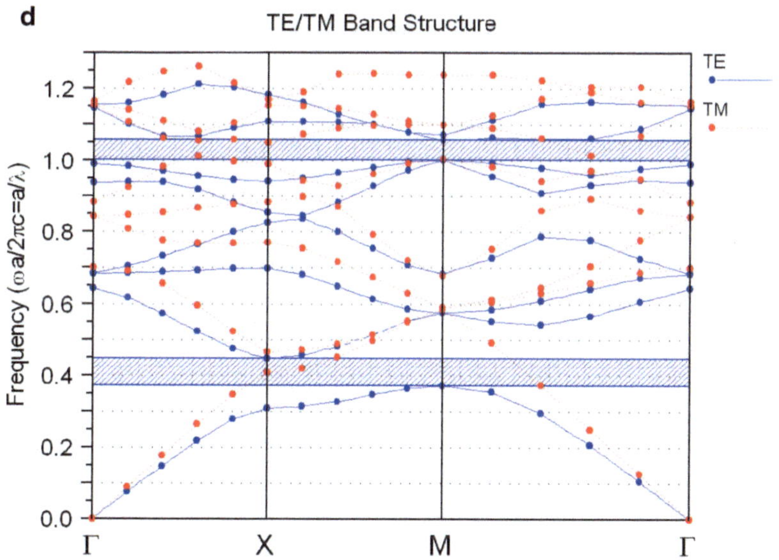

Fig. 8.9 (continued)

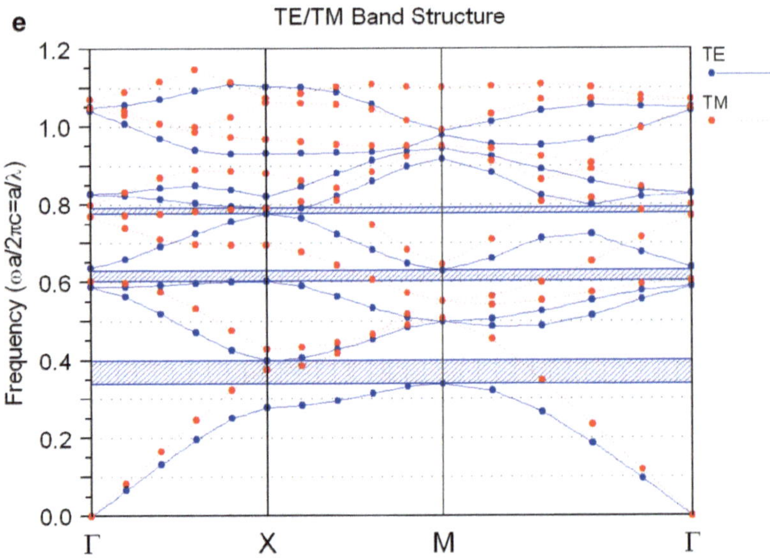

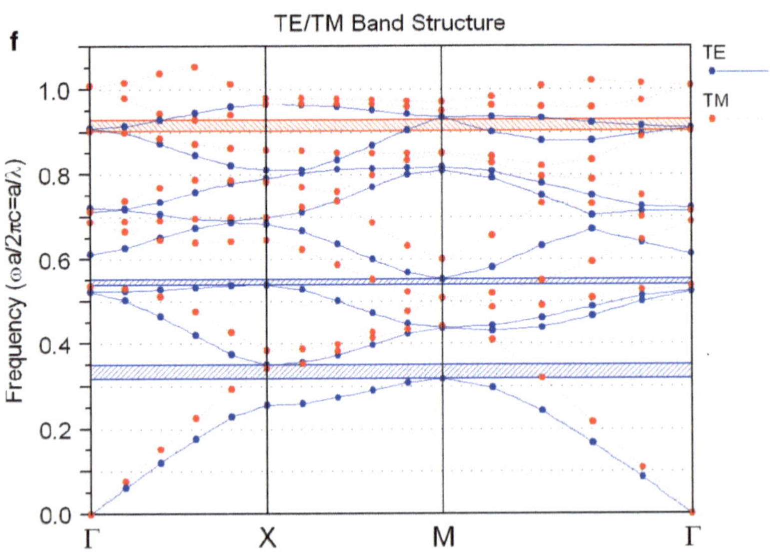

Fig. 8.9 (continued)

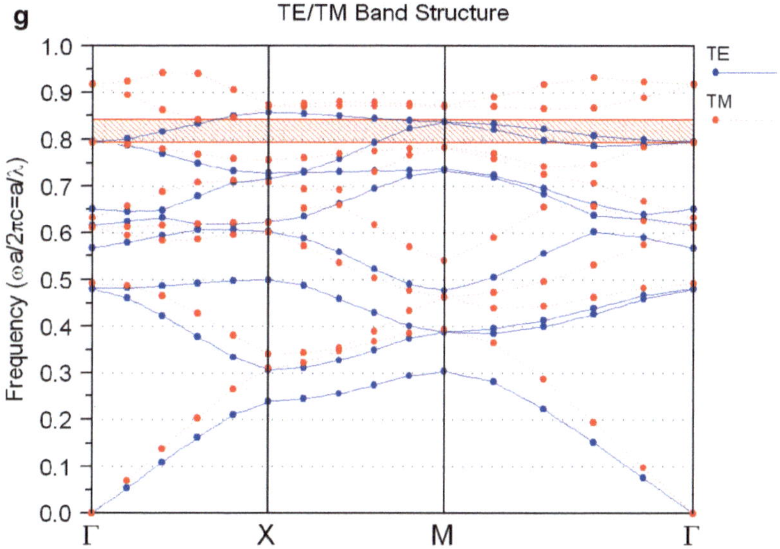

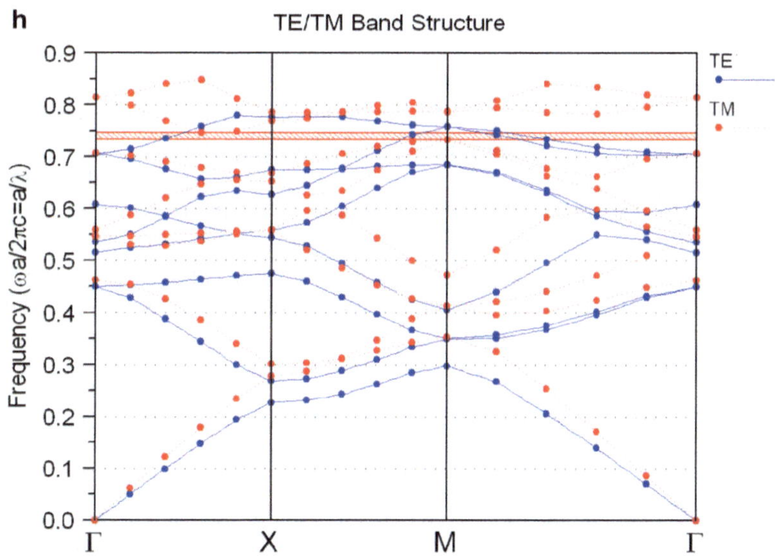

Fig. 8.9 (continued)

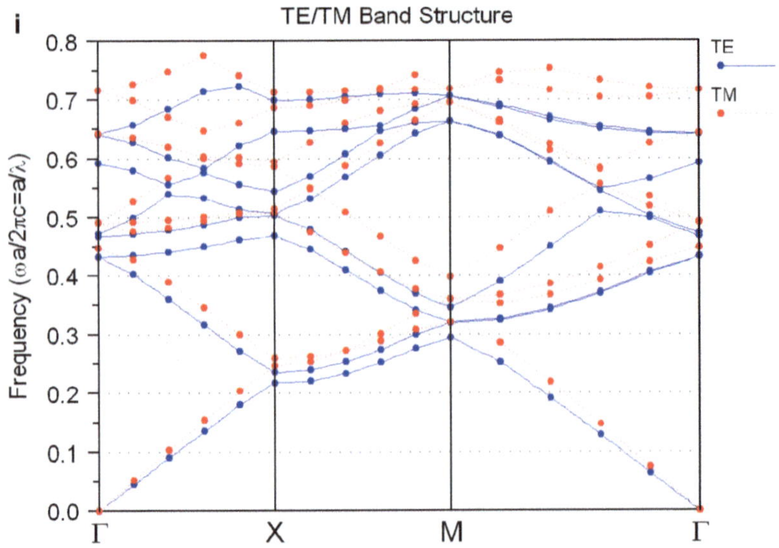

Fig. 8.9 (continued)

Besides, the transmittance of photonic crystals is shown in Fig. 8.12. When the pitch of photonic crystal is 600 nm, the etch depth is 110 nm, the transmission is higher, and the selectivity of transmission at 460 nm is better.

The above simulation is only for the p-GaN layer, ignoring the ITO layer effect. Hexagonally arranged photonic crystals are better than that of square arrangement. Photonic crystal was transferred to the ITO layer into the p-GaN layer, as schematically shown in Fig. 8.13. The etched depth is between 250 and 550 nm.

The enhancement factor of extraction efficiency (E_r) of LED with PC structure is defined as the ratio of light extraction efficiency (E) of LED with PC structure and without PC structure (E_o):

$$E_r = \frac{E}{E_o}. \tag{8.15}$$

Figure 8.14 shows the relation of etch depth and light extraction efficiency. When pitch $a = 0.2, 0.3, 0.6$, and 1 μm and etch depth is all 375 nm, the E_r is 0.3, 0.5, 0.33, and 0.2, respectively. If $a = 0.3$ μm, $R_p = 0.5$, and etch depth is 375 nm, extraction efficiency of LED is enhanced five times. If $a = 0.8$ μm, $d = 0.3$ μm, and etch depth is 475 nm, it is enhanced 2.8 times. With an increment of etch depth, light extraction efficiency is also increased. But if the etch depth is more than the value of the wavelength of guide modes in p-GaN types (187.5–195.8 nm, $\sim\lambda/n$), light extraction efficiency begins to decrease. If light extraction efficiency is maximum,

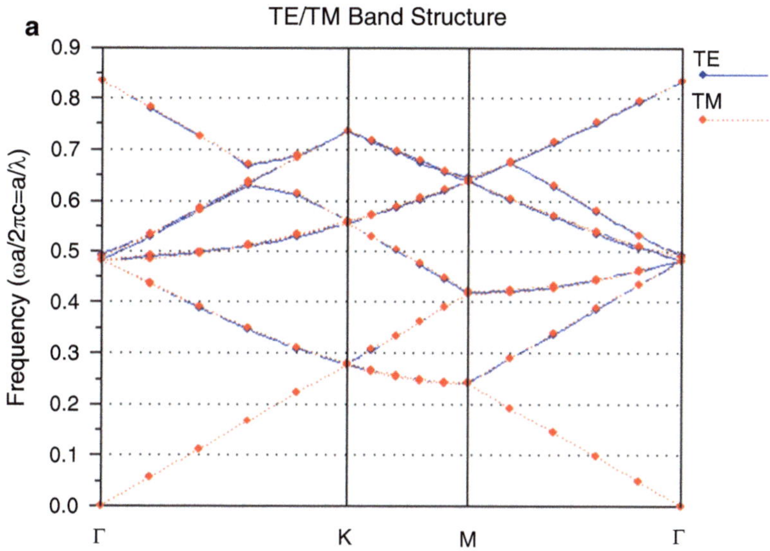

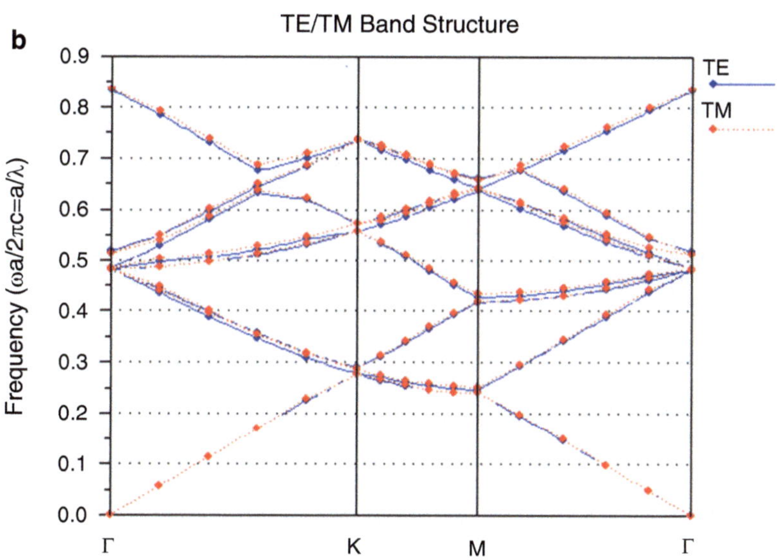

Fig. 8.10 Band diagram for hexagon-type photonic crystals made of air holes in GaN. (**a**) $R_p = 0.1$. (**b**) $R_p = 0.2$. (**c**) $R_p = 0.3$. (**d**) $R_p = 0.4$. (**e**) $R_p = 0.5$. (**f**) $R_p = 0.6$. (**g**) $R_p = 0.7$. (**h**) $R_p = 0.8$. (**i**) $R_p = 0.9$. (**j**) $R_p = 1$

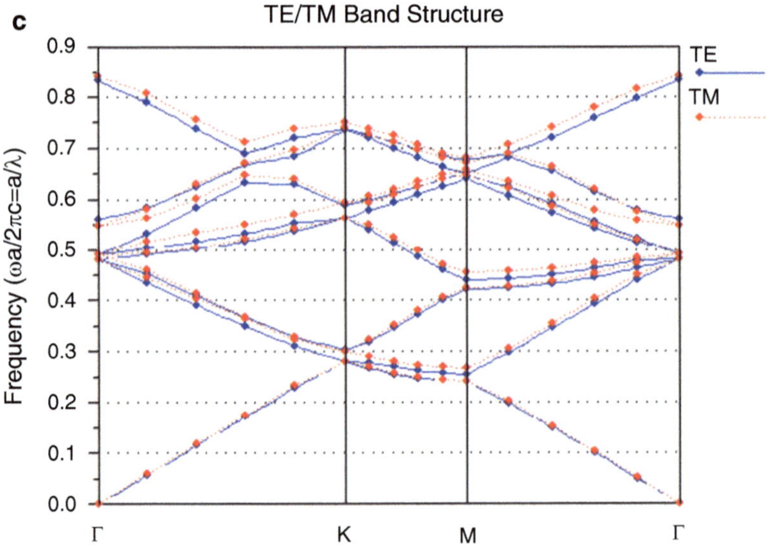

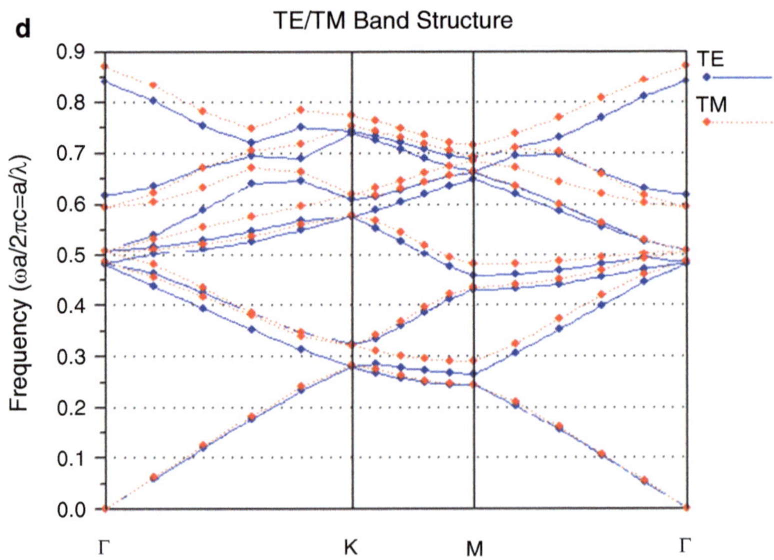

Fig. 8.10 (continued)

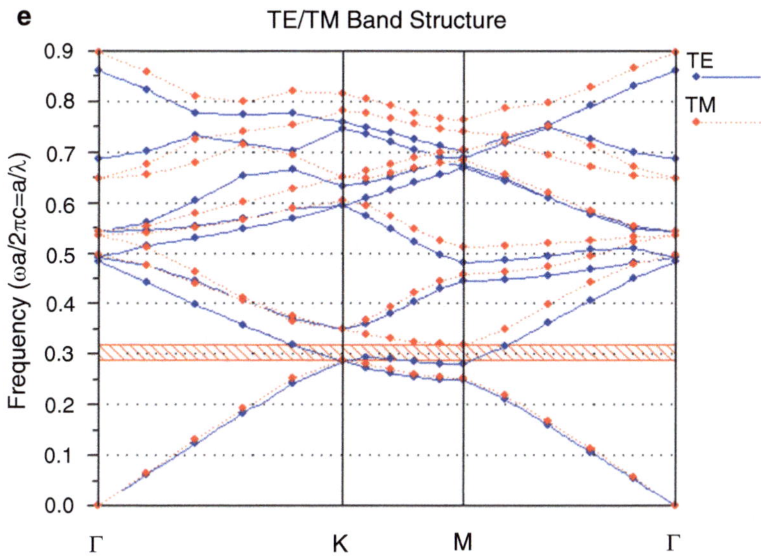

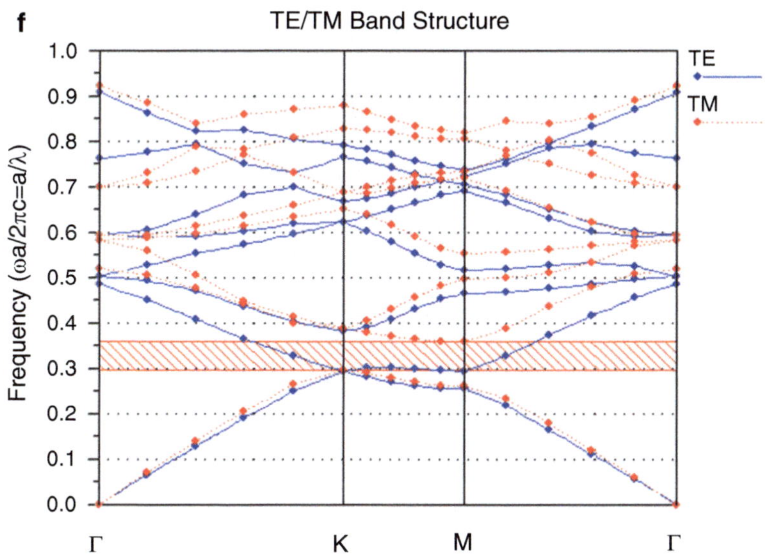

Fig. 8.10 (continued)

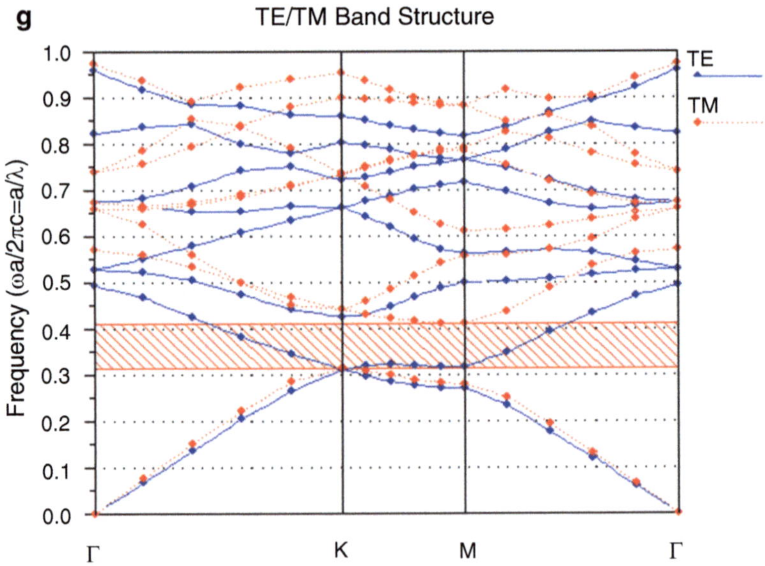

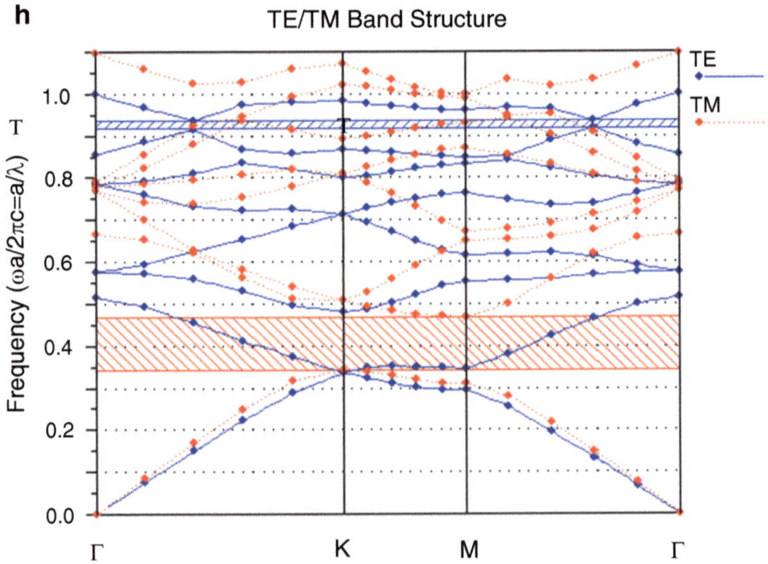

Fig. 8.10 (continued)

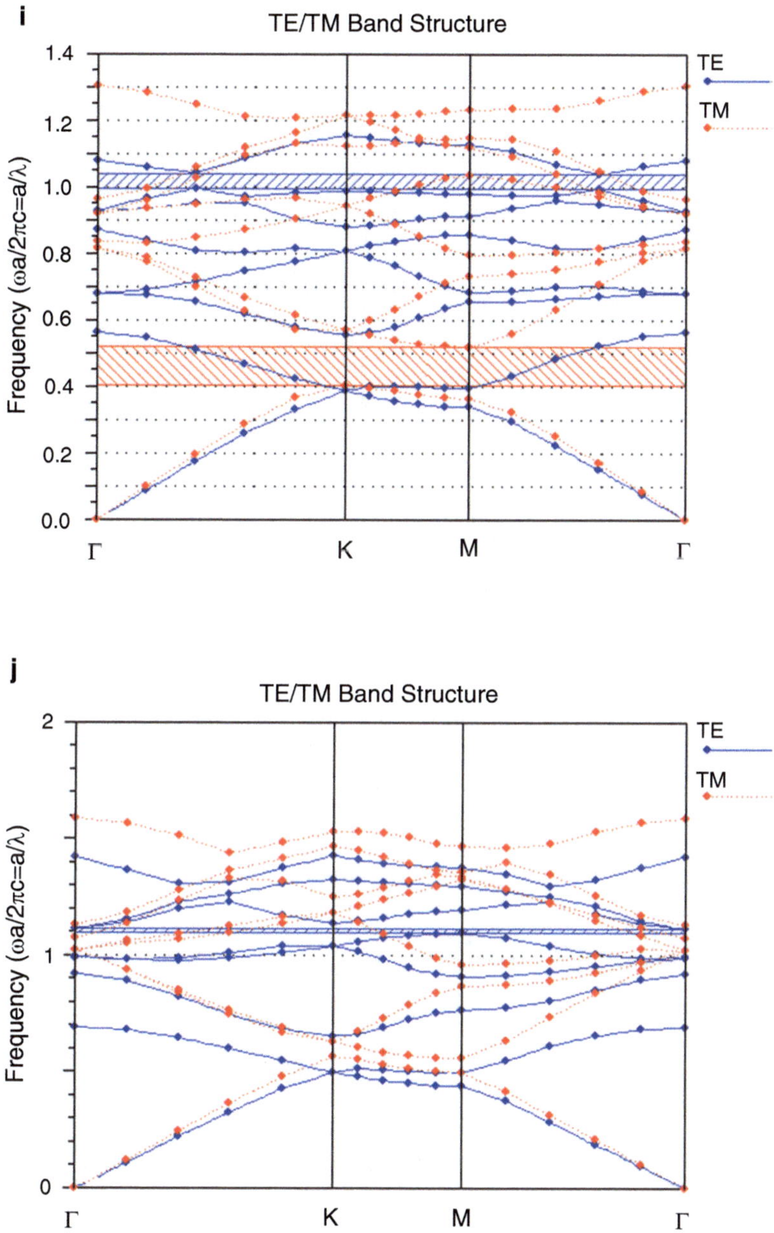

Fig. 8.10 (continued)

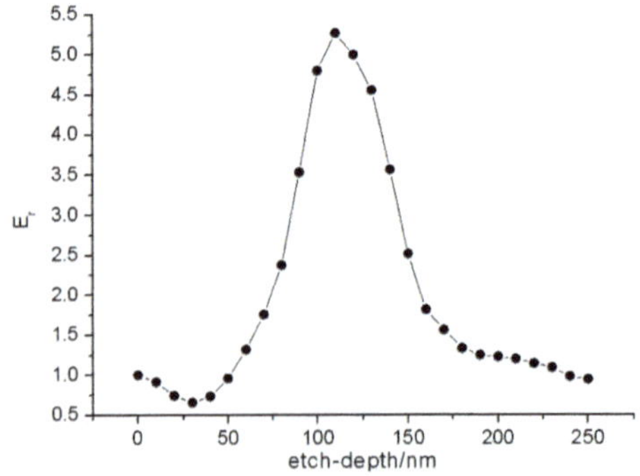

Fig. 8.11 E_r as a function of etch depth with $a = 430$ nm when R_p is 0.8 and $r = 172$ nm

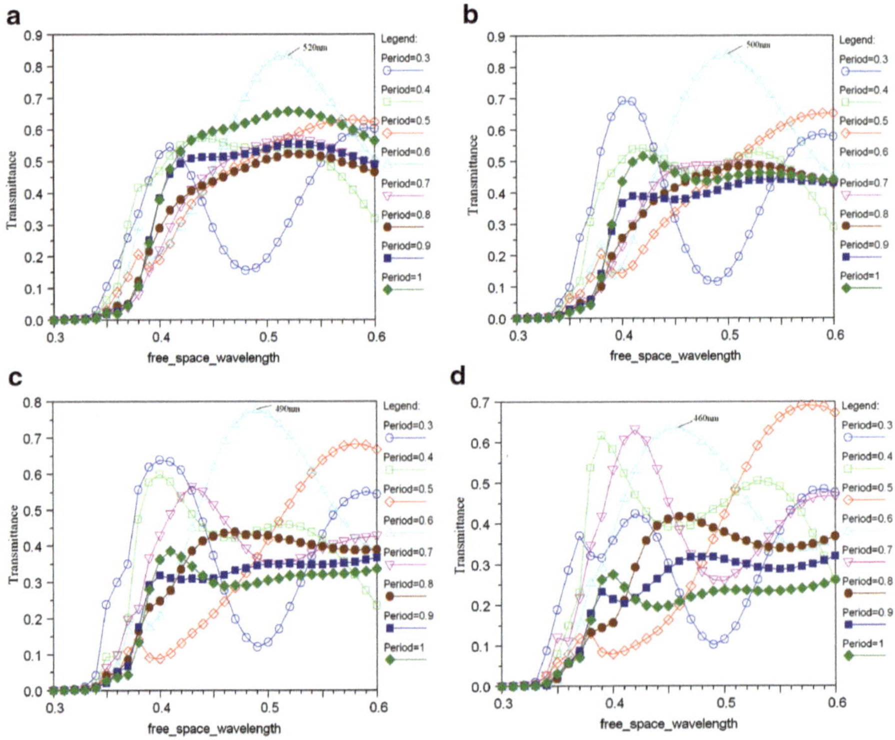

Fig. 8.12 The transmittance of photonic crystals. (**a**) $R_p = 0.5$, (**b**) $R_p = 0.6$, (**c**) $R_p = 0.7$, (**d**) $R_p = 0.8$

the reflection must be minimum, only $f = 0.15$ and $R_p = 0.41$. It is also confirmed from Fig. 8.14 that the maximum extraction efficiency R_p is approaching 0.41.

Etch depth is given by

$$h = (2m + 1)\lambda_o/(4n_1), \quad n_1 = 1.548 \tag{8.16}$$

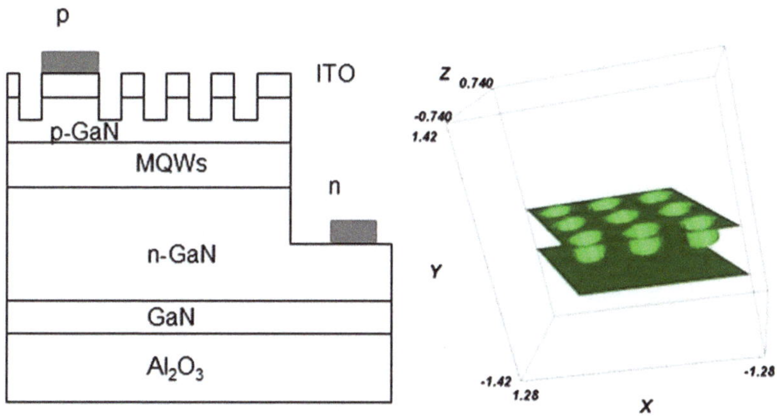

Fig. 8.13 The schematic of GaN photonic crystal LED

When $m = 1$, $h = 222.8 < 250$ nm and when $m = 2$, $h = 371.4$–375 nm. The calculating result is agreeable with the data presented in Fig. 8.14.

The transmittance curves have been demonstrated in Fig. 8.15. When $a = 300$ and 600 nm, the transmission is higher and the selectivity of transmission at 460 nm is better.

When $a = 600$ nm and etch depth = 375 nm, the transmittance as a function of R_p is shown in Fig. 8.16. When the optional parameter is $a = 0.3$ μm, $R_p = 0.5$, and etch depth is 375 nm, the reflection of photonic crystals is $3.42 \times 10^{-4}\%$ and the extraction efficiency is improved to about five times.

8.4.4 *Fabrication Tolerances and Defects*

The geometric parameters of photonic crystals deviate the designed value, taking into account fabrication tolerances. Using FDTD (finite-difference time-domain) method, the geometric parameters of random disturbance have an impact on the extraction efficiency of PC-LEDs.

Actual radius (r) and designed radius (R) of photonic crystals often occur because of the fabrication process. If $r/R \neq 1$, there is a defect. In the calculated region, the number of photonic crystals having defects is n and the ratio of defect is $\frac{n}{25} \times 100\%$. In the research, the effect of the number of defects and r/R on the light extraction efficiency of photonic crystal LED is explained.

The decreased degree of extraction efficiency is defined by

$$\Delta\eta\% = \frac{\eta_0 - \eta}{\eta_0}, \tag{8.17}$$

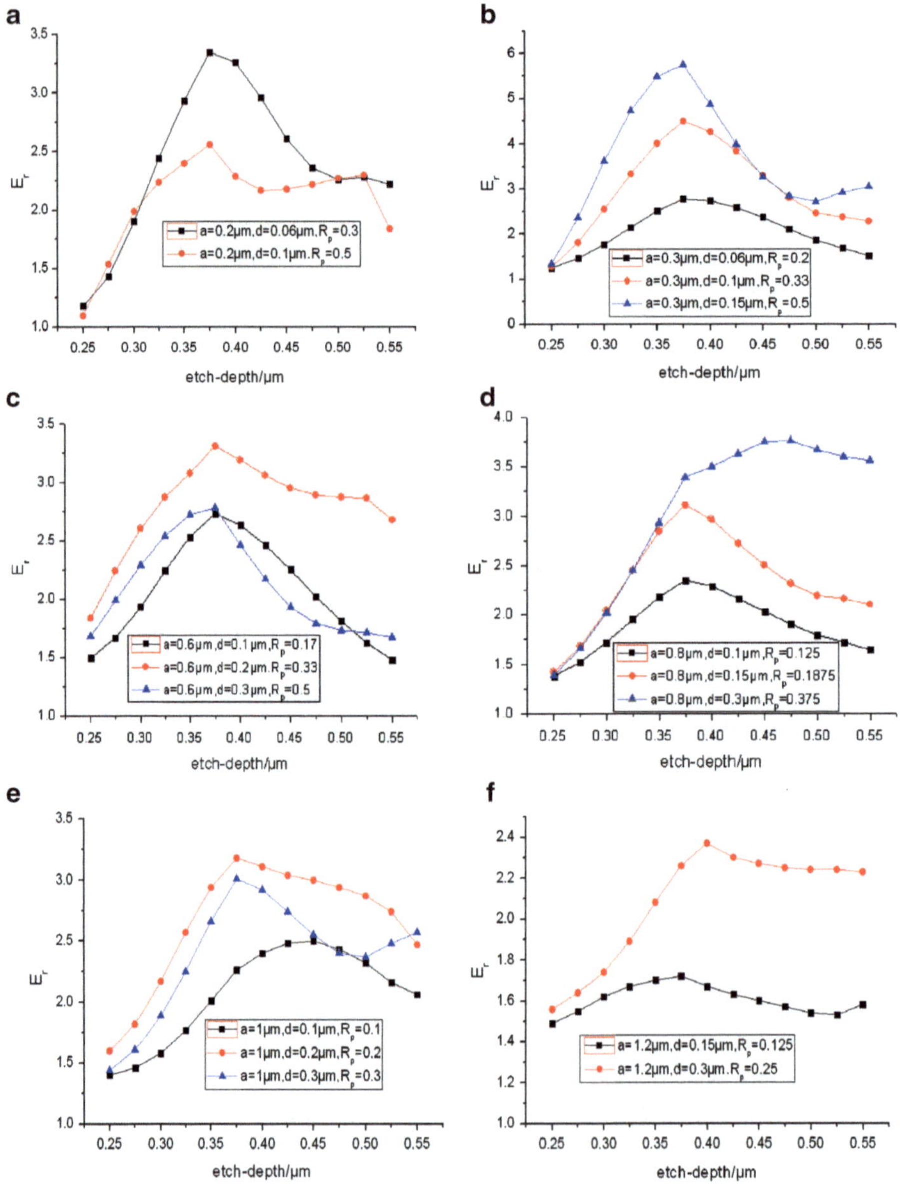

Fig. 8.14 E_r as a function of etch depth. (**a**) $a = 200$ nm, $d = 60$ and 100 nm. (**b**) $a = 300$ nm, $d = 60$, 100, and 150 nm. (**c**) $a = 600$ nm, $d = 100$, 200, and 300 nm. (**d**) $a = 800$ nm, $d = 100$, 150, and 300 nm. (**e**) $a = 1$ μm, $d = 0.1$, 0.2, and 0.3 μm. (**f**) $a = 1.2$ μm, $d = 0.15$ and 0.3 μm

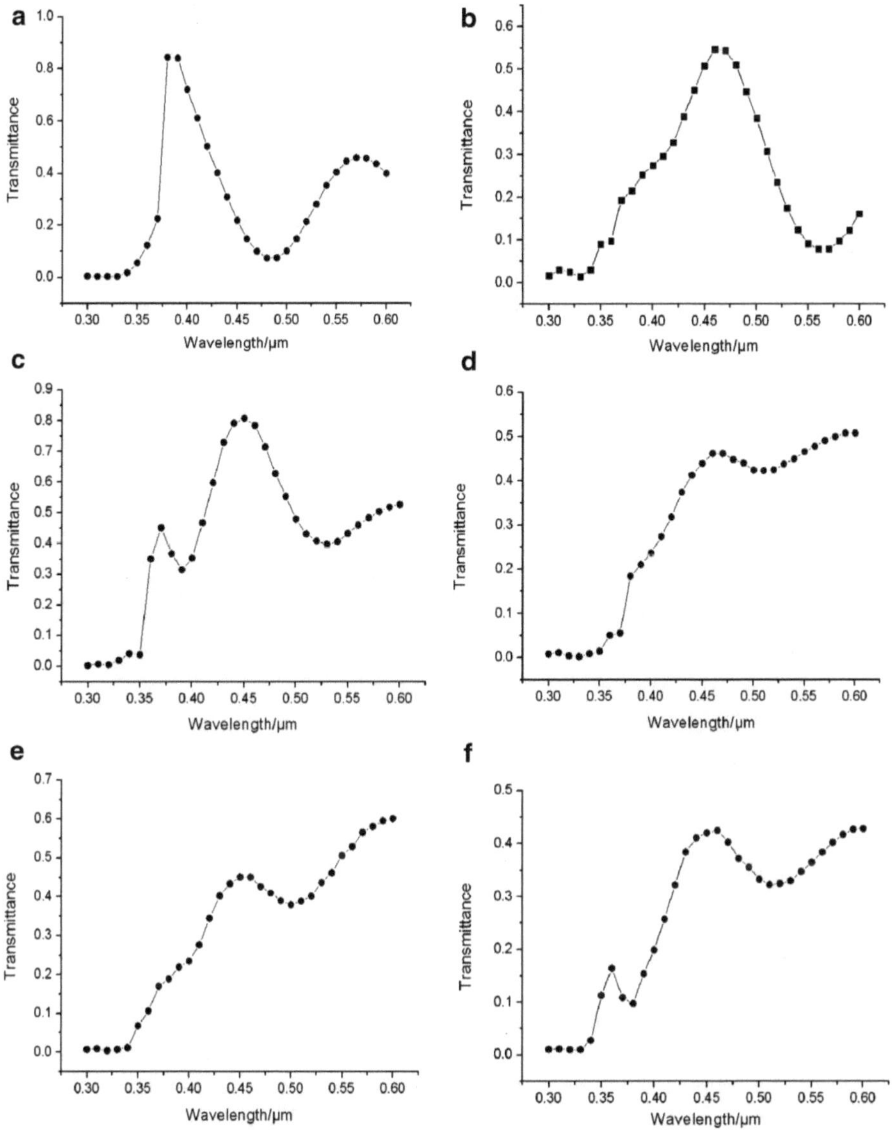

Fig. 8.15 The transmittance curves. (**a**) $a = 200$ nm. (**b**) $a = 300$ nm. (**c**) $a = 600$ nm. (**d**) $a = 800$ nm. (**e**) $a = 1{,}000$ nm. (**f**) $a = 1{,}200$ nm

where η_0 and η are light extraction efficiency of photonic crystals without defects and with defects, respectively. The calculated unit is limited to a space of $2 \times 2.5 \times 2$ μm, as shown in Fig. 8.17. In order to accurately restore, the periodic boundary condition (PBC) was set around the x–z plane, and perfectly matched layer (PML) is set on the y-axis direction, as shown in Fig. 8.17.

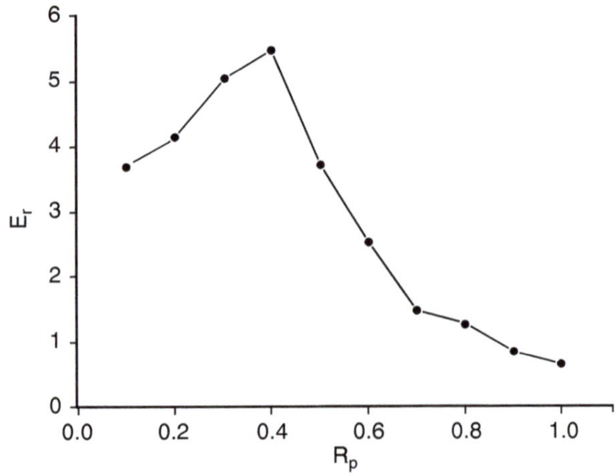

Fig. 8.16 The efficiency as a function of R_p

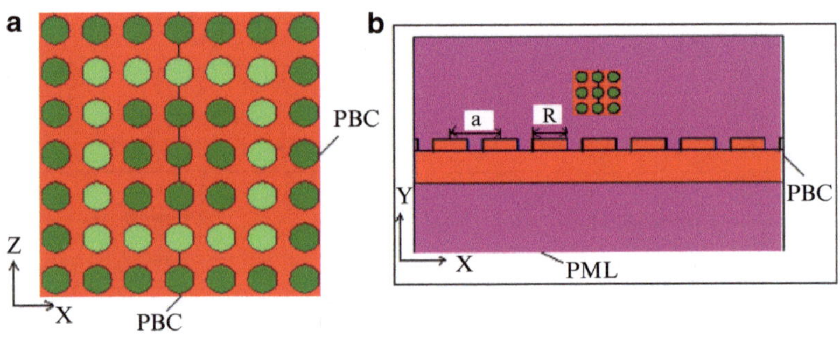

Fig. 8.17 Boundary conditions of FDTD (**a**) of *x–z* plane and (**b**) of *x–y* plane

8.4.4.1 Square Lattice

Figure 8.18 shows light extraction efficiency as a function of the number of defects and r/R. It can be seen that the light extraction efficiency is maximum when defects are zero. The decreased degree of extraction efficiency depends on the value of r/R. When n is 1, 2, 3, 4, and 5, the maximum $\Delta\eta\%$ is 17.5, 33.5, 39.6, 47.6, and 49.2%, respectively.

Figure 8.19 shows bandgap as a function of r/R when n is 1. When r/R is 1, or without fabrication tolerances, the number of bandgaps is maximized and a maximum light extraction efficiency is obtained. When r/R is 0.6, only one bandgap occurs. When r/R is less than 0.4, there is no bandgap. When r/R is more than 1, one bandgap appears. The results are agreeable with the aforementioned conclusion.

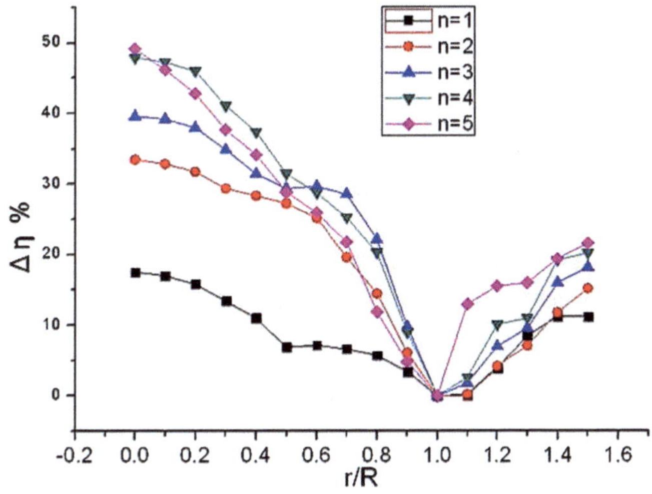

Fig. 8.18 $\Delta\eta\%$ as a function of r/R, $n = 1$–5

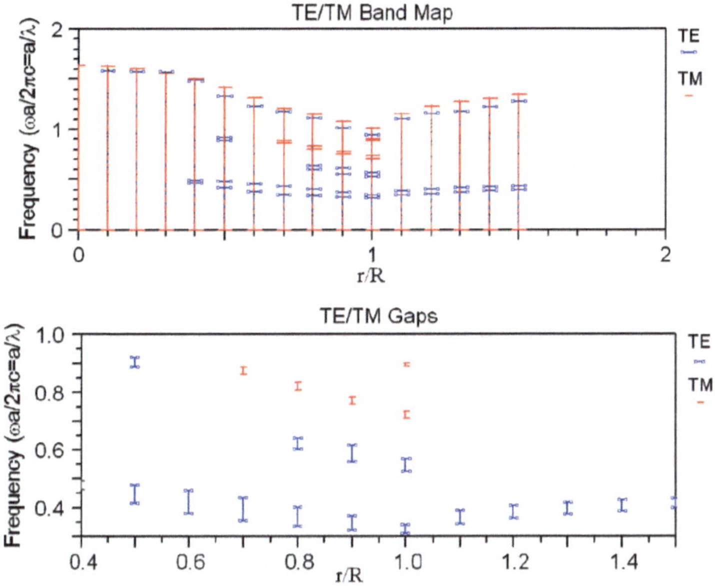

Fig. 8.19 Bandgap numbers and widths of square lattice holes of 2D photonic crystals with different r/R

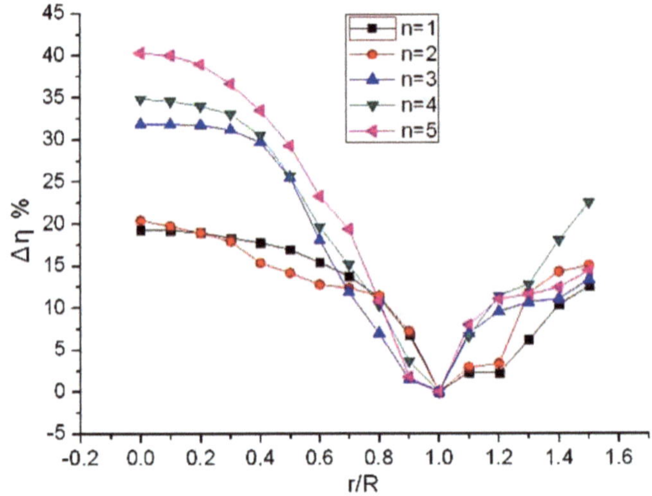

Fig. 8.20 $\Delta\eta\%$ as a function of *r*/*R*, $n = 1$–5

8.4.4.2 Hexagonal Lattice

Figure 8.20 shows light extraction efficiency as a function of the number of defects and *r*/*R*. The decreased degree of light extraction efficiency depends on the value of *r*/*R*. When *n* is 1, 2, 3, 4, and 5, the maximum $\Delta\eta\%$ is 19.3, 20.4, 31.9, 34.5, and 40.4%, respectively. Furthermore, the property of hexagonal lattice is superior to that of square lattice.

To further demonstrate the issue, the emitted light energy distributions are given in Fig. 8.21. It can be seen that the energy distribution is not uniform. Energy distribution in the vertical direction of guided wave propagation is relatively uniform, and the energy amplitude is small, as shown in Fig. 8.21a. When square lattice holes were introduced into photonic crystal structure, the peak of energy forms and the energy was greatly enhanced (Fig. 8.21b). As seen from Fig. 8.21c, the energy distribution is greatly improved. Therefore, the structure of PC-LED has an impact on light extraction efficiency. The hexagonal lattice is more excellent than the square lattice in light extraction efficiency.

8.4.5 Pitch Designing of Photonic Crystals

The pitch of photonic crystals is the other issue to be considered in the designing of photonic crystals. If the pitch is large, the diffraction distance is also largex, and the light cannot emit outside of the LED chip. By designing the pitch, the light can escape from LED. Figure 8.22 shows the influence of photonic crystal pitch on the diffraction direction. Figure 8.22a shows that some of the originally trapped guided modes with $|k| > k_0$ become Bloch modes with $||k| + G| < k_0$. Some of the originally extracted modes are coupled to the reciprocal lattice and turn back as

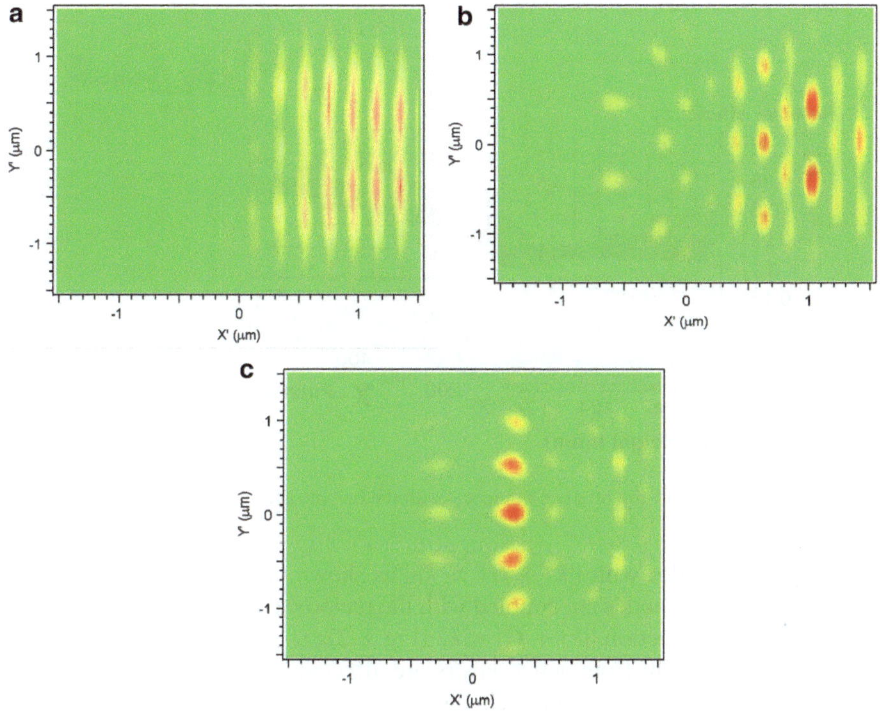

Fig. 8.21 Emitted light energy distribution. (a) No photonic crystals. (b) Square arrays of circular photonic crystals. (c) Hexagonal arrays of circular photonic crystals

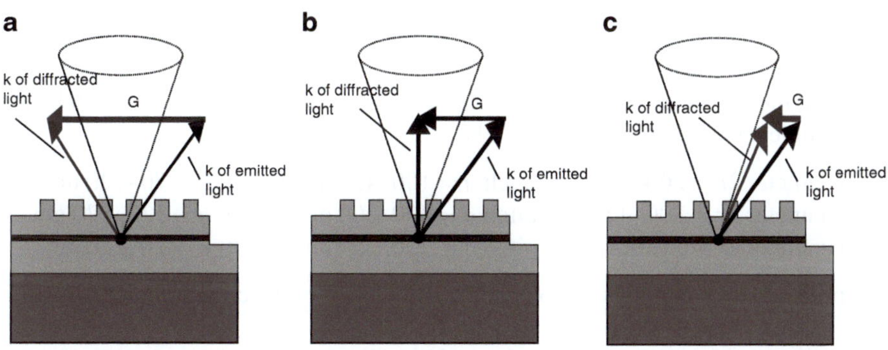

Fig. 8.22 Influence of pitch on the diffraction direction. (a) the originally trapped guided modes with $|k| > k_0$ become Bloch modes with $|k|| + G| < k_0$. (b) the originally extracted modes are coupled to the reciprocal lattice and turn back as trapped guided modes with $|k|| + G| > k_0$. (c) the originally extracted modes are coupled with the reciprocal lattice and still extracted with the changed direction $|k|| + G| < k_0$

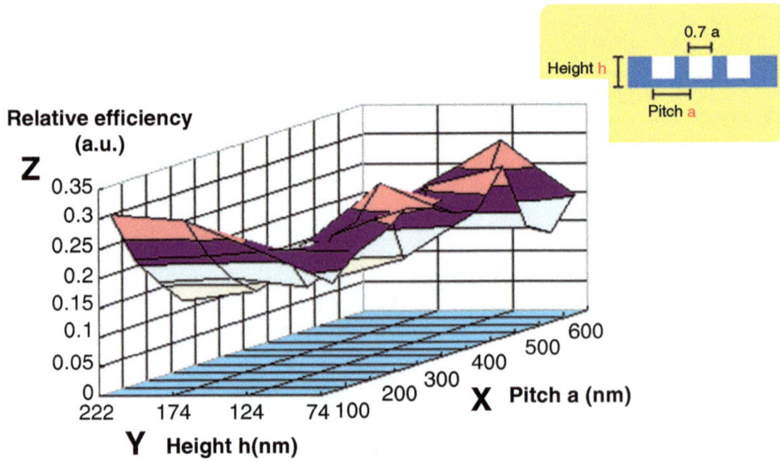

Fig. 8.23 Relationship of the relative efficiency with the heights and pitches of photonic crystals

trapped guided modes with $||k| + G| > k_0$, as shown in Fig. 8.22b. Besides, the originally extracted modes are coupled with the reciprocal lattice and still extracted with the changed direction $||k| + G| < k_0$ (Fig. 8.22c).

In fact, not all light from active layers are affected by diffraction effect. Light extraction efficiency depends on pitch, duty ratio, depth of holes, and so on.

Figure 8.23 shows the relationship of the relative efficiency and height and pitch of photonic crystals. As can be seen, two approaches are possible to obtain higher light extraction efficiency: one is small pitch and large ratio of depth/width. For example, pitch is below 150 nm, and height is more than 179 nm. The other is larger pitch and low ratio of depth/width. For example, the pitch is in a range of 350–600 nm, and height is less than 120 nm. Obviously, the latter approach is suitable for fabrication requirement.

In all, the following parameters can be selected for design parameters to enhance light efficiency of photonic crystal LED:

(a) Diameter $2r = 200$–315 nm, pitch $a = 100$–450 nm, height $h = 70$–120 nm
(b) Diameter $2r = 300$–420 nm, pitch $a = 450$–600 nm, height $h = 120$–170 nm
(c) Diameter $2r = 175$–245 nm, pitch $a = 250$–350 nm, height $h = 70$–120 nm

where $R_p = 0.7$, r is radius of hole, and a is the pitch of the circular hole lattice.

8.5 Nanoimprint LED Process

8.5.1 Compatibility with LED Process

LED chip process includes epitaxy growth, chip fabrication, and packaging. Nanoimprint lithography process should have good compatibility with conventional

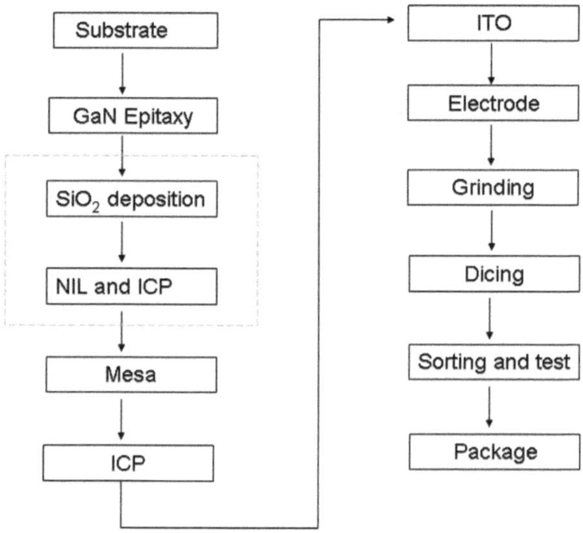

Fig. 8.24 Compatible process design of PC-LED fabrication by combining NIL technology and semiconductor processes

LED chip process. The key issue is to transfer the imprinting patterns into substrate surface and not to change the assembly line. The photonic crystals can be fabricated on the surface of ITO layer or GaN layer or sapphire.

Figure 8.24 shows a compatible process design of PC-LED fabrication by combining NIL technology and semiconductor processes. The process is as follows:

(a) GaN wafer substrate was made by conventional LED process, such as MOCVD technique.
(b) ITO/GaN surface was coated a layer of SiO_2, which acts as etching layer. In order to avoid damage of GaN wafer, a low imprint force is necessary, resulting in shallow hole patterns. If only resist was used and the deep holes cannot be transferred onto the surface of the substrate. SiO_2 film was coated by PECVD process. And the reaction gas mixtures are SiH_4 and N_2O.
(c) The UV-imprinting resin was coated on the surface of SiO_2 layer.
(d) Alignment of nanoimprint lithography.
(e) UV cured and separation of stamp and substrate.
(f) The residual layer was removed by plasma.
(g) The dry etch was used to transfer the patterns into the surface.
(h) The next step is performed by a conventional LED process.
(i) The LED chip was packaged, tested, and analyzed.

8.5.2 *Imprint Patterning Solutions*

As we know, nanoimprint lithography can fabricate micro-/nanostructures with low cost and large area. Photonic crystal structures can enhance extraction efficiency

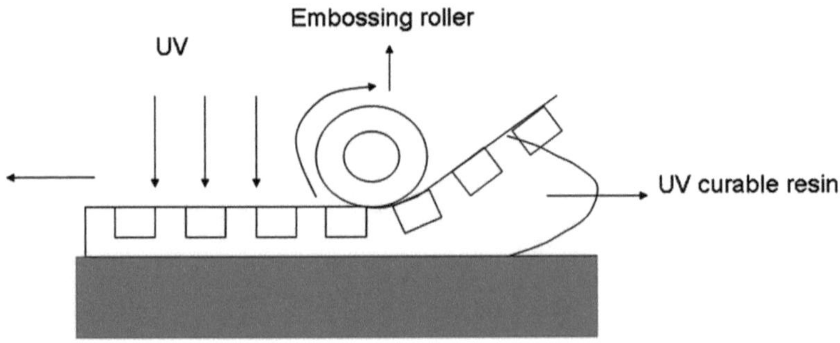

Fig. 8.25 Schematic of pattern transfer with a disposable stamp film. A roller is shown to press the patterned film into a UV-curable polymer layer on a harder substrate. Photo-curing of the imprinted polymer allows clean separation of the master

of LED. Often, GaN surface is non-flatness, and optical lithography is not reliably used for the fabrication of photonic crystals. Electron beam lithography is too slow and expensive. Nanoimprint lithography is an idea patterning method to fabricate photonic crystals of LED.

As for wafer non-flatness, soft nanoimprint lithography is a good choice. The flexible mold makes it suitable for the nonplanar nanoimprint process, thus having wider application. Several vendors have proposed some full imprinting techniques. EVG uses a two-layer plastic mold. MII and Nanonex use a conformal hard thin glass mold. Obducat uses a "one-use" plastic mold in a two imprinter [24]. SUSS MicroTec recommends the substrate conformal imprint lithography (SCIL) technique. The aforementioned methods are full-wafer imprint, and each time a wafer can be patterned. Roller nanoimprint lithography is a continuous process and can be applied in the fabrication of photonic crystals of LED. The schematic of roller imprint used for photonic crystal pattern is shown in Fig. 8.25.

8.6 Photonic Crystal Types by Nanoimprint Lithography

Great progress was made in photonic crystal LED in recent years. Photonic crystals can be formed on the surface of ITO layer or GaN layer or sapphire or both layers. The lattice is either square/hexagonal holes or rods. The light output of photonic crystal LEDs is increased by two times, showing a potential application in enhancing the extraction efficiency of LED.

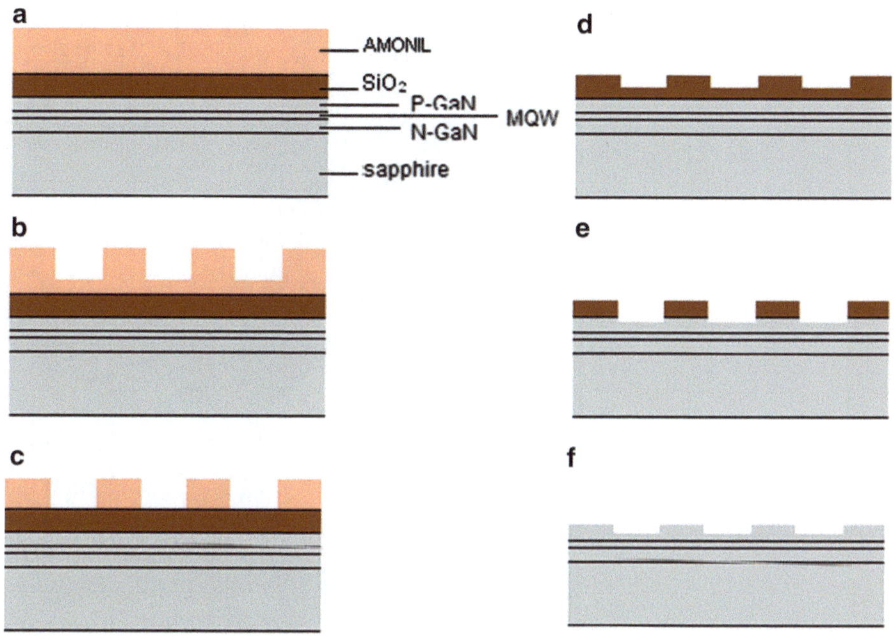

Fig. 8.26 A Schematic of the fabrication flow of nanopatterned LEDs using nanoimprint lithography. (**a**) SiO_2 deposition and AMONIL resin coating. (**b**) After UV nanoimprint lithography process. (**c**) Removing residual layer. (**d**) Etching SiO_2 and removing AMONIL resin. (**e**) The p-GaN RIE etching. (**f**) Removing the residual SiO_2 layer by buffer oxidation etchant

8.6.1 ITO Photonic Crystals

ITO photonic crystals were fabricated by hot embossing technique. Under 20-mA current injection, it has been shown that forward voltages were 3.24, 3.25, and 3.24 V, while the LED output powers were 11.7, 12.6, and 13.3 mW for the conventional ITO LED, ITO LED patterned with 1.75-um holes, and ITO LED patterned with 0.85-um holes, respectively. Twelve percent enhancement in LED output power by patterning ITO layer with 0.85-um holes was achieved [25].

8.6.2 p-GaN Photonic Crystals

We fabricated p-GaN photonic crystals with quartz and soft stamp, respectively [26]. Figure 8.26 is the schematic of the process flow for nanopatterned GaN layer of LEDs. A SiO_2 layer (50 nm) was deposited on the GaN wafer surface by PECVD, which acts as the transfer layer of etching. UV-curable resin AMONIL-MMS4

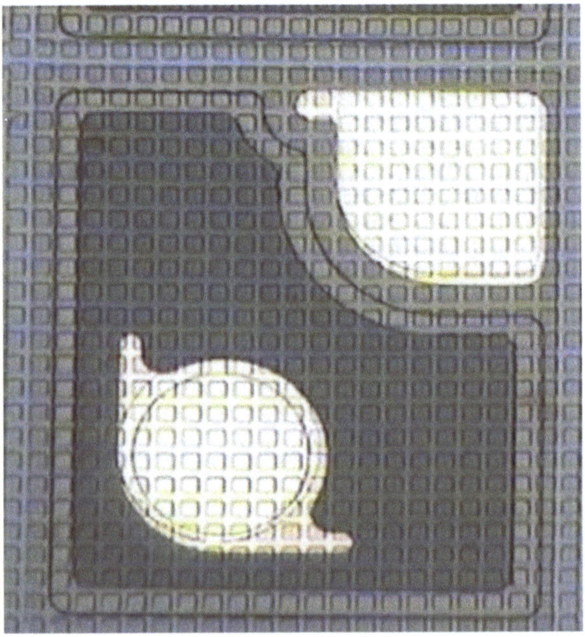

Fig. 8.27 A typical LED chip

(AMO GmbH, Germany) was spun on top of the SiO_2. Nanoimprint lithography was performed using an EVG620 machine (EVG Group, Austria) with a pressure of 300 mbar and 300 s UV exposure. A reactive-ion etching (RIE) system (Nextral 860R) is used for the removing of the residual polymer layer through O_2 plasma with a bias RF power of 100 W and a chamber pressure of 100 mTorr. Afterward, $Cl_2/Ar/CH_4$ mixtures were introduced to transfer the pattern to the SiO_2 layer and the p-GaN layer. A buffer oxidation etchant was used to remove the residual SiO_2 layer. Finally, the LEDs were fabricated by standard process with a mesa area. Figure 8.27 shows an optical photo of LED chip. An image of GaN photonic crystal is shown in Fig. 8.28.

Figure 8.29 shows measured current–voltage (I–V) characteristics of the fabricated LEDs. Under 20-mA current injections, it was found that forward voltages were 3.29, less than 3.0 V for the conventional GaN-LED, GaN-LED patterned with different diameters of holes. In Fig. 8.29, the 1a, 1b, ... denote different diameters of holes.

8.6.3 *Sapphire Photonic Crystals*

The substrate is an important factor to grow GaN wafer. Often, the sapphire and SiC are first considered. GaN, Si, and ZnO substrates are still in research phase.

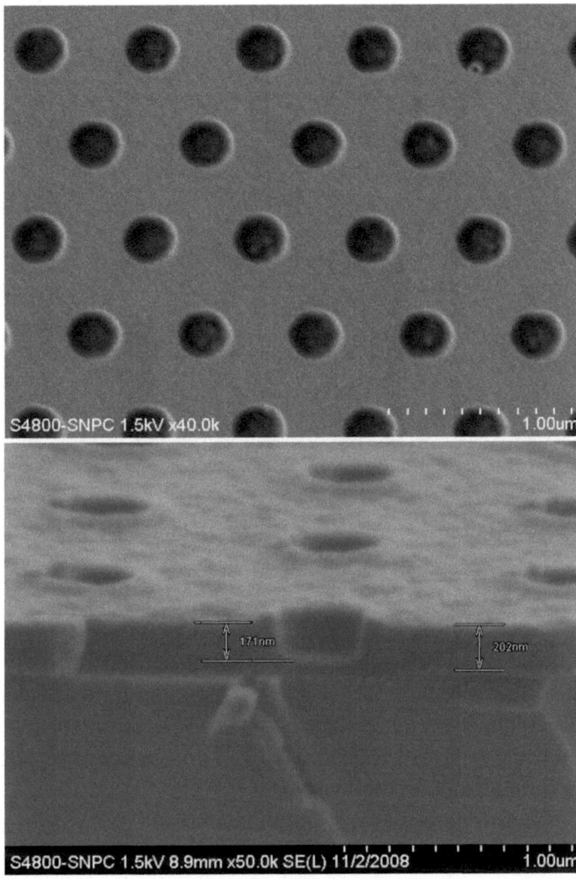

Fig. 8.28 SEM images of GaN photonic crystals

Sapphire substrate has some advantages for growing GaN wafer: fabrication technique is relatively mature, and quality can be guaranteed; the stability of sapphire is relatively higher, and it can be used at high temperatures; mechanical strength is high; and is easy to clean. These advantages render it as growing substrate. However, the lattice mismatch between GaN and sapphire exists, resulting in a large number of threading dislocations (10^9–10^{12} cm^{-2}) at the interface. Therefore, the dislocation could lead to unwanted outcomes for lifetime and quality of LED chip.

Patterned sapphire substrate technique refers to first form pattern on sapphire substrate, and epitaxy grows GaN material. It can change interface properties between GaN and sapphire and decreases defects and enhances internal quantum efficiency of LED. Meanwhile, the roughened surface can enhance light extraction efficiency for multiple reflections at the GaN/sapphire interface, resulting in the enhancement of efficiency of LED.

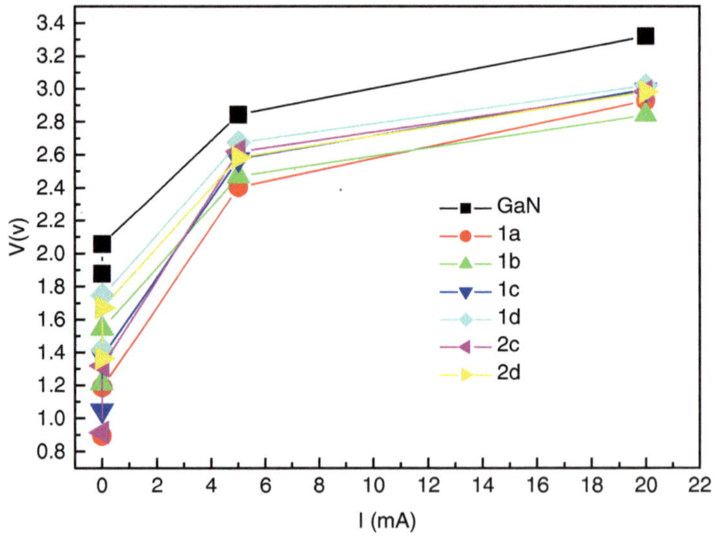

Fig. 8.29 Measured I–V characteristics of the fabricated LED

The patterned substrate is microstructures in the production. Some results show that defects can be decreased by 25% and light efficiency of LED was increased by 43%. The wet etching process is used for fabrication of patterned sapphire substrate, and it can avoid the damage by dry etching. Figure 8.30 is an AFM image of patterned sapphire substrate.

Nanoscale patterned sapphire substrate (NPSS) is very expensive by conventional lithography. Nanoimprint lithography is one of the most promising techniques for nanoscale pattern fabrication due to its capability. The sapphire substrate is fabricated by nanoimprint lithography in Luxtaltek Corporation. The patterned substrates are nanohole patterned sapphire substrate (NHPSS) and a double 12-fold photonic quasicrystal (PQC). The schematic diagram is shown in Fig. 8.31.

At a driving current of 20 mA, light output power of an LED with NHPSS and an LED with PQC structure is enhanced by 34 and 61%, respectively, compared with the conventional LED. The SEM images of NPSS are not shown here [27].

8.6.4 Flip Chip

The conventional chip has also a disadvantage of heating and static electricity and some light that cannot escape. The flip chip was used for higher power LED. The patterned sapphire substrate is also used for the flip chip of LED, as shown in Fig. 8.32.

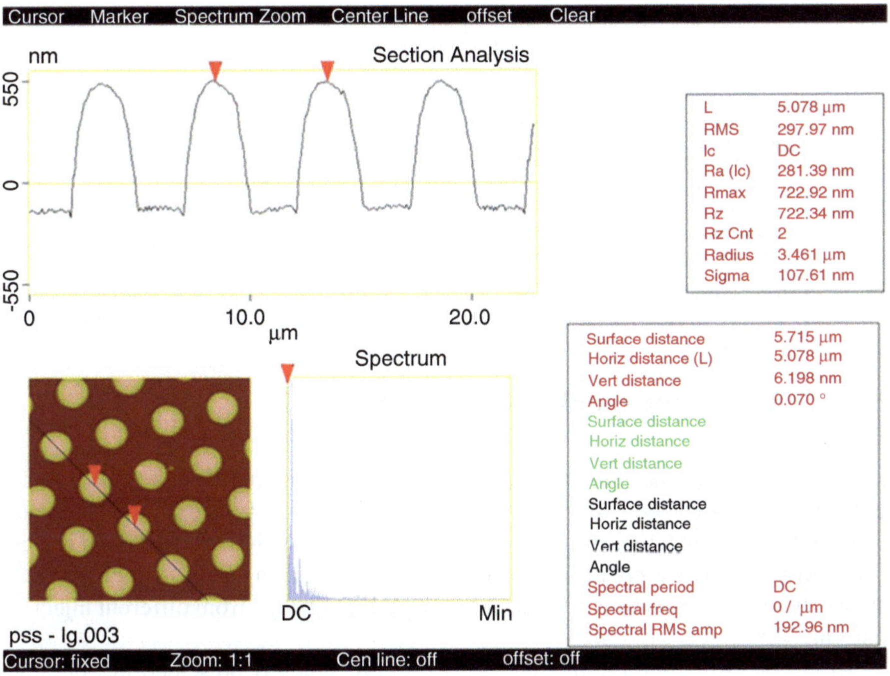

Fig. 8.30 AFM image of patterned sapphire substrate

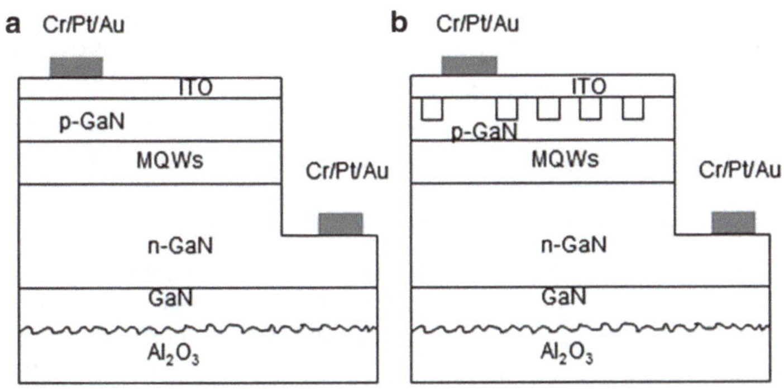

Fig. 8.31 The schematic diagram of LED. (**a**) LED with NHPSS. (**b**) LED with double PQC structure

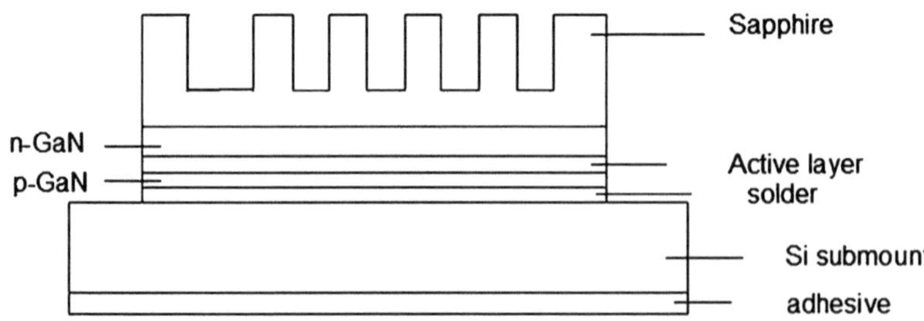

Fig. 8.32 Patterned sapphire used for flip chip

8.7 Encapsulation and Testing of LED

LED encapsulation can not only protect the LED chip but also enhance output efficiency. It is involved with electronic, thermal, photonic, and structural characteristics. After package, LED property was measured by PMS-50. At a driving current of 20 mA, electronic characteristic of LED was given in Table 8.1. Figure 8.33 shows the room-temperature PL intensity mapping obtained from different injection current.

After encapsulation, the relative extraction efficiency is presented in Fig. 8.34. The 1a, 1b, 1c,... denote different dimensions of photonic crystals on the p-GaN surface.

According to simulation and experimental results, it has been shown that the higher the ratio of air/GaN (diameter/pitch), the higher the extraction efficiency. Besides, if the ratio of air/pitch is the same, the extraction efficiency depends on the pitch and etching depth of photonic crystals. The light energy distribution of the same R_p, but different pitch is given in Fig. 8.35.

It has been shown that the smaller pitch has a higher penetration, resulting in higher extraction efficiency. The simulation data agrees well with the experiment results.

It should be pointed that defects of photonic crystals, transparent electrodes, and package materials all have an impact on extraction efficiency of LEDs. The factors are not considered in the simulation process. The maximum extraction efficiency is relative. In the further research works, the simulation should be optimized and further provides guidance for photonic crystals used in extraction efficiency of LEDs.

Table 8.1 The characteristic of LED after package

	Luminous flux (Lm)	Luminous flux efficiency (Lm/W)	Wavelength (nm)	Radiant power (mW)	Voltage (V)	Current (mA)
Without nanopatterns	0.923529	14.04	474.1857	10.22114	3.296286	20
1a	0.94185	15.9275	479.625	8.06475	2.94575	20
1b	0.88882	15.008	477.94	8.272	2.9574	20
1c	1.027386	17.24571	477.4	9.770286	2.979857	20
2c	0.668267	11.38778	475.7	6.933	2.933333	20
1d	1.044029	17.63714	477.7143	9.764	2.958714	20
2d	1.134089	19.18556	475.7889	11.67556	2.954778	20

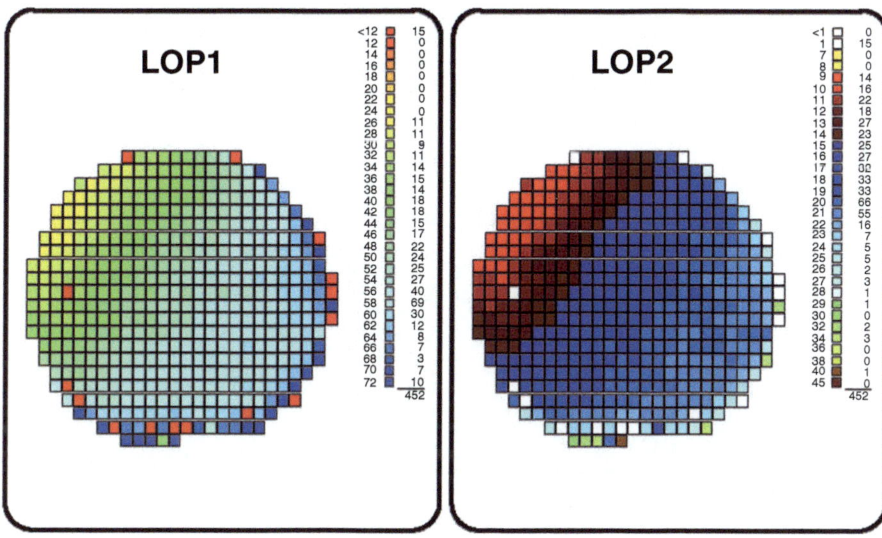

Fig. 8.33 Room-temperature photoluminescence intensity mapping of photonic crystal LED

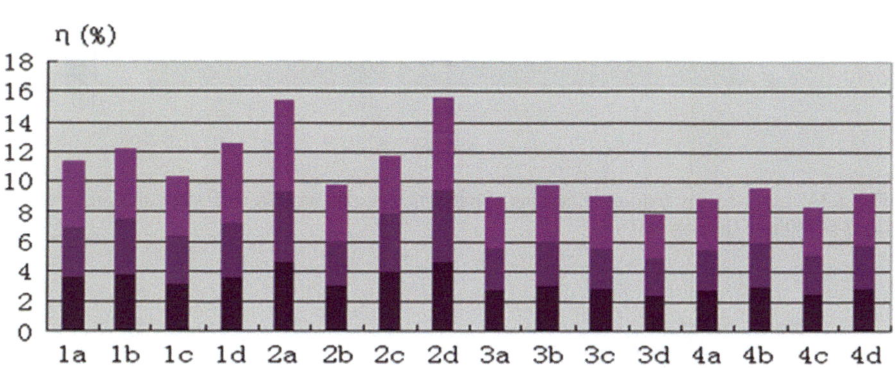

Fig. 8.34 The relative extraction efficiency of photonic crystal GaN-LED

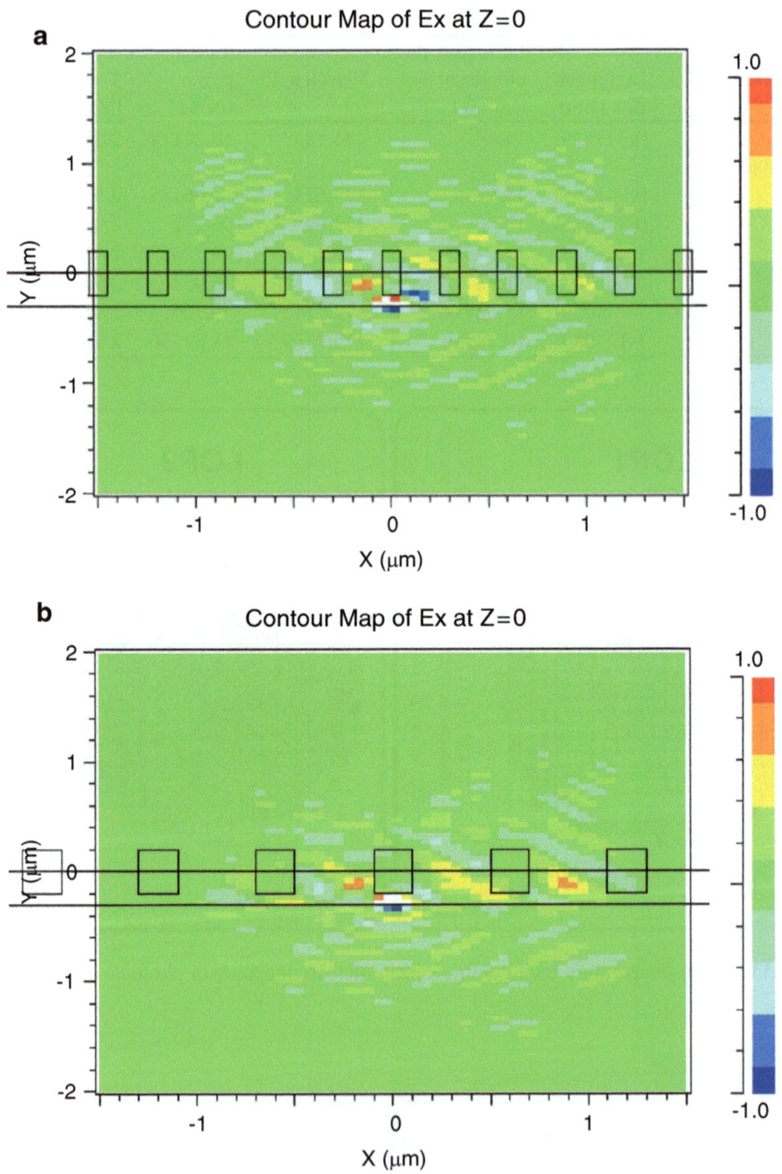

Fig. 8.35 Light energy field distributions of photonic crystals with $R_p = 1/3$ but different pitches. (**a**) $a = 300$ nm. (**b**) $a = 600$ nm

References

1. Schubert EF, Kim JK (2005) Solid-state light sources getting smart. Science 308:1274–1278
2. Huang HW, Chu JT, Kao CC, Hseuh TH, Lu TC, Kuo HC, Wang SC, Yu CC (2005) Enhanced light output of an InGaN/GaN light emitting diode with a nano-roughened p-GaN surface. Nanotechnology 16:1844–1848
3. Hsieh MY et al (2008) Improvement of external extraction efficiency in GaN-based LEDs by SiO_2 nanosphere lithography. IEEE Electr Device L 29:658–660
4. Pan SM, Tu RC, Fan YM, Yeh RC, Hsu JT (2003) Improvement of InGaN–GaN light-emitting diodes with surface-textured indium-tin-oxide transparent ohmic contacts. IEEE Electr Device L 15:649–651
5. Fan S, Villeneuve PR, Joannopoulos JD, Schubert EF (1997) High extraction efficiency of spontaneous emission from slabs of photonic crystals. Phys Rev Lett 78:3294–3296
6. Kim DH, Cho CO, Roh YG (2005) Enhanced light extraction from GaN-based light-emitting diodes with holographically generated two-dimensional photonic crystal patterns. Appl Phys Lett 87:203508–203510
7. Yablonovitch E (1987) Inhibited spontaneous emission in solid-state physics and electronics. Phys Rev Lett 58:2059–2062
8. John S (1987) Strong localization of photons in certain disordered dielectric superlattices. Phys Rev Lett 58:2486–2489
9. Armenise MA, Campanella CE, Ciminelli C, Olio FD, Passaro VMN (2010) Phononic and photonic band gap structures: modeling and application. Phys Prodeida 3:357–364
10. Leung KM, Liu YF (1990) Full vector wave calculation of photonic band structures in face-centered-cubic dielectric media. Phys Rev Lett 65:2646–2649
11. Zhang Z, Satpathy S (1990) Electromagnetic wave propagation in periodic structures: Bloch wave solution of Maxwell's equations. Phys Rev Lett 65:2650–2653
12. Pendry JB, MacKinnon A (1992) Calculation of photon dispersion relations. Phys Rev Lett 69:2772–2775
13. Kuzmiak V, Maradudin AA, Pincemin F (1994) Photonic band structures of two-dimensional systems containing metallic components. Phys Rev B 50:16835–16844
14. Yee K (1966) Numerical solution of initial boundary value problems involving Maxwell's equations in isotropic media. IEEE T Antenn Propag 14:302–307
15. Oder TN, Kim KH, Lin JY, Jiang HX (2004) III-nitride blue and ultraviolet photonic crystal light emitting diodes. Appl Phys Lett 84:446–448
16. Lee HC, Park JB, Bae JW, Thuy P, Yoo MC, Yeom GY (2008) Effect of the surface texturing shapes fabricated using dry etching on the extraction efficiency of vertical light-emitting diodes. Solid State Electron 52:1193–1196
17. Wu ML, Lee YC, Lee PS, Kuo CH, Chang JY (2008) III-nitride-based microarray light-emitting diodes with enhanced light extraction efficiency. Jpn J Appl Phys 47:J6757–J6759
18. Kim JY, Kwon MK, Park IK, Cho Y, Park SJ, Jeon DM, Kim JW, Kim YC (2008) Enhanced light extraction efficiency in flip-chip GaN light-emitting diodes with diffuse Ag reflector on nanotextured indium-tin oxide. Appl Phys Lett 93:93021121–93021123
19. Liu WD, Gan ZZ (2009) Analysis of the light extraction efficiency of light-emitting diodes with photonic crystals combined with the generalized Rouard method. J Opt Soc Am A Opt Image Sci Vis 26:289–296
20. Na SI, Ha GY, Han DS, Kim SS, Kim JY, Lim JH, Kim DJ, Min KY, Park SJ (2006) Selective wet etching of p-GaN for efficient GaN-based light-emitting diodes. IEEE Photonic Tech L 18:1512–1514
21. Krauss TF, Richard M, Rue DL (1999) Photonic crystals in the optical regime-past, present and future. Prog Quant Electron 23:51–96
22. Niu PJ, Li XY, Liu HW, Tian HT, Gao TC, Yang GH (2007) Enhancing the light extraction efficiency of GaN-based LEDs. Proc SPIE 6828:682801–682811

23. Wuu DS, Wang WK, Wen KS, Hang SC, Lin SH, Horng RH, Yu YS, Pan MH (2006) Fabrication of pyramidal patterned sapphire substrates for high-efficiency InGaN-based light emitting diodes. J Electrochem Soc 153:G765–G777
24. Michael PC (2007) The value, solution, and costs of patterning LED's. Proc SPIE 6462:64620N1-12
25. Chang SJ, Shen CF, Chen WS (2007) Nitride-based light emitting diodes with indium tin oxide electrode patterned by imprint lithography. Appl Phys Lett 91:013504-1-3
26. Zhou WM, Min GQ et al (2010) Enhanced efficiency of light emitting diode with nano-patterned gallium nitride surface by soft UV nanoimprint lithography. Nanotechnology 21:205304–205309
27. Huang HW, Lin CH, Huang ZK, Lee KY, Yu CC, Kuo HC (2009) Improved light output power of GaN-based light-emitting diodes using double photonic quasi-crystal patterns. IEEE Electr Device L 30:1152–1154

Chapter 9
Application of NIL in Memory Devices

9.1 Introduction

Semiconductor memories together with logic devices have played critical roles in the IC industry in terms of providing cost-effective solutions for ever-increasing need for data storage [1]. Among these nonvolatile products, flash memory is very popular especially in portable application fields. To enhance its density and high programming throughput, flash devices must be continued scaling to lower dimension. However, the fabrication techniques are difficult to meet the needs of scaling. Researchers are trying to develop the next-generation nonvolatile memory. Some of those emerging memory technologies, such as ferroelectric, magnetic, and chalcogenide, are under development. These memories are evolutional, but some challenges must be overcome to realize small cell sizes, high density, and low power consumption.

Nowadays, Ferroelectric Random Access Memory (FeRAM), Magnetic Random Access Memory (MRAM), Resistive Random Access Memory (RRAM), and Phase-Change Random Access Memory (PCRAM) have been developed intensively, and there is a body of interesting and exciting achievements.

A FeRAM cell consists of one capacitor and one transistor, a so-called 1T1C device. In a FeRAM cell, a ferroelectric material, typical lead zirconate titanate (PZT) is used. The ferroelectric material has a nonlinear relationship between the applied electric field and the apparent charge, in the form of a hysteresis loop [2].

MRAM cell consists of a magnetic tunnel junction structure in which two ferromagnetic electrodes are separated by a thin insulating barrier. The tunneling current flowing through the junction depends on the relative orientation of the magnetization in both electrodes. The resistance change takes place between the low resistance parallel alignment and the high resistance antiparallel alignment [3].

RRAM is a two-terminal device where the switching medium material is sandwiched between top and bottom electrodes. The resistance of the middle medium material can be changed by applying electrical current or voltage.

W. Zhou, *Nanoimprint Lithography: An Enabling Process for Nanofabrication*,
DOI 10.1007/978-3-642-34428-2_9, © Springer-Verlag Berlin Heidelberg 2013

Table 9.1 Nonvolatile memory technology requirements extracted from the ITRS 2011 edition [5]

	FeRAM		MRAM		PCRAM	
Cell elements	1T1C		1T1R		1T1R or 1D1R	
Year	2009	2024	2009	2015	2009	2024
Feature size F (nm)	180	65	130	65	65	8
Cell area F^2	22	12	45	45	5	4
Read time (ns)	45	<20	20	<0.5	60	<60
W/E time	10	1	20	<0.5	50/120	<50
Retention time (years)	>10	>10	>10	>10	>10	>10
Write cycles	1.0E + 14	>1.0E + 16	>3.0E + 16	>3.0E + 16	1.0E + 9	1.0E + 15
Write operating voltage (V)	0.9–3.3	0.7–1	1.5	<1.5	3	<3
Read operating voltage (V)	0.9–3.3	0.7–1	1.5	<1.5	3	<3
Write energy (J/bit)	3.0E−14	5.0E−15	7.0E−11	2.0E−11	5.0E−12	<1.0E−13

Many materials such as perovskite oxide ($SrZrO_3$, $LiNbO_3$, $SrTiO_3$), binary metal oxide (NiO, CuO_2, TiO_2, HfO_2), and solid electrolytes (AgGeS, CuSiO) have resistance properties [4].

Phase-change memory (also known as PCME, PRAM, PCRAM, Ovonic Unified Memory, Chalcogenide RAM, and C-RAM) is one of several new memory technologies competing in the nonvolatile memory. The phase-change material switches crystalline and amorphous state by heat produced by the passage of an electric current. The following Table 9.1 lists their characteristics to reveal their advantages and disadvantages.

Among these nonvolatile memories, PCRAM is an excellent candidate for flash memory. Its advantages of fast read and write, low operating voltage, low energy operation, high intrinsic endurance and retention, smaller feature size, and high density offer the most promising candidate for nonvolatile memory. High-density PCRAM can be realized by nanofabrication approaches. Among the manufacturing processes, the nanoscale PCRAM cell device fabrication is vital. To improve its properties, two ingredients are necessary: one is nanoscale PCRAM cells, and the other is nanoscale phase-change materials. In terms of nanofabrication, a large number of techniques are used. However, nanoimprint lithography is most suitable for mass production.

9.2 Phase-Change Random Access Memory

9.2.1 Principle of PCRAM

PCRAM is based on the theory of phase-change material whose resistance between the amorphous and the crystalline states is switched by heating. The amorphous

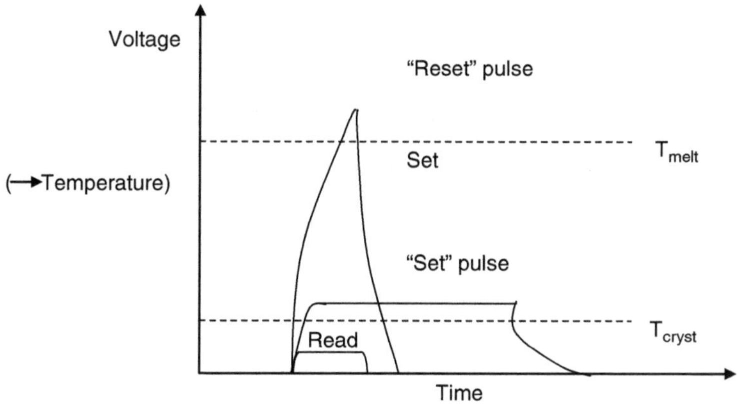

Fig. 9.1 Programming of a PCRAM device

phase has high electrical resistivity, while the crystalline phase reveals a low resistivity. A write process is realized by sending an electric pulse to heat a portion of the cell above the crystallization temperature of the phase-change material so that it is on a low resistance state. The pulse is very short, usually shorter than 10 ns. A read process is that a pulse is reset on the phase-change material and cut off abruptly and the molten material quenches into amorphous phase and behaves a high resistance state. The read operation is performed by measuring the cell devices at low voltage [6]. The operations are given in Fig. 9.1.

Since the finding of Ovshinsky, phase-change material and device fabrication developed slowly last century because of the lack of these techniques. The PCRAM has been paid little attention. In 1990s, phase-change material was widely used in re-writable optical media (such as CD-RW and DVD-RW) and reached the commercialization stage for optical media. With the development of IC, especially the nanofabrication, PCRAM has been becoming a focus for consumer electronic devices and an alternative nonvolatile memory. The memory material is a chalcogenide alloy of germanium, antimony, and tellurium (GeSbTe), called GST. Other composite materials, such as SiSbTe (SST) and GeSb, have also been used. Many semiconductor companies have been working on the PCRAM, such as Ovonyx, Intel, Samsung, STMicroelectronics, Hitachi, and British Aerospace. In September 2006, Samsung announced a prototype 512-Mb device using diode switches. Intel and STMicroelectronics demonstrated their own PCM devices at the 2006 Intel Developer Forum in October and showed a 128-Mb device. In February 2008, Intel together with STMicroelectronics revealed their first multilevel (MLC) PCM array prototype. At the same time, Intel and STMicroelectronics began releasing their first PCM products to customers. The products were 90 nm and 128 Mb. In April 2010, Numonyx announced the Omneo line of 128-Mb NOR-compatible phase-change memories. Samsung announced its shipment of 512-Mb PCRAM in a multi-chip package (MCP) for use in mobile handsets.

Table 9.2 Development of PCRAM by years since 1966

Year	Events
Sept, 1966	Stanford Ovshinsky files first patent on phase-change technology
Jan, 1969	Charles H. Sie published a dissertation on chalcogenide phase-change memory device
Jun, 1969	US Patent 3,448,302 licensed to Ovshinsky claims first reliable operation of PCM device
Sept, 1970	Gordon Moore publishes research in Electronics Magazine
Nov, 1999	Lockheed Martin works with Ovonyx on PCRAM for space applications
Feb, 2000	Intel invests in Ovonyx, licenses technology
Dec, 2000	STMicroelectronics licenses PCRAM technology from Ovonyx
Mar, 2001	Macronix found laboratory for PCRAM research
Nov, 2001	Intel, Ovonyx, and Azalea work on 4-Mb PRAM technology
Mar, 2002	Macronix files a patent application for transistor-less PRAM
July, 2003	Samsung begins work on PCRAM technology
Aug, 2004	Nanochip licenses PCRAM technology from Ovonyx for use
Aug, 2004	Samsung announces successful 64-Mb PCRAM chip
Feb, 2005	Elpida licenses PRAM technology from Ovonyx
Sept, 2005	Samsung announces successful 256-Mb PRAM chip
Oct, 2005	Intel increases investment in Ovonyx
Dec, 2005	Renesas and Hitachi announce 1.5-V PCRAM with 100 μA programming current
Dec, 2005	Samsung licenses PRAM technology from Ovonyx
July, 2006	BAE Systems begins selling the first commercial PCRAM
Sept, 2006	Samsung announces 512-Mb PCRAM chip
Oct, 2006	Intel and STMicroelectronics show a 128-Mb PRAM chip
Dec, 2006	IBM Research Labs demonstrate a prototype chip 3*20 nm
Jan, 2007	Qimonda licenses PRAM technology from Ovonyx
Feb, 2008	Intel and STMicroelectronics begin shipping samples to customers
Jun, 2009	Samsung's PCRAM goes into mass production
Sept, 2009	Samsung announces mass production of 512-Mb PCRAM device
Dec, 2009	Numonyx announces 1-Gb 45-nm product
April, 2010	Samsung releases 512-Mb PCRAM with 65-nm process
Feb, 2011	Samsung showed 58-nm 1.8-V 1-Gb PCRAM
Feb, 2012	Samsung showed 58-nm 1.8-V 1-Gb PCRAM

After phase-change material had been proposed, some major progress was listed in a chronological sequence, as shown in Table 9.2 [7].

In China, many proposals had been adopted by China government, and great progress has been made. The research proposals deal with the fabrication of phase-change material, electrode material, insulating material, memory mechanism, design and simulation, test system, chip designing, and device integration. Some universities and institutes in China are involved in these related research fields, such as Shanghai Institute of Microsystem and Information Technology, Shanghai Jiaotong University, and Fudan University.

9.2.2 Critical Issue in PCRAM

The properties of phase-change materials making PCRAM as one of the candidates for nonvolatile memories are high switching speed, high density, and low power consumption. Although PCRAM is promising, it is still in its infancy and limited by various technical difficulties. Some stringent issues are modification of phase-change materials, crystallization and electroconductibility mechanism, structure designing of devices, material thermodynamics, memory mechanism, method of reducing current, repeatability of devices, stability and failure, nanoscale cell devices, and high-density chip integration and fabrication.

Nanoscale PCRAM depends on the fabrication technique. How to realize a smaller region of the phase-change nanoscale structure is a sign of lower voltage, lower power consumption, and compatibility with CMOS technique. A great challenge remaining for PCRAM is the continued scaling down of memory cells for power efficient device operation. In nanoscale PCRAM, the instability of resistance and threshold voltage in the amorphous phase of phase-change material limits its property. In nanoscale region, phase change energy is also reduced. How to select an appropriate voltage is to match the structure feature of memory cell. Therefore, a high-density array is a fast method to test the structures and properties of PCRAM.

With shrinking dimensions, phase-change rate of phase-change material is improved and the switching power will be reduced. The cell device has a good compatibility with nanoscale CMOS process, resulting in low voltage, low operation power, and fast read/write speed. Phase-change nanowire can also be used in the PCRAM. It has been demonstrated that melting temperature of phase-change nanowire can be decreased and switching current and powers in PCRAM devices will also be reduced. Furthermore, the incubation time for recrystallization will be shortened for melt-quenched amorphous nanowires [8].

But phase-change nanowire cannot easily be manipulated for cell fabrication. In terms of nanoscale PCRAM cell devices, NIL can fabricate the nanoscale cell devices with low cost, high throughput, and high fidelity of dimension [9–13].

NIL is considered as an alternative technique for selection of novel material and device structure. We fabricated PCRAM cell, which provides a fast and simple method to test the properties of emerging phase-change material and the cell structure of PCRAM [14–16].

9.3 PCRAM Fabricated by Nanoimprint Lithography

9.3.1 Microstructure Array of Phase-Change Cell Devices

The schematic of the UV-NIL process for phase-change material (SST) array was given in Fig. 9.2. First, the substrate was coated with a layer of AMONIL resist, and the imprinting process was performed on an EVG620.

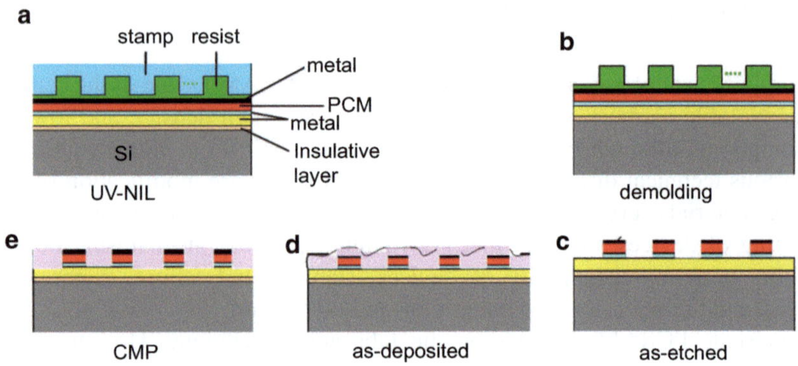

Fig. 9.2 PCRAM array was fabricated by UV-NIL

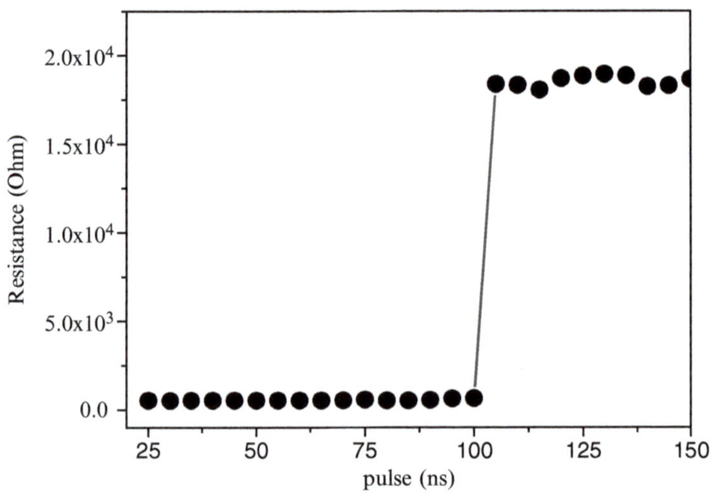

Fig. 9.3 R-TRESET characteristic curve of SST PCRAM

The imprint stamp, imprinted patterns, and as-etched SST array are fabricated by EBL, imprint, and dry etching, respectively (not shown here). It had been demonstrated that as-imprinted pattern had a good fidelity of original mold dimension. By calculation, the density of the SST was 18 $M^2/in.^2$.

Determining appropriate pulse characteristics for the RESET and SET operations is very important for PCRAM. RESET pulses of a fixed voltage (10 V) and of different lengths were applied to SST PCRAM. The maximum resistance obtained from the memory cell was plotted on the curve made of round dots in Fig. 9.3. As a result of this experiment, a RESET pulse width of 100 ns was chosen for the RESET operation. When the pulse height is constant, the heating energy increased with pulse width until the temperature is over the melting temperature of SST material. The transition from an amorphous phase to crystalline phase and RESET process was completed.

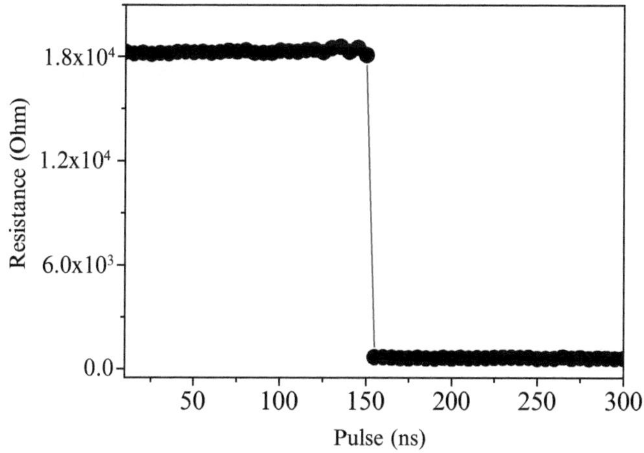

Fig. 9.4 R-TSET characteristic curve of SST PCRAM

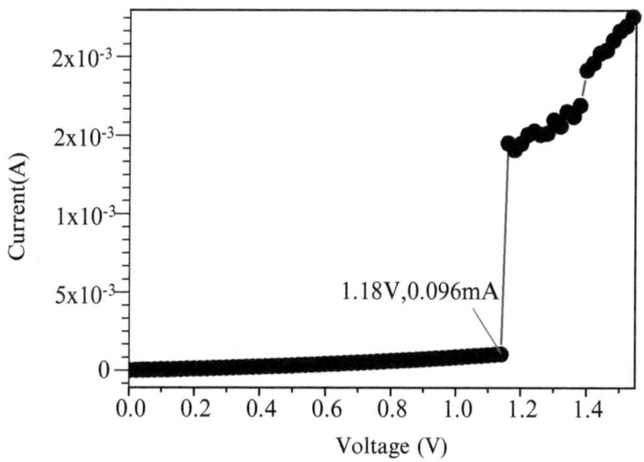

Fig. 9.5 V-I characteristic curve of SST PCRAM

The R-TSET curve is given in Fig. 9.4. There was a sharp decrease in the resistance for the shortest pulse (155 ns), which meant that it was approaching the minimum crystallization time. Note that the 155-ns SET pulse had a larger programming current than the equivalent RESET pulse. As for this pulse, enough current was therefore applied to remelt the amorphous region. However, due to the long trailing edge of the pulse, the device recrystallized instead of quenching into the amorphous phase [17].

A typical I–V curve of PCRAM cell device is shown in Fig. 9.5. The threshold voltage and threshold current of PCRAM cell were 1.18 V and 9.6 μA, respectively. By joule heating, the cell device changed from high resistance to low resistance,

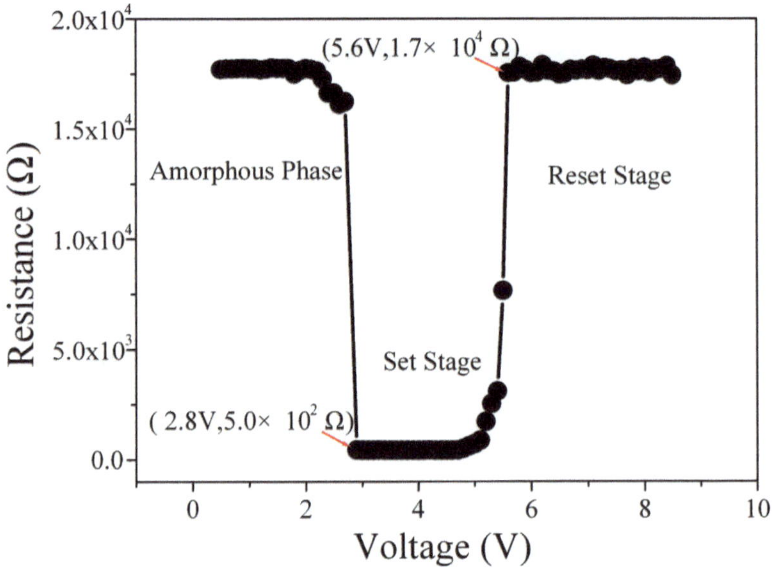

Fig. 9.6 V-R characteristic curve of SST PCRAM

corresponding to the transition from an amorphous state to crystalline state. It had been demonstrated that SST material PCRAM had a lower threshold voltage and reduced operation power consumption.

Read and write properties of PCRAM cell device are key parameters for PCRAM designing. By applying a voltage pulse of 20 ns, the R-V curve of the cell device was given in Fig. 9.6. It can be seen that the ratio between the high resistance and low resistance was up to 33. The cell device changed from high resistance to low resistance at a voltage of 2.8 V and then switched back to amorphous state when the voltage was 5.6 V.

The endurance characteristic of the memory cell is also a key parameter for memory. The measured result was shown in Fig. 9.7. 3×10^5 cycles with a resistance ratio of 30 were achieved. RESET pulse (6.0 V, 20 ns) and SET pulse (3.0 V, 20 ns) were used in the endurance measurements.

9.3.2 Nanostructure Array of Phase-Change Cell Devices

Figure 9.8 shows SEM images of the mold, imprinted pattern, and as-etched array. The SST array was formed on the TiN layer. The feature of SST array was 198 nm and the pitch was 400 nm. A density of 4 Gbit/in.^{2}. PCRAM cell was achieved by UV-NIL. By calculation, the contact area between SST and electrodes of memory cell was about 0.0392 μm^2.

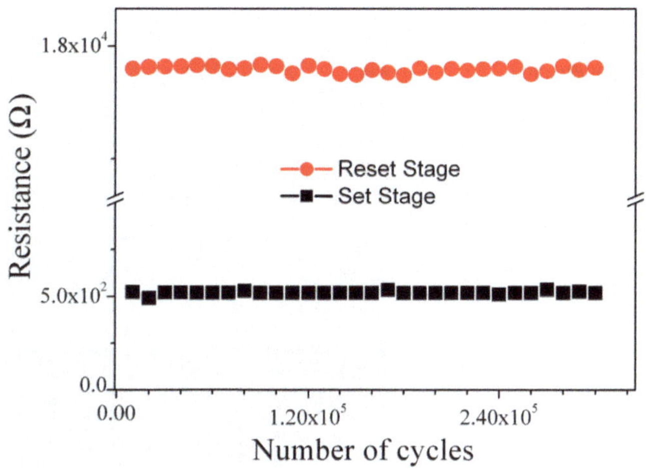

Fig. 9.7 *R* as a function of number of cycle curves of SST PCRAM

In order to test the electrical characteristic of the cell, conductive AFM (C-AFM) setup was used. C-AFM uses a voltage between a conductive tip and the sample surface to measure the current. Current flows through the metal-coated tip of the microscope and the conducting sample. Besides, it can simultaneously map topography and current distribution of a sample. It is a powerful current measurement technique that can detect current signals from sub pA to mA.

The schematic of SST memory cell electrical probe storage system was presented in Fig. 9.9. The SST array was sandwiched between AFM tip and bottom electrode. First, tapping mode was used to scan on the surface of the sample and detected the SST location. Then, the contact tip was placed on top of the SST array to obtain an I–V curve.

Figure 9.10 shows a typical I–V characteristic of one single array measured by C-AFM. During the first sweep, the current was very low because of high resistance. When the voltage was 1.6 V, the resistance reduced dramatically, which resulted in high current and switched to crystalline phase. The second sweep showed that the I–V relationship between SST array and TiN electronic was linear and the contact was a good Ohmic contact. The threshold voltage for the SST cell device was 1.6 V.

9.3.3 Further Fabrication of PCRAM

As far as memory application is concerned, it is believed that flash memory will approach the end of scaling within one decade or so. Hence, one of the most urgent challenges in semiconductor industry is the need of an emerging NVRAM fabrication technique which has high density, nonvolatile, fast read/write, and low

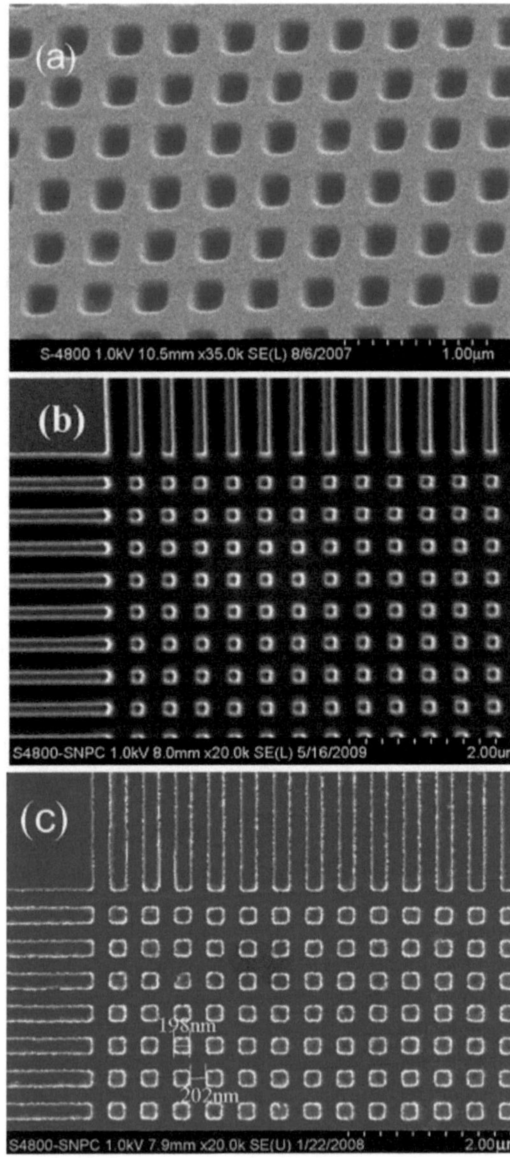

Fig. 9.8 SEM images (**a**) of Quartz stamp, (**b**) of imprinted resist pattern, and (**c**) of as-etched SST

power consumption, and compatible with CMOS process. PCRAM is regarded as one such alternative memory technique, which has received ever increasing interest from researchers both in industry and in academia. The latest Non-Volatile Memory Technology Symposium (NVMTS 2011), held in Shanghai on November 7–9, 2011, was presented great progress in the related fields of PCRAM.

Just as mentioned hereinbefore, NIL is a fast and simple fabrication method to test the properties of emerging phase-change material and the cell structure of

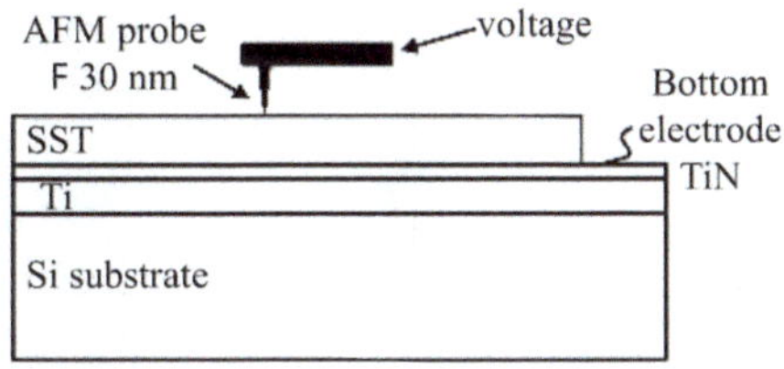

Fig. 9.9 Schematic of the SST-based electrical probe storage system

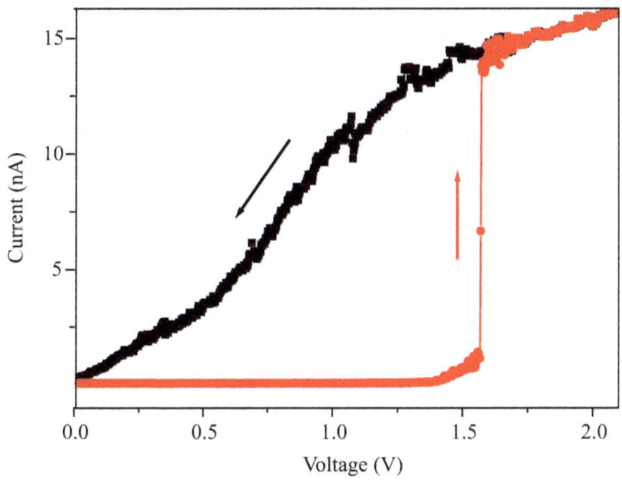

Fig. 9.10 I–V characteristic of one single array measured by C-AFM

PCRAM. It can reduce the process of an emerging memory cell by conventional lithography. The crossbar architecture, which was a 17-nm-width and a 34-nm-pitch nanowire electrode, was fabricated by UV-NIL. The density of the cell memory was 86.5 Gbits/cm^2. This process of NIL for array fabrication can be used to construct high-density memory arrays [18].

1T1R (one transistor, one resistor) memory devices are being widely studied as a key technology in high-density nonvolatile memories [19]. As shown in Fig. 9.11, the switching element (e.g., phase-change material (PCM)) is connected with a select transistor in series. The 1T1R method can be readily integrated with CMOS technique. However, the storage density has still been limited by one transistor. The 1T1R structure is at the expense of large cell size (6F^2) and high thermal budget. It is unfavorable to the high-density memory array where the cell size and low-temperature 3D stacking are of concern [20, 21].

1D1R (one diode, one resistor) structure is also used in memory cell. 1D1R cell can cut off the parasitic cross talk. The ratifying PN junction provides sufficient on-off ratio of the current densities. A cell size of 1D1R is 4F^2 and smaller than that of a 1T1R cell [22]. The schematic of a 1D1R is given in Fig. 9.12.

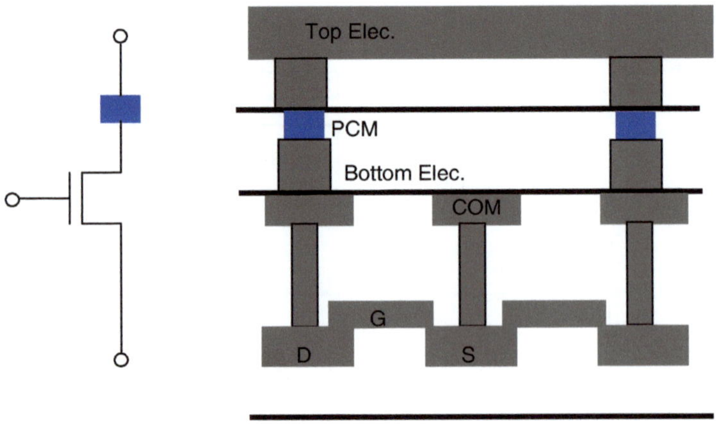

Fig. 9.11 PCRAM cell with a transistor

Fig. 9.12 Schematic of 1D1R memory cell

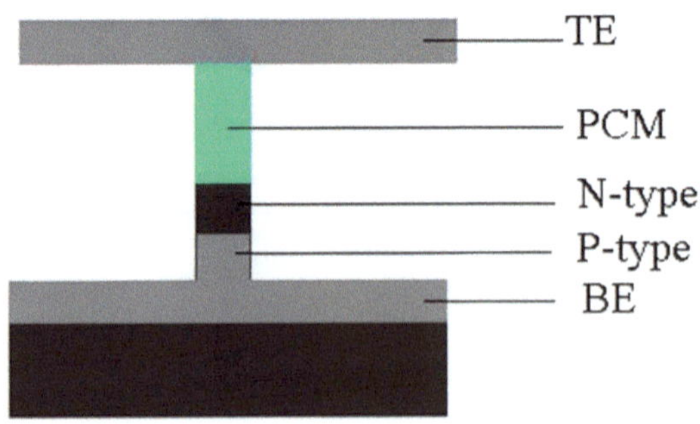

Crossbar array is another approach to improve the density of PCRAM. In crossbar array structure, each cross point at the intersections of the top and bottom electrodes is an individual PCRAM cell. The crossbar array structure offers a simple structure, ease of fabrication, and good defect tolerance for high-density memory. Figure 9.13 shows the schematic of a crossbar array cell.

Today, there is an ultimate limitation to further devices scaling for device fabrication technology. The CD of semiconductor devices is already comparable to that of nanoscale structure materials. Novel fabrication technologies come out to develop better electronic devices. One of the most promising candidates for memory fabrication is nanoimprint lithography with the ability to pattern at the sub-10 nm. However, some important requirements are to meet: large-area patterning with uniformity and the compatibility of nanoimprint lithography with current semiconductor processes. In the long run, nanoimprint lithography can replace several steps to pattern the critical features of memory devices, for example, the metal line fabrication of crossbar structured nanodevices.

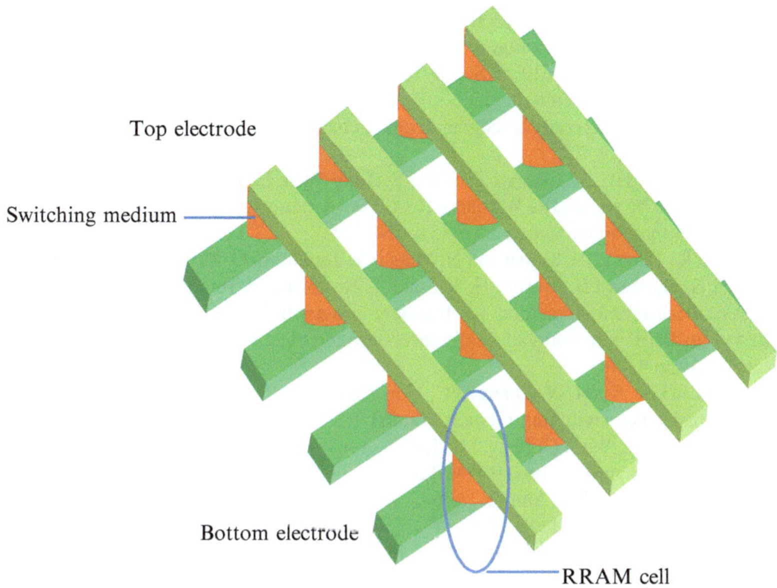

Fig. 9.13 Schematic of crossbar array structure

References

1. Nishi Y (2001) Challenges and opportunities for future non-volatile memory technology. Curr Appl Phys 11:e101–e103
2. Scott JF, Paz-de-Araujo CA (1989) Ferroelectric memories. Science 246:1400–1405
3. Sousa RC, Prejbeanu IL (2005) Non-volatile magnetic random access memories (MRAM). C R Phys 6:1013–1021
4. Jo SH (2010) Nanoscale memristive devices for memory and logic application. Dissertation, University of Michigan
5. The ITRS International Technology Working Groups, International Technology Roadmap for semiconductors. http://www.itrs.net/Links/2011ITRS/Home2011.htm. Accessed 18 Dec 2011
6. Burr GW, Breitwisch MJ, Franceschini M (2010) Phase change memory technology. J Vac Sci Technol B 28:223–262
7. Song ZT (2010) Phase change random access memory. Science Publisher, Beijing
8. Raoux S, Wutting M (2008) Phase change material science and application. Springer, Heidelberg
9. Chou SY, Kelmel C (2002) Ultrafast and direct imprint of nanostructures in silicon. Nature 417:835–838
10. Lee H, Hong S-H, Yang K-Y et al (2007) Fabrication of Ge2Sb2Te5 based PRAM device at 60 nm scale by using UV nanoimprint lithography. Microelectron Eng 84:573–576
11. Chou SY, Krauss PR, Renstrom PJ (1996) Imprint lithography with 25-nanometer resolution. Science 272:85–87
12. Ito T, Okazaki S (2000) Pushing the limits of lithography. Science 406:1027–1031
13. Lee JJ, Choi KB, Kim GH et al (2007) The UV-nanoimprint lithography equipment with multi-head imprinting unit for sub-50 nm half-pitch patterns. Microelectron Eng 84:963–966
14. Liu YB, Li XL, Zhou WM (2008) Fabrication of the Si2Sb2Te5 phase change cell structure for PCRAM by using UV nanoimprint lithography. In: 2008 2nd IEEE international nanoelectronics conference, INEC 2008, Shanghai, 2008, pp 807–810

15. Liu YB, Song ZT, Zhang T (2009) Fabrication, constructions and electrical property of $Si_2Sb_2Te_5$ electrical probe storage system. Microsyst Technol 15:1389–1393
16. Liu YB, Zhang T, Niu XM (2009) $Si_2Sb_2Te_5$ phase change material studied by an atomic force microscope nano-tip. J Semiconduct 30:063003–063008
17. Gibby AM (2008) A layered chalcogenide phase change memory device. Dissertation, Stanford University
18. Jung G, Johnston HE, Wu W, Yu Z, Wang S, Tong WM, Li Z, Green JE, Sheriff BA, Boukai A, Bunimovich Y, Heath JR, Williams RS (2006) Circuit fabrication at 17 nm half-pitch by nanoimprint lithography. Nano Lett 6:351–354
19. Kurotsuchi K, Takaura N, Matsuzaki N, Matsui Y, Tonomura O, Fujisaki Y, Ki-tai N, Takemura R, Osada K, Hanzawa S, Moriya H, Iwasaki T, Kawahara T, Terao M, Matsuoka M, Moniwa M (2006) Measurement method for transient programming current of 1T1R phase-change memory. In: IEEE international conference on microelectronic test structures, 2006. Austin, Texas, USA, pp 43–46
20. Huang JJ, Lin GL, Kuo CW, Chang WC, Hou TH (2009) Room-temperature TiOx oxide diode for 1D1R resistance-switching memory. In: ISDRS 2009, College Park, 2009
21. Chen A, Haddad S, Wu YC, Fang TN, Lan Z et al (2005) Non-volatile resistive switching for advanced memory applications. IEDM Tech Dig 746–749
22. Lee MJ, Park Y, Kang BS, Ahn SE, Lee C, Kim K, Xianyu W, Stefanovich G, Lee JH, Chung SJ, Kim YH, Lee CS, Park JB, Baek IG, Yoo IK (2007) 2-stack ID-IR cross-point structure with oxide diodes as switch elements for high density resistance RAM applications. IEDM Tech Dig 771–774

Chapter 10
Application of NIL in Solar Cell

10.1 Introduction

The problems of environmental and energy crisis in the world are serious and pressing, which cause many countries to spend billions of dollars on research to produce clean energy. Among these novel energies, solar energy is a kind of clean energy and a solution to the energy crisis. The International Energy Agency said that "the development of affordable, inexhaustible and clean solar energy technologies will have huge long-term benefits. It will increase countries' energy security reliance on an indigenous, inexhaustible and mostly import-independent resource" [1]. Many nations have put forward proposals to use solar energy. For example, the so-called Solar Roofs Plan in the USA and Germany began in 1996. Japan especially paid attention to the real applications of solar energy. "New Sunshine Project" started in 1993, and the incentive program "Residential PV System Dissemination Program" began in 1994 [2]. China has also taken effective measures to promote the development of solar cell.

Although solar energy is an inexhaustible and clean energy technology, the cost of solar cell is relatively expensive compared to other sources. Enhancing the efficiency of solar cell is one way of decreasing the cost of solar cell. With the development of nanotechnology, it has emerged as one promising approach of enhancing the efficiency of solar cell. Two approaches are included: one is nanomaterials, and the other is nanoscale structures.

10.2 Solar Cell

10.2.1 History of Solar Cell

Solar cell (also called photovoltaic cell) is an electrical device that converts light directly into electricity by the photovoltaic effect, which was first recognized in

W. Zhou, *Nanoimprint Lithography: An Enabling Process for Nanofabrication*,
DOI 10.1007/978-3-642-34428-2_10,

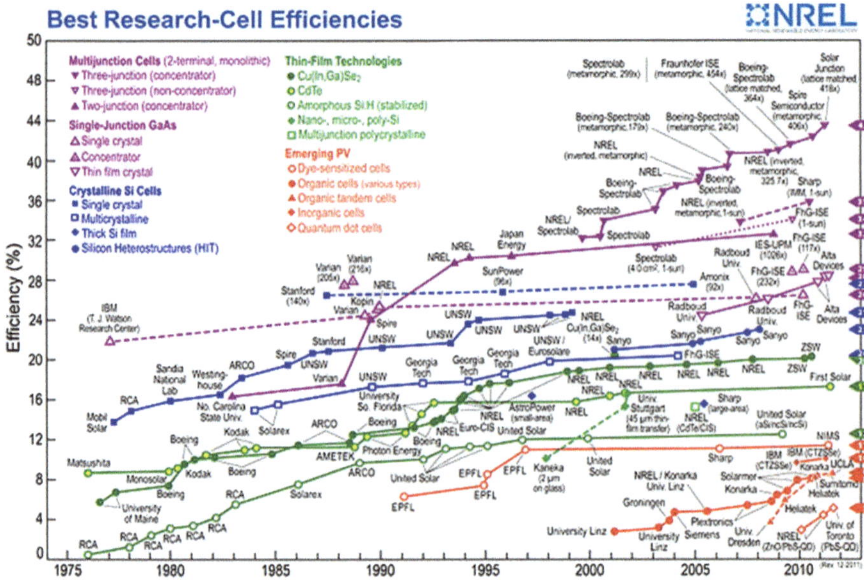

Fig. 10.1 Evolution of solar cell energy conversion efficiencies in the past 30 years (From National Renewable Energy Laboratory (USA). The dates are the maximum efficiencies)

1839 by French physicist. It was not until 1883 that the first photovoltaic cell was realized. In 1954, a modern photovoltaic cell with a diffused silicon p-n junction was developed in Bell Laboratories, reaching around 6% efficiency. It has since developed very quickly. As a novel energy, it was used for satellites in 1958. With the development of material and processes, its price has decreased and been used for civil application. Nowadays, due to the environmental and energy crisis, solar cell gains attention again, and many nations have proposed the related plans of solar cell. The requirements of solar cell with low cost, long lifetime, high efficiency, and excellent stability are becoming hot topics. The efficiency is one of the key factors for solar cell. Figure 10.1 shows solar cell energy conversion efficiencies for the reported various materials and processes in the past 30 years.

10.2.2 Prospects for Solar Cell

The first-generation solar cell was based on silicon wafers, and the second generation was thin-film solar cell technology. In 2004, Prof. Green proposed the concept of the third-generation photovoltaic cell, which was able to overcome the Shockley-Queisser limit of 31–41% power efficiency [3]. It has all the benefits of the second-generation solar cell, more than the limit of Shockley-Queisser, stable properties, and abundance in nature. In short, it is a kind of solar cell that has high performance, low cost, long lifetime, and high stability and approaches to an ideal photovoltaic cell.

The development of solar cell aims to enhance the efficiency and decrease the cost of solar cell. The third-generation solar cell is designed to combine the benefits of both the first- and second-generation solar cell. There are several attempts to improve the efficiency of solar cell through the following four methods: multi-junction cells, intermediate-band cells, hot carrier cells, and spectrum conversion [4].

For fabrication of solar cell, two factors are mattered. One is to widen the absorbing range of solar wavelength, and the other is to decrease energy loss of solar cells. Multi-junction solar cells consist of several p-n junctions, which is by separated the absorption of the polychromatic solar spectrum into semiconductors with different bandgaps. The wide bandgap junction absorbs high-energy photons, and the narrow bandgap junction absorbs lower energy photons. Therefore, a large portion of the solar spectrum is to be absorbed. However, the electrical characteristics of each layer must be matched, which is especially difficult for multi-junction solar cell production. Moreover, carrier transport through multilayers remains an obstacle which would decrease the thickness of layers.

Research and development of solar cells with low cost and high-conversion efficiency often result in nanostructures. Nanosolar cells refer to solar cells taking advantage of nanomaterials or nanostructures, which usually have excellent absorbing properties, quantum effect, or multi-excitons. By being nanostructured, it can improve the performance of conventional solar cells, obtaining relatively high-conversion efficiency from inexpensive materials with low cost and low-energy consumption, or obtaining conversion efficiency higher than the theoretical limit of conventional p-n junction solar cell [5].

10.3 Nanostructured Solar Cell

10.3.1 Physics of Solar Cell

The solar cell theory is based on the photovoltaic effect, as shown in Fig. 10.2.

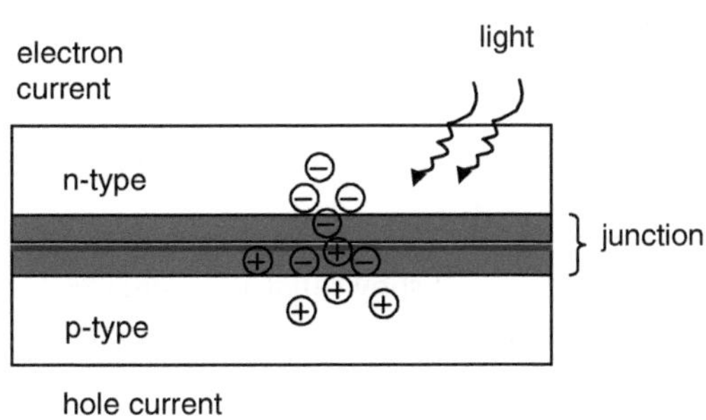

Fig. 10.2 Simplified physical model of a solar cell

In response to sunlight, a voltage difference is generated at the junction of two different materials. Three steps are involved: absorption of light and generation of charge carriers, separation of charge carriers, and collection of the carriers at the electrodes. When a photon hits semiconductor material, there are three possibilities to happen: (a) photon passing straight through the semiconductor, (b) photon reflected by the material surface, and (c) photon absorbed by the material. If the value of photon energy is higher than that of the semiconductor bandgap, electron–hole pairs are generally generated. Before the electrons and the holes recombine, the internal electric field at the p-n junction of the solar cell drives the electrons to the n-region and the holes to the p-region. The diffusion of carrier charges builds up on both sides of the junction and creates an electric field. This is equivalent to a p-n junction with a forward bias voltage, which can generate a current. When the cell is open-circuited, the voltage across the output terminals is defined as open-circuit voltage. Similarly, being short-circuited, the current through the terminals is defined as short-circuit current.

In the dark condition, most solar cells behave like diodes. The dark current density is given by [6]

$$J_{\mathrm{dark}}(V) = J_0(e^{qV/K_B T} - 1), \tag{10.1}$$

where J_0 is a constant, K_B is Boltzmann's constant, and T is temperature in degrees Kelvin.

Under illumination, the current of solar cell consists of three components and is given by

$$J_1 = J_p + J_n + J_d, \tag{10.2}$$

where J_p and J_n are photogenerated currents by illumination and J_d is drift current of photogenerated current in the depletion region.

When the circuit is shortened, the short-circuit current density J_{sc} is given by

$$J_{sc} \approx I_L. \tag{10.3}$$

I_L is a photogenerated current.

When the circuit is open, the difference of electrical potential between two terminals of solar cell has its maximum value. The open-circuit voltage V_{oc} is given by

$$V_{oc} = \frac{kT}{q} \ln\left(\frac{J_{sc}}{J_0} + 1\right), \tag{10.4}$$

where J_{sc} is short-circuit current density, J_0 is saturated current density, and q is electron charge. It shows that V_{oc} increases logarithmically with light intensity.

The fill factor (FF) of solar cell depends on the open-circuit voltage V_{oc} and short-circuit current density J_{sc}. FF is given by

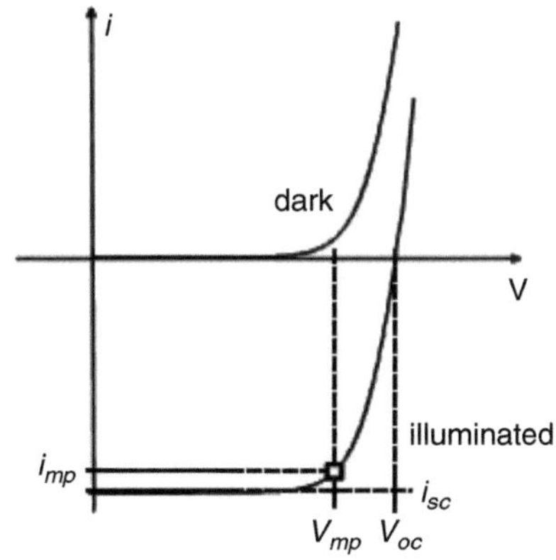

Fig. 10.3 Typical dark and illuminated solar cell I–V curve

$$FF = \frac{I_{mp}V_{mp}}{V_{oc}J_{sc}}, \quad (10.5)$$

where I_{mp} and V_{mp} are the current density and the voltage for maximum power, respectively.

The FF is the ratio of the two rectangular areas, as shown in Fig. 10.3.

The efficiency (η) of solar cell is given by

$$\eta = \frac{I_{mp}V_{mp}}{P_{in}} = \frac{V_{oc}J_{sc}FF}{P_{in}}, \quad (10.6)$$

where P_{in} is the incident light power density.

In all, these four parameters, J_{sc}, V_{oc}, FF, and η, are the key performance characteristics of a solar cell.

10.3.2 Nanopatterning Scheme for Nanoimprint Lithography

In this book, we mainly deal with nanopatterning rather than nanomaterials to enhance the efficiency of photovoltaic solar cell. The nanopatterns have close relations with light management of solar cell. Using light-trapping schemes, the efficiency of photovoltaic solar cells can be increased. There are three main approaches to pattern the nanostructures by nanoimprint lithography:

(a) Imprinting structures on top contact: it can increase the incident angle and decrease the reflection of sunlight.
(b) Imprinting period structures of p/n layers and forming a p-n junction: it can increase power conversion efficiency
(c) Imprinting structures on the back contact: it can increase the internal light reflection.

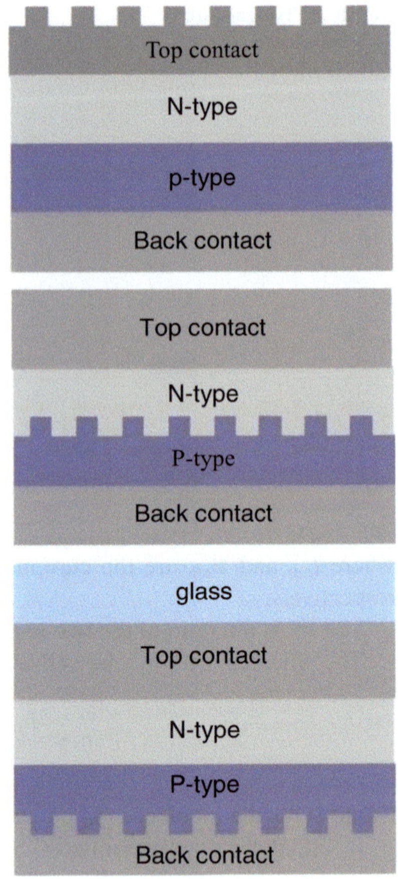

Fig. 10.4 Schematic of three main approaches by imprinting structures

The schematic of three main approaches is shown in Fig. 10.4. Light-trapping techniques have been demonstrated in the form of antireflection coating, photonic crystal pattern, and enhancing the interface area for junction and surface plasmon.

10.4 Antireflection Coating

10.4.1 Textures by Wet Etching

Decreasing light reflection to improve absorption of photovoltaic devices is of importance. In industrial field, texturing is used. Two roles are involved: the first is to decrease the reflectivity over the relevant spectrum. Secondly, the incoming

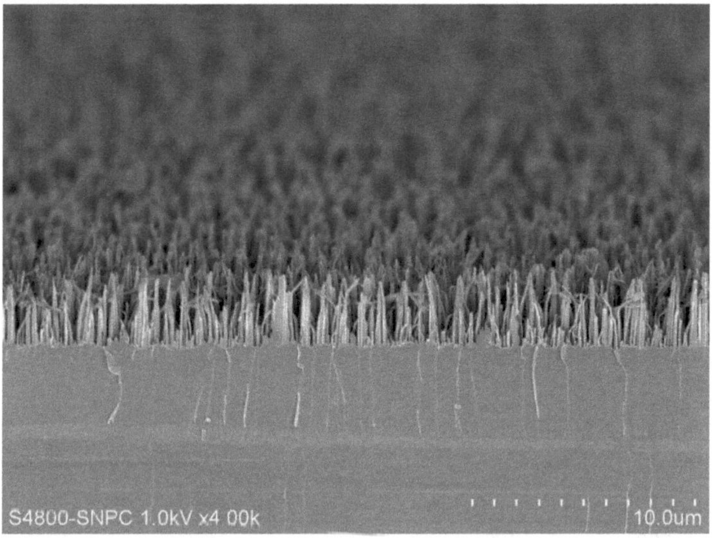

Fig. 10.5 Low-resolution and high-resolution SEM images of typical nanowire film

light is multi-reflected within the cell and the increased path length can enhance absorption. This textures approach makes use of silicon properties. Wet etching is used for obtaining the nanostructure.

Among these nanostructures, Si nanowire (SiNW) arrays are widely considered as an important class of antireflection nanostructures, along with their compatibility with modern integrated circuits. It is being explored as an alternative to nanosolar cell for photovoltaic application [7–9].

The ordered SiNWs were synthesized by electroless etching technique. Firstly, the silicon wafer was cut to 2×2 cm^2 and ultrasonicated in acetone and ethanol at room temperature for 20 and 10 min, respectively. Then, the silicon substrate was heated in boiling piranha solution (3:1(V/V) H_2SO_4/H_2O_2) and RCA solution (1:1:5(V/V/V) $NH_4OH/H_2O_2/H_2O$) for 2 h. Subsequently, the silicon substrate was rinsed several times with deionized water. The substrate was then immersed in HF (5 M)/$AgNO_3$(0.02 M) for 2 h at room temperature and was washed by HNO_3 thereafter. Finally, the substrate was rinsed with deionized water and dried with N_2 gas.

The as-synthesized SiNWs had a black nonspecular appearance showing that a large fraction of incident light was reflected by scattering. A cross-sectional SEM image of the vertically aligned SiNW array was shown in Fig. 10.5. The length of SiNWs was $\sim$3.85 μm, and the diameter was $\sim$178 nm. Figure 10.6 showed typical optical reflectance spectra of the SiNWs and silicon bulk counterparts, and a clear reduction of the reflectance was observed compared to the silicon bulk surface. The reflectance of the SiNWs film was less than 7% over the entire

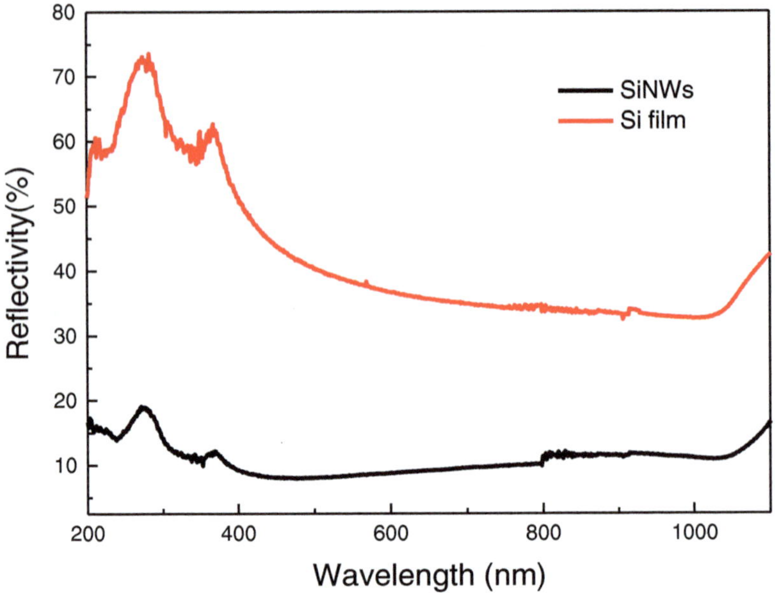

Fig. 10.6 Reflection measurements of silicon substrates with nanostructure, solid silicon substrate

spectral range (300–800 nm). The low reflectivity of SiNWs was attributed to the nanostructure surface of the arrays—the subwavelength-structure surface and a change in refractive index with depth. Therefore, the synthesized SiNWs surface resembled a multilayer antireflection surface.

To examine the effect of structural parameters of the nanowire in antireflective film on the reflectivity, a theoretical analysis is desirable. Using time-domain FDTD method [10, 11], when the Si plate height h_{Si} is infinity, TM and TE reflection spectra of the two models are shown. It can be seen that the structure of surface has greater reflectance and the reflectance is less than 5% within wavelength range from 320 to 1,100 nm. At the simulation process, we are equivalent to set h_{Si} as Infinity, so the reflection at the interface above entirely comes from free space and the smooth surface' reflection, the corresponding time-domain simulation results are shown in Fig. 10.7b, c, the time-domain simulation results and theoretical analysis results of the previous section are basically consistent, it also can be seen as the structure achieves better impedance matching between free space and silicon space, which makes reflectance of the structure close to zero near at a wavelength of 700 nm. Also important to note is that in the actual simulation, we have not set the height of Si plate to infinity based on the linear light source, which is at 1a position above the Si board, and the Dft flux plane, which is at 1a position on the top of the line light source. By setting a reasonable lighting time of source, Si plate height h_{Si}, and the time of time-domain simulation, we can make the light entered the Si plate will not be detected by Dft flux in the simulation process.

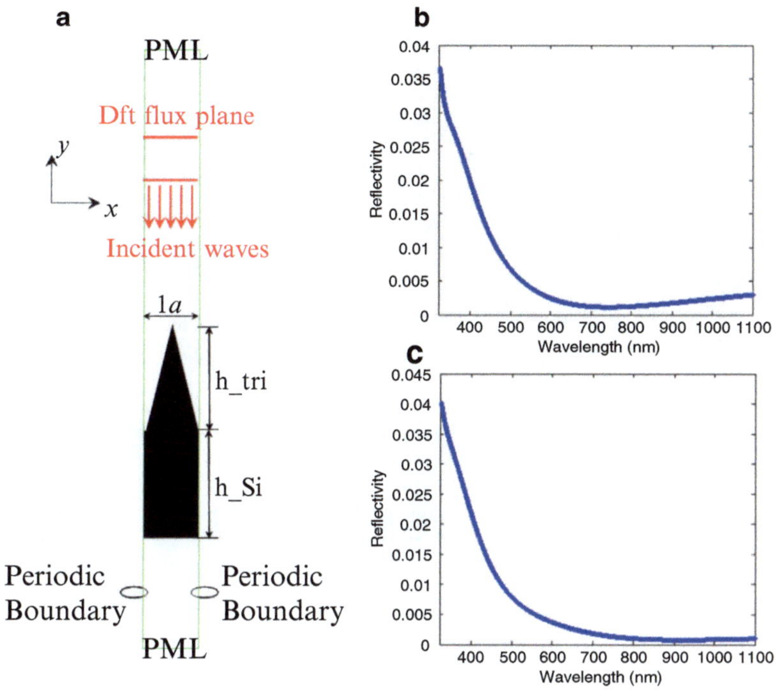

Fig. 10.7 (**a**) Periodic surface is antireflective film constructed by the triangular silicon plate. a is periodic length in x direction in this structure, and h_{Si} is the height of silicon plate; h_tri is the height of the triangle. At the same time, it can be seen from the diagram that when using the FDTD time-domain simulation in two dimensions, we use the PML boundary conditions in the y direction and the periodic boundary conditions in the x direction. (**b**) when $a = 178$ nm, $h_{\mathrm{Si}} = \infty$ and $h_{\mathrm{tri}} = 21.63a$, the calculated TM mode reflection spectrum. (**c**) when $a = 178$ nm, $h_{\mathrm{Si}} = \infty$, and $h_{\mathrm{tri}} = 21.63a$, the calculated TE mode reflection spectrum

10.4.2 Textures by Nanoimprint Lithography

Monocrystalline silicon can be easily textured, but multicrystalline silicon is prone to inefficient textures by chemical etching. Hauser et al. investigated that it is industrially feasible for the UV nanoimprint lithography to replace substitute photolithography to define textures [12].

The master mold was designed by calculation, and the hexagonal patterned photoresist was created via three-beam interference lithography on a glass substrate. The replication imprinted PDMS mold was used as working stamp. It allowed a full-wafer process to pattern the desired structures and the resist pattern on top of the multicrystalline silicon substrate. After etching, an SEM image of inverted pyramid Si was obtained.

The results of reflection measurements of different patterns Si were tested. It had been demonstrated that the NIL honeycomb texture of Si has excellent

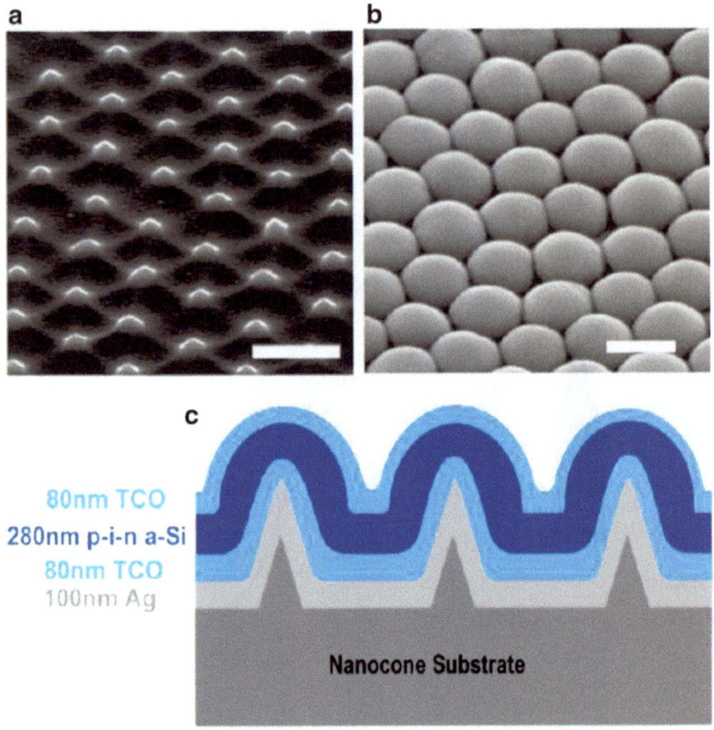

Fig. 10.8 Nanodome a-Si:H photovoltaic cell structure. (**a**) SEM image of nanocone quartz substrate. (**b**) a-Si:H nanodome solar cell after depositing multilayers of materials on nanodome. (**c**) Schematic of the cross-sectional structure of nanodome solar cell (Reprinted with permission from [13]. Copyright (2010) American Chemical Society)

reflection properties among these four different Si structures. The roller nanoimprint lithography used in the research may have a potential to generate an etching mask on multicrystalline silicon in the batch fabrication process.

The nanodome solar cell consisting of the periodic nanoscale layers from the bottom substrate, to the active absorber, and to the top transparent contact, was fabricated. The SEM images and schematic of the solar cell was given in Fig. 10.8. The patterned solar cell can absorb 94% of the light with wavelengths of 400–800 nm, 65% higher than the absorption of the thin-film cell. Besides, after modifying with hydrophobic molecules, the cell had a superhydrophobic surface and self-cleaning effect.

10.4.3 Moth-Eye Antireflection Coating

Mimicking nature is a good way to obtain excellent properties of antireflection. Moth-eye nanostructures have been developed in solar cell, which could reduce the reflection from solar cells. The eco-friendly production is suitable for the fabrication of high-efficiency photovoltaic cells.

Moth-eye coatings have the same effective graded index profile from superstrate to substrate on single material, which is achieved by pattern transfer and plasma-reactive ion etching (RIE) [14, 15].

Nanoimprint lithography is an excellent technique to realize the moth-eye nanostructures. Subwavelength moth eye on Si film was generated by soft-roller nanoimprint. An average reflection of 1% in the spectral range from 400 to 1,000 nm at normal incidence had been demonstrated. Low reflection (<12%) had been achieved with up to 60° incident angle [16].

Ordered arrays of subwavelength hydrogen silsesquioxane (HSQ) nanopillars on glass substrate were fabricated using anodized aluminum oxide membranes (AAO) mold nanoimprint lithography. It had been demonstrated that moth-eye nanorod arrays have superior omnidirectional antireflection property in visible wavelengths. Figure 10.9 showed SEM images of AAO stamp and imprinted nanorods and the schematic diagram of fabrication steps. The diameter of as-fabricated nanorods was about 80 nm, its lattice constant 130 nm, and the height was between 200 and 400 nm.

Figure 10.10 shows the specular reflectance of thin-film HSQ and nanopatterned HSQ. When the incidence angle is 90°, HSQ thin-film coated glass is better than plain glass in the reflection, and HSQ was best in performance over a large range of incidence angles (up to 70°).

10.5 Photonic Crystals

Photonic crystals are materials that have periodic variation in the permittivity with lattice periods on the order of the wavelength of light. The photonic bandgap is designed to forbid the propagation of electromagnetic waves, thereby enabling the manipulation of light. The ability of photonic crystals to control the flow of light has been widely used for improving the property of optoelectronic devices such as light-emitting diodes (LED), as mentioned in Chap. 8. In this section, photonic crystals used to enhance the property of photovoltaic cells will be included.

Nanoimprint lithography is used for organic photovoltaic (OPV) cells process. Especially the roller nanoimprint lithography meets the need of batch fabrication and exhibits good compatibility with organic materials.

10.5.1 Organic Photovoltaics

OPV cell is a photovoltaic cell that uses semiconducting polymers for light absorption and charge transport. Since the discovery of conductive polymer, the OPV cell has developed quickly [18].

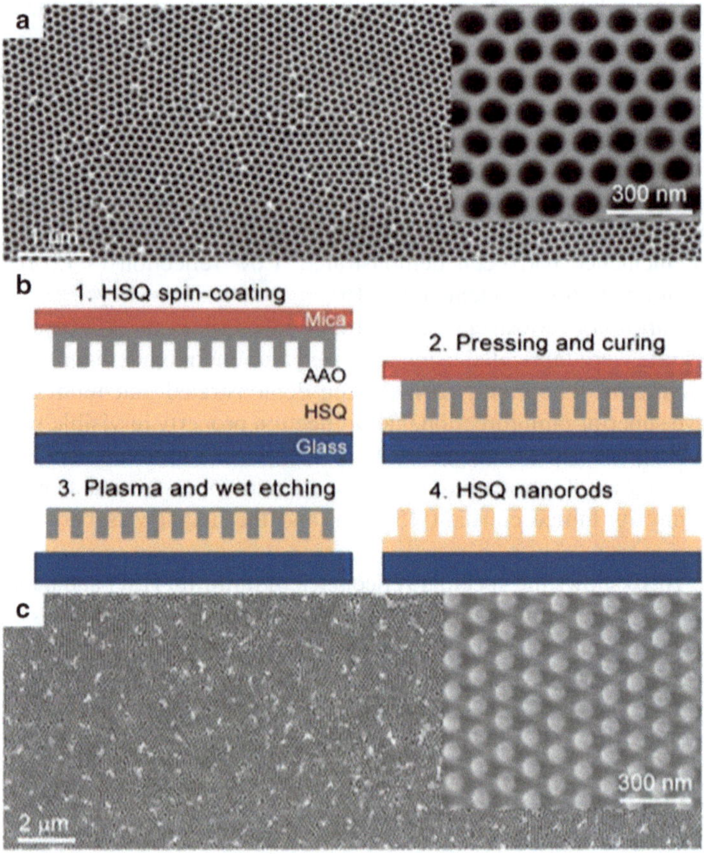

Fig. 10.9 (**a**) SEM image of AAO stamp. (**b**) Schematic diagram of fabrication processes of HSQ ordered nanorod arrays. (**c**) SEM images of HSQ nanorod arrays (Reprinted with permission from [17]. Copyright (2011) American Institute of Physics)

OPV cell offers several advantages over inorganic solar cell: (1) easier manufacturing and low cost, (2) less toxic and not easy to cause pollution, (3) simple material preparation, and (4) diversified solar cell structure and suitable for the fabrication of large-area flexible photovoltaic devices. Therefore, OPV cells are being intensely studied and have more and more markets because of low cost, ease of fabrication, flexible devices with large area.

In terms of the structure of OPV, the photon absorption layer is in the middle of two electrodes. One electrode as anode-collecting hole is semitransparent material, usually ITO. The other is a low work function metal electrode collecting electrons, such as Al, Ca, and Mg.

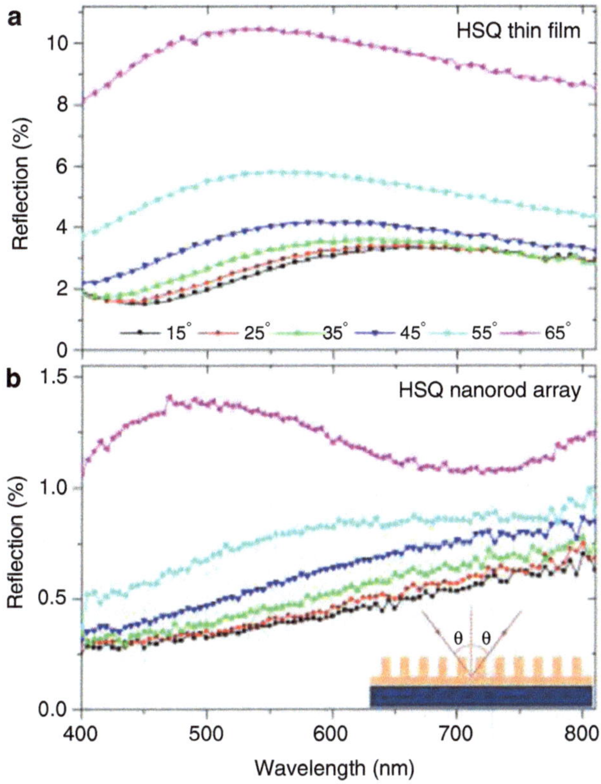

Fig. 10.10 (**a**) Reflectance properties at different incidences from HSQ film. (**b**) Reflectance properties at different incidences from HSQ nanorod surface (Reprinted with permission from [17]. Copyright (2011) American Institute of Physics)

PEDOT: PSS has emerged as a hole transport material, assisting in band bending for anode and enhancing the performance of the cell. It can be used for buffer layer of anodes for OPV to enhance holes extraction. The structure was shown in Fig. 10.11. LiF is a commonly used transport layer material of electrons and can enhance open-circuit voltage V_{oc} and the fill factor (FF) of OPV, resulting in the enhancement of conversion efficiency.

The active layer material consists of a donor and an acceptor, which forms the junction of OPV. The materials of donor include Ppy, PVK, PPV, MEH-PPV, PAn, PTh, P3OT, PEDOT, PF, P3HT, etc. Their molecular formulas are shown in Fig. 10.12.

The acceptor materials include C60, PCBM, PCBB, and C70. Their molecular formulas are shown in Fig. 10.13.

Fig. 10.11 The molecular formula of PEDOT: PSS

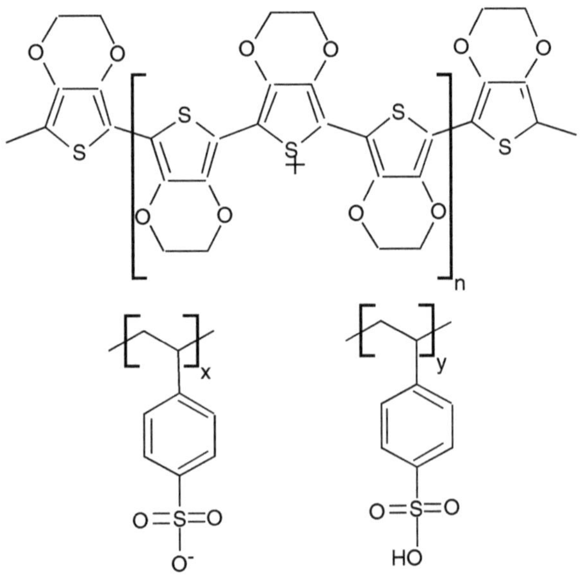

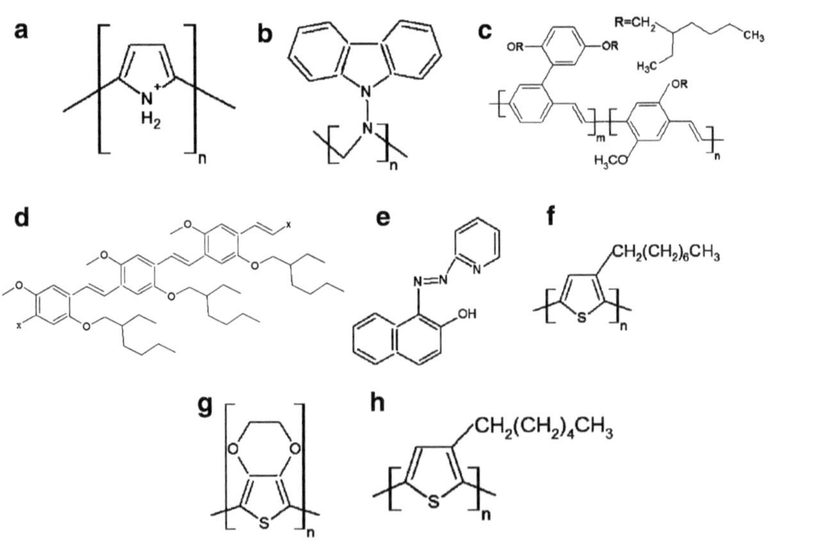

Fig. 10.12 Molecular formulas of donor materials. (a) PPy. (b) PVK. (c) PPV. (d) MEH-PPV. (e) PAn. (f) P3OT. (g) PEDOT. (h) P3HT

10.5.2 Different Heterojunction for OPV

The structure of OPV cell is simple and is made by sandwiching organic materials between two metallic conductors. There are three types of junctions for OPV, as shown in Fig. 10.14.

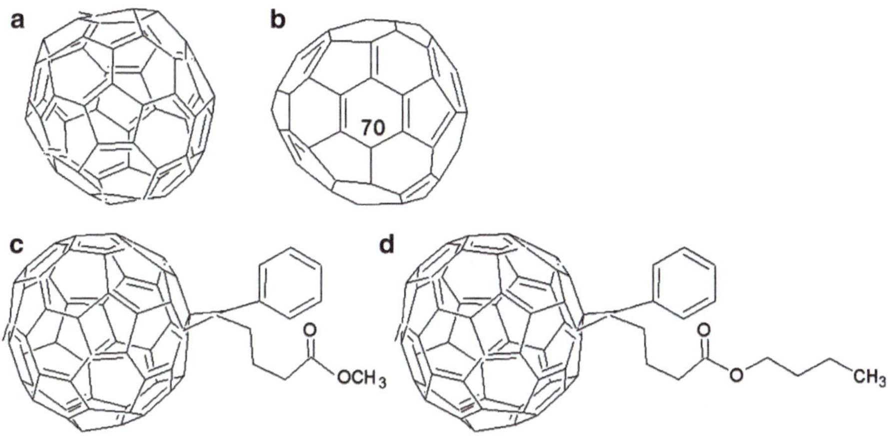

Fig. 10.13 Molecular formulas of acceptor materials. (**a**) C60. (**b**) C70. (**c**) PCBM. (**d**) PCBB

Fig. 10.14 Different heterojunction for OPV: Sketch of a single-layer organic photovoltaic cell, bilayer organic photovoltaic cells, and bulk heterojunction photovoltaic cells

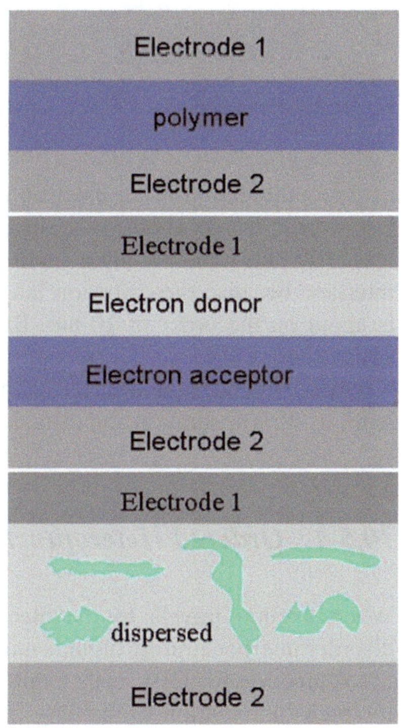

Single-layer OPV cell is a layer of organic electronic material, which is a close analogy to the structure of metal-insulator-metal (MIM) devices. However, it has a low conversion efficiency (<0.1%). The recombination of electrons and holes often occurs before they reach the electrodes. Bilayer heterojunction OPV cell consists

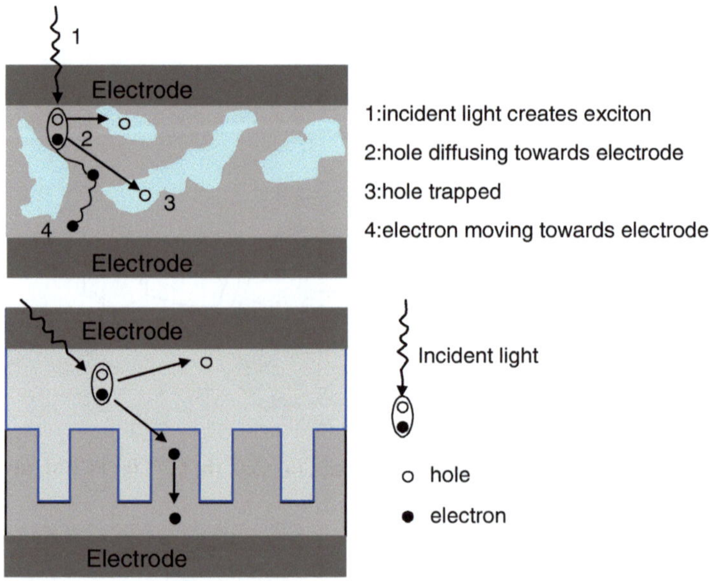

Fig. 10.15 Heterojunction for OPV: highly folded heterojunction and ordered heterojunction

of the combination of an electron donor and an electron acceptor material layer. However, it brings about some disadvantages: the thickness of bilayer material (at least 100 nm) renders only a small fraction of exciton to reach the heterojunction interface because the diffusion length of excitons in organic electronic materials is about on the order of 10 nm. Bulk heterojunction (BHJ) is the most common structure for OPV cell. The electron donor and acceptor are mixed together to form a polymer blend, which has a nanoscale interpenetrating network. This structure helps exciton separation and diffusion [19].

10.5.3 Ordered Heterojunction OPV

When exciton travels to the electrodes, it can be trapped/recombined in the disorder interpenetrating organic material, resulting in lower efficiency. The ordered heterojunction for OPV cells exhibits small distance between donor and acceptor for high dissociation probability and is used as high charge carrier collector at electrodes, preventing competing processes of interfacial recombination, as shown in Fig. 10.15 [20].

The nanoscale-ordered patterns require advanced pattern process. There are many methods to form nanoscale-ordered patterns, such as electron beam lithography, focused ion beam lithography, block copolymer templates, nanoimprint

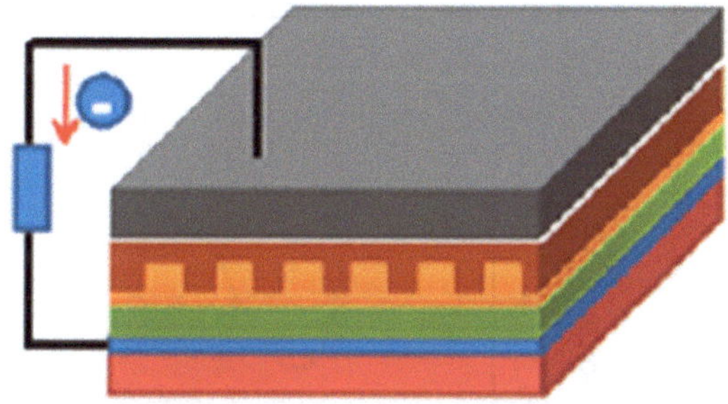

Fig. 10.16 General schematic of the OPV

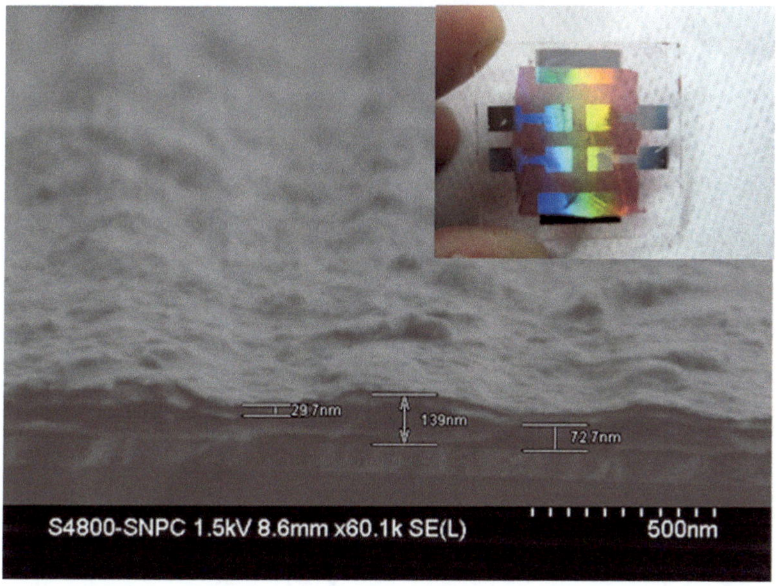

Fig. 10.17 SEM image of the cross-sectional solar cell, the inset was a photo of solar cell

lithography, and so on. Nanoimprint lithography is a preferable approach to generate these nanoscale structures.

Solar cells were fabricated using P3HT in the following structure of OPV: indium tin oxide (ITO)/PEDOT: PSS/P3HT: PCBM/LiF/Al, and the general schematic of OPV was shown in Fig. 10.16. The ITO-coated glass substrate was treated with ultraviolet ozone for 5 min, and then P3HT was spin-coated on the ITO surface. P3HT was imprinted by soft nanoimprint lithography. The cavity of P3HT nanostructures was filled by PCBM. Finally, 100-nm Al was thermally evaporated on the top as the cathode.

Figure 10.17 showed an SEM image of the cross-sectional solar cell. The thickness of active layer was 70–110 nm. The Al electrode had a thickness of 30 nm. The insert on Fig. 10.17 can be found color fringe.

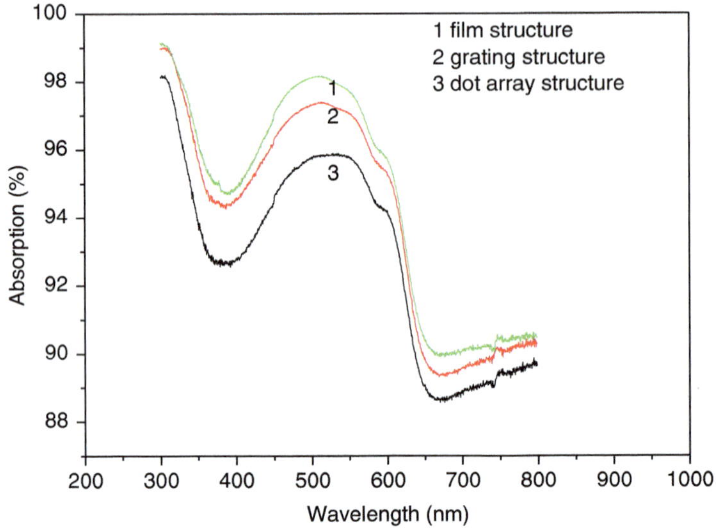

Fig. 10.18 Measured absorption for solar cell

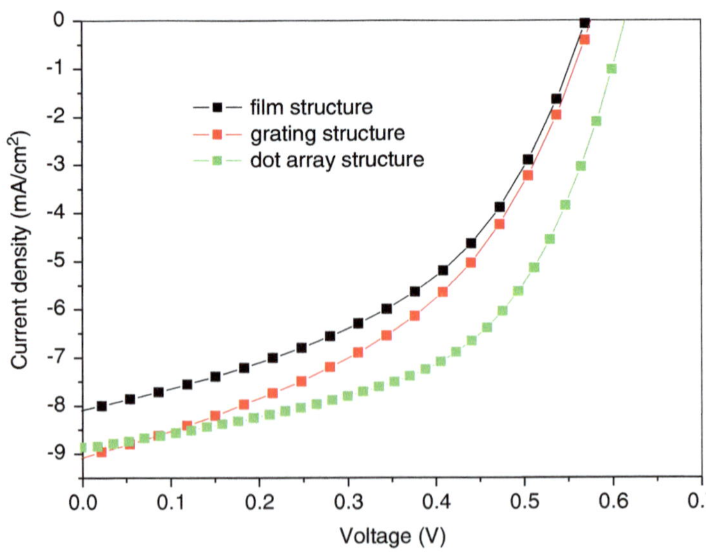

Fig. 10.19 Measured I–V characteristics for three different solar cells

As shown in Fig. 10.18, a clearly discernible absorption enhancement in the patterned solar cells over the planar cells was detected. The nanoarray structure changed the light direction in the active layer. The measured I–V characteristics for three different solar cells were shown in Fig. 10.19. The short-circle current density

(J_{sc}) of the nanograting devices was 9.05 mA/cm^2, but that of the film structure was 8.07 mA/cm^2. A 7.9% increase in PCE over the non-imprinted planar device was observed. Furthermore, the overall PCE of the dot structure of solar cell was enhanced about 37.5% than that of the planar devices.

10.6 Nanoarray Structures

When a material is down to the nanometer scale, its properties such as electrical conductivity, ductility, and energy band structure change, which are the so-called nanosize effects. For solar cells, nanorods or nanopore arrays are often used in high-performance solar cells. It is used to improve the ratio of the p-n junction of semiconductor materials and enhance the transport efficiency of carriers.

10.6.1 Nanorod Array

Fan et al. from University of California, Berkeley, created single crystal n-type CdS nanorod array by means of AAO substrate. They then deposited a layer of p-type CdTe light absorption layer embedded CdS nanorods, as shown in Fig. 10.20 [21]. The cell structure was more conductive to light absorption and helped to effectively collect carriers.

10.6.2 Nanohole Array

Compared to nanorods array, nanohole array exhibits superior mechanical robustness and is not fragile and easily cracked. Peng et al. reported that p-type silicon planar structure with the nanopore array structure was fabricated by deep ultraviolet lithography and metal-assisted Si etching. Then, p-type holes were filled with a liquid $POCl_3$ source at 910 for 20 min, to form the radial p-n junctions for Si solar cell. The schematic of the fabrication process of the radial p-n junction of Si solar cell is shown in Fig. 10.21.

A typical Si nanohole solar cell has hole depths of $\sim$2 um, and the thicknesses of the hole walls are 280 and 880 nm for the first and second neighbor holes, respectively. It has been demonstrated that the properties of the nanoholes array solar cell are superior to those of SiNWS, planar Si, and pyramid-textured Si. The nanohole array reveals a potential for high-efficiency energy conversion.

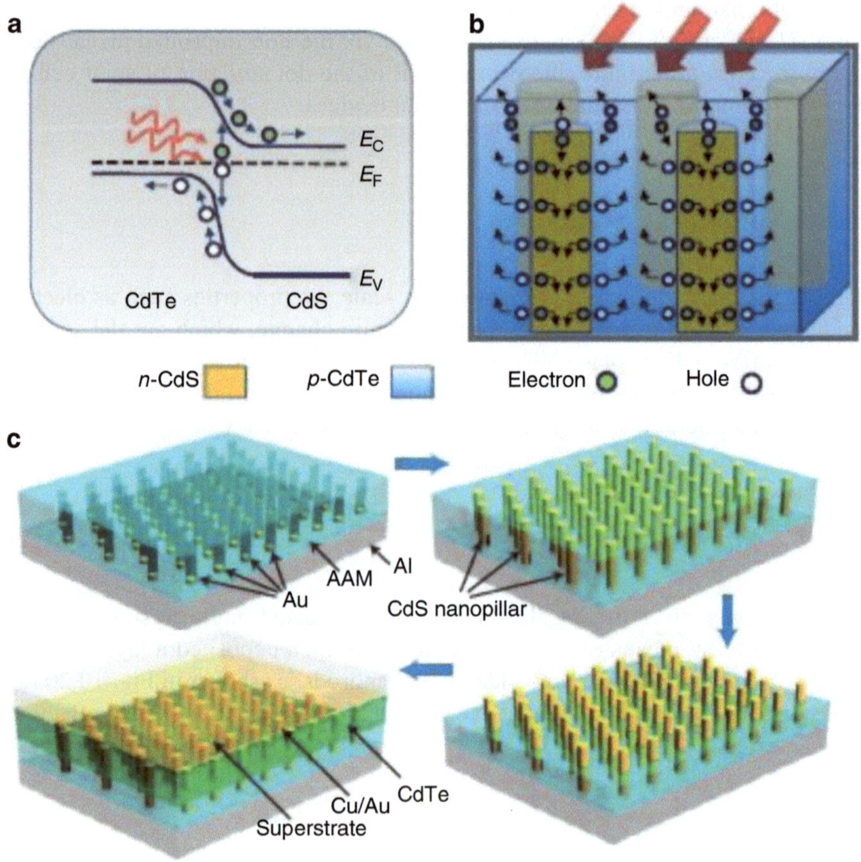

Fig. 10.20 CdS/CdTe SNOP cells. (**a**) Energy band diagram of a CdTe/CdS photovoltaic. (**b**) Cross-sectional schematic diagram of a SNOP cell, illustrating the enhanced carrier collection efficiency. (**c**) SNOP-cell fabrication process flow (Reprinted with permission from [21])

10.6.3 Optical Simulation of Nanoarray Structures

Si nanowire array has excellent optical and electronic properties, differing from those of Si bulk and thin films. The parameter of Si geometry has an impact on the optical properties. Chen's group reported the simulation analysis of the effects of nanowire diameter (d), length, and filling ratio on the optical absorption in silicon nanowires arrays by the transfer matrix method (TMM). The schematic of the period nanowire structure is given in Fig. 10.22a. Figure 10.22b showed the reflectance and transmittance for a nanowire structure ($d = 50$ nm and $a = 100$ nm) and a thin film. It had been demonstrated that the reflectance of nanowires was significantly lower than that of the thin film in the entire spectral range due to the reduced density of the nanowire structures.

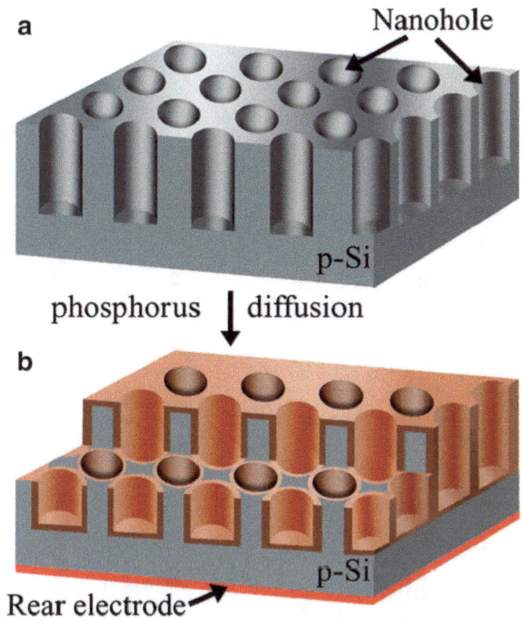

Fig. 10.21 The fabrication process of Si nanohole solar cell with radial p-n junctions via thermal phosphorus dopant diffusion. The front electrode was not shown here (Reprinted with permission from [22]. Copyright (2010) American Chemical Society)

The filling ratio was also studied, as shown in Fig. 10.22c, d. It can be seen that Si nanowire arrays with smaller-filling ratios occurs lower reflection and higher absorption in the high-frequency region, while Si nanowire arrays with larger filling ratios yielded higher absorption in the low-frequency region for the same wire length.

10.7 Surface Plasmon Effect for Light Trapping

Surface plasmon effect has been introduced to trap light on the surface of solar cell. It is that the confined light waves propagate along the metal and dielectric interface, which need metal nanoparticles of gold or silver implanted on surface or imbedded in dielectric materials.

10.7.1 Surface Plasmon Effect

Metal nanoparticles are known to exhibit different optical characteristics compared to the bulk form of metals [24, 25]. For example, surface-enhanced Raman scattering (SERS) is widely used for biomolecular manipulation, labeling, and detection [26–28].

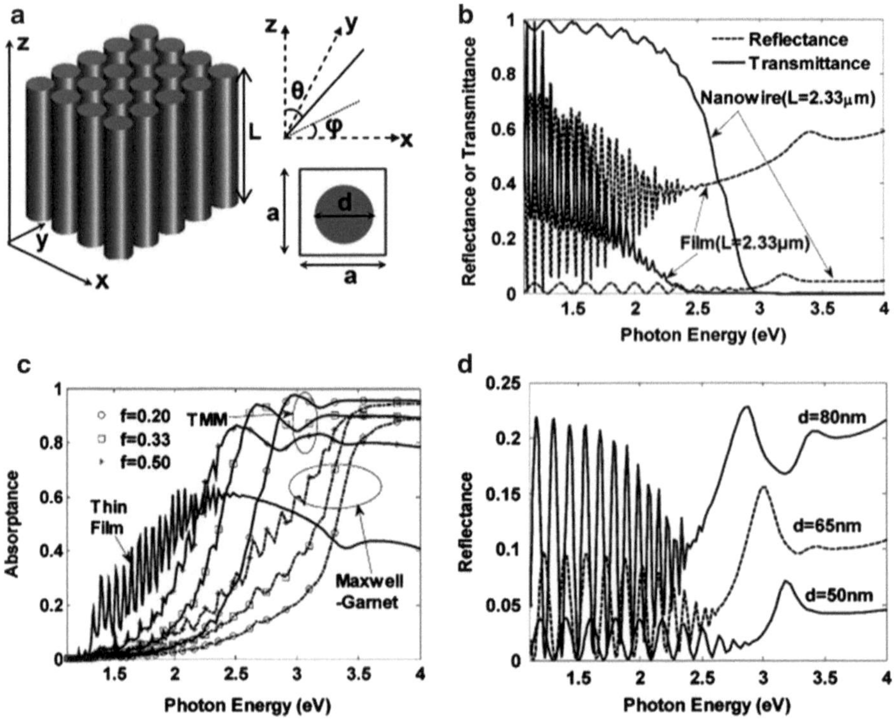

Fig. 10.22 (**a**) Schematic drawing of the periodic silicon nanowire structure. The parameters are the length *L*, the period *a*, and the diameter *d*. In the figure, θ and ψ are the zenith and azimuthal angles, respectively, (**b**) of reflectance of the nanowires and the thin-film, (**c**) of absorptance obtained by TMM and Maxwell-Garnet approximation, and (**d**) of reflectance of nanowires with diameter of 50, 65, and 80 nm (Reprinted with permission from [23]. Copyright (2010) American Chemical Society)

The performance depends on the surface plasmon absorption. Surface plasmons are electromagnetic surface waves that propagate along the boundary between a metal and dielectric. The surface plasmon mode needs a relatively free electron cloud which can move easily by an electromagnetic wave to sustain a coupled propagation of plasmons and photons. Both surface plasmon polaritons (SPPs) propagate at the metal/semiconductor interface. Localized surface plasmons excited in metal nanoparticles are common types and receive interest.

There are at least three methods of improving the efficiency of solar cell. First, metallic nanoparticles can be used as subwavelength-scattering elements and enhance light absorption of an absorbing layer. Second, metallic nanoparticles can be used as subwavelength antennas where the near-field plasmon is coupled to the semiconductor, increasing its effective absorption cross section. Third, light trapping by the excitation of surface plasmon polaritons at the metal/semiconductor interface can enhance the efficiency. The schematic of plasmonic light trapping for solar cell is shown in Fig. 10.23.

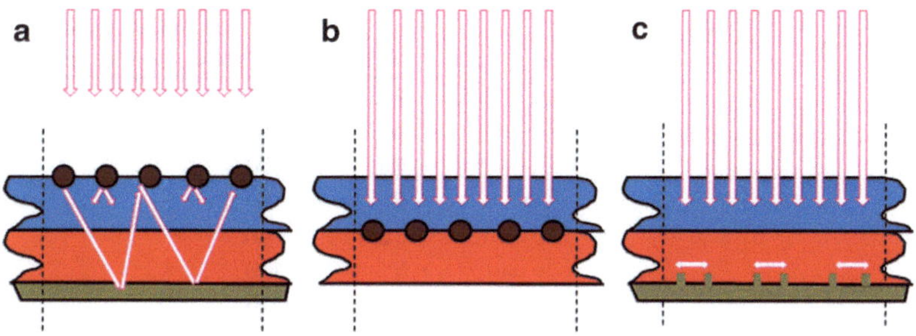

Fig. 10.23 Plasmonic light-trapping geometries for solar cell. (**a**) Light trapping by scattering from metal nanoparticles at the surface of the solar cell. (**b**) Light trapping by the excitation of localized surface plasmons in metal nanoparticles embedded in the semiconductor. (**c**) Light trapping by the excitation of surface plasmon polaritons at the metal/semiconductor interface. (Reprinted with permission from [29]. Copyright (2010) Nature Publishing Group)

(a) Light trapping by scattering from metal nanoparticles at the surface of the solar cell
(b) Light trapping by the excitation of localized surface plasmons in metal nanoparticles embedded in the semiconductor
(c) Light trapping by the excitation of surface plasmon polaritons at the metal/semiconductor interface

As for device configuration of plasmonic solar cell, there are three types: (1) outward excitation on the photoactive layer, (2) excitation of the embedded metal nanoparticles in the photoactive layer, and (3) excitation via metallic gratings [30].

10.7.2 *Plasmonic Solar Cells with Metal Nanoparticles*

Metal nanoparticles can collect the incident light with the extinction cross section near the surface plasmon resonance and then scatter the incident light at an angle greater than the critical angle for reflection to increase the optical patch length in the absorption layer. For nanoparticles in the quasi-static limit, the scattering and the absorption cross sections are presented by [31]

$$C_{\mathrm{abs}} = \frac{2\pi}{\lambda}\,\mathrm{Im}[\alpha], \quad \text{and} \quad C_{\mathrm{sca}} = \frac{1}{6\pi}\left(\frac{2\pi}{\lambda}\right)^4 |\alpha|^2, \tag{10.7}$$

where α is the polarizability of particles. For a small spherical particle in vacuum, α is given by

$$\alpha = 3V\left[\frac{\varepsilon/\varepsilon_{\mathrm{m}} - 1}{\varepsilon/\varepsilon_{\mathrm{m}} + 2}\right], \tag{10.8}$$

where V is the volume of the particle and and $_{\mathrm{m}}$ are the permittivity of the metal and the surrounding medium, respectively.

Table 10.1 Some plasmon wavelengths [32]

Metal	Localized surface plasmon (resonance) wavelength (nm)
Pd	~250
Ag	~390
Ba	~400
Eu	~380
Ca	~500
Y	~430
Au	~525
Pt	~230
Cu	~210
Cs	>700

The scattering efficiency is given by

$$Q_{\text{sca}} = C_{\text{sca}}(C_{\text{sca}} + C_{\text{abs}}). \tag{10.9}$$

Absorption and scattering are dependent on nanoparticles, geometry, and medium. If the dimension of particle is less than wavelength of incident light, absorption is of relative importance, and the energy is losing in the form of heat. Otherwise, scattering is a main factor. Table 10.1 is a listing of metal nanoparticle sizes, in which light at or near the wavelengths can couple into the plasmon scattering mechanism.

Metal nanoparticles are usually formed on the surface of thin-film solar cell by deposition. However, large-area fabrication of metal nanoparticles is still difficult for conventional lithography. For example, electron beam lithography is used for the fabrication of photovoltaic module production in laboratory, but the cost is expensive. Silver nanoparticles on a substrate are fabricated by thermal evaporation and annealing; however, the Ag nanoparticles are random and result in low-efficiency light trapping [33, 34].

A regular array of metal Ag nanoparticles on 6-in.-diameter Si wafer was fabricated using substrate conformal imprint lithography, followed by Ag evaporation and lift-off process. Figure 10.24a showed a SEM image of hexagonal array of Ag nanoparticles using substrate conformal imprint lithography, deposition, and lift-off technique. It can be seen that the nanoparticle had a diameter of 300 nm. Figure 10.24b was a photograph of a 6-in. Si wafer patterned with a regular array of Ag metal nanoparticles.

10.7.3 Plasmonic Solar Cells with Metallic Electrode

The metallic layer at the bottom of solar cell layer is schematically given in Fig. 10.23c. Incident light can couple into surface plasmons propagating at the

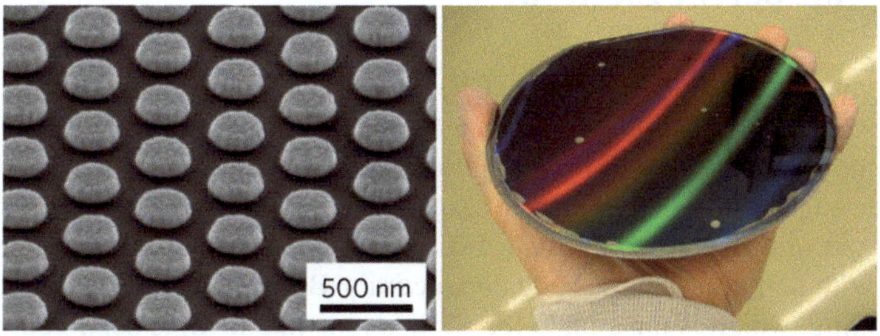

Fig. 10.24 Ag nanoparticles, (**a**) of SEM image of hexagonal array of Ag nanoparticles, and (**b**) of a photograph of a 6-in.-diameter Si wafer patterning with Ag nanoparticles arrays (Reprinted with permission from [29]. Copyright (2010) Nature Publishing Group)

metal/semiconductor interface via some subwavelength-size pattern. In this geometry, the energy flux is changed from normal to lateral direction of photovoltaic cell, which is an order of dimension larger than the optical absorption length.

The triggering of surface plasmons at the interface of a metal grating and a photovoltaic bulk heterojunction blend of alternating polyfluorenes and a fullerene derivative has been demonstrated by Tvingstedt et al. It has been revealed that an increased absorption results from surface plasmon resonances, which influence the extracted photocurrent from solar cell devices [35].

Na et al. investigated that the efficiencies of P3HT/PCBM-based polymer photovoltaic cells were increased by means of surface relief gratings (SRG) on the active layers as a light-trapping scheme. Soft nanoimprint lithography was used to realize the organic solar cell with surface relief gratings. The effect of periodic gratings (period, depth, and dimension) on solar cell performance had been investigated [36]. The effect of 1D and 2D gratings on the performance of organic solar cell was presented. Under 1 sun with AM 1.5-G illumination, a V_{oc} value of 0.61 V, J_{sc} of 9.5mAcm^{-2}, FF of 62%, and of 3.6% were decided for the reference cell. However, the two cells with 1D and 2D SRG were as follows: V_{oc} of 0.62 V, J_{sc} of 10.5 and 10.9 mA cm^{-2}, FF of 63 and 64%, and of 4.1 and 4.3%, respectively. It had been demonstrated that SRG triggered more photogenerated charge carriers by stronger absorption of incident light because of an increase of the optical path length and light trapping.

Kang et al. demonstrated organic solar cell with transparent Ag nanowires electrodes acting as anode, which can enhance the conversion efficiency by surface plasmon resonance and waveguide effects. The device had the layer structure of Ag nanowire electrode anode, PEDOT: PSS, CuPc, C60, BCP, and cathode (thick Ag film). The Ag device was enhanced by about 35% in comparison with that of the ITO device under unpolarized light illumination [37].

10.8 Transparent Electrodes

Transparent conducting electrodes (TCE) are optically transparent and electrically conductive. It is an important component in photosensitive optoelectronic devices such as photovoltaic cell and light-emitting diodes. TCE act as a window for light to pass through to the active layer of solar cell and an ohmic contact for carrier transport out of the photovoltaic cell. ITO (tin-doped indium oxide) is an excellent material with both transparency and conductivity. But it is brittle and tends to crack upon bending, which is unsuitable for applications like flexible solar cell. Besides, the scarcity of indium also helps the attempt to find alternatives to ITO, which are fluorine-doped tin oxide (FTO), doped zinc oxide, carbon nanotube networks, and graphene. Alternatives of ITO will have the following characteristics [38]:

- Highly transparent and high conductive
- Suitable surface morphology
- Resistant to chemical etching, thermal
- Nontoxic
- Durable films with low cost
- Low resistance on contact
- Material compatibility (wetting, diffusion)

10.8.1 Transparent Semiconductor Electrodes

Transparent conducting oxide semiconductor is a special class of semiconductors, such as impurity-doped ZnO, In_2O_3, SnO_2, as well as multicomponent oxides consisting of combinations of ZnO, In_2O_3, and SnO_2, including some ternary compounds existing in their systems [39].

Fluorine-doped tin oxide (FTO), Al-doped ZnO (AZO), and Ga-doped ZnO (GZO) exhibit promise for transparent electronic application. ZnO is low cost and low toxicity and can be doped with Al or Ga. A lower resistivity of AZO can be achieved compared to that of ITO.

A nanomolding process that provides exactly such a platform, enabling precise, large-area, nanoscale patterning of functional zinc oxide films at low cost, has been established. With nanomolded process, zinc oxide films as transparent front electrodes in amorphous silicon solar cells have been achieved. The high-efficiency light-harvesting scheme was responsible for excellent initial conversion efficiencies of 10.1%. Figure 10.25 presented SEM images of cross sections of two solar cells milled with focused ion beam. From these images, it can be seen that fine-grained structures of ZnO front electrodes will help the nucleation and growth of amorphous silicon. The parameters of the solar cell were summarized in the Table 10.2.

ITO is fragile and tends to break. Besides, high transmittance and conductivity require high temperature of annealing. These disadvantages are not compatible with a flexible plastic solar cell. Therefore, developing new TCE is necessary.

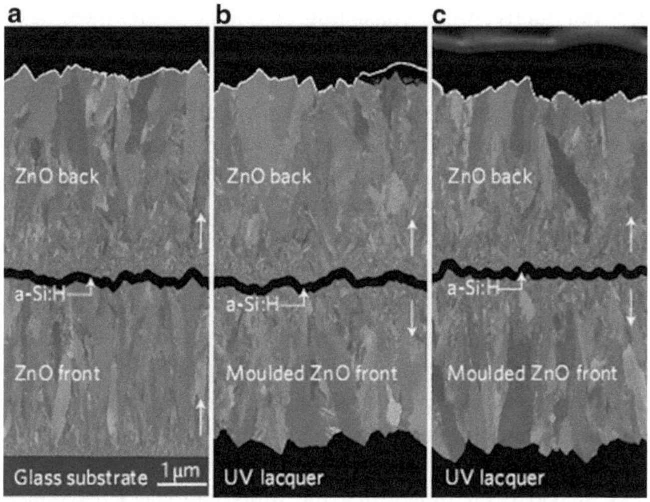

Fig. 10.25 Amorphous silicon solar cells on nanomolded ZnO. (**a**) FIB cuts across the solar cells on the master ZnO. (**b**) The nanomolded ZnO replica. (**c**) The nanomolded TA replica. Growth direction of the ZnO layers is indicated by *white arrows* (Reprinted with permission from [40]. Copyright (2011) Nature Publishing Group)

Table 10.2 Characteristics of amorphous silicon solar cells grown on the ZnO master, its ZnO replica, flat ZnO, and replicated textured aluminum (TA)

	J_{sc} (mA cm^{-2})	V_{oc} (mV)	FF (%)	Efficiency (%)
ZnO master	15.2	880	76	10.1
ZnO replica	15.6	879	74	10.1
Flat ZnO	11.6	896	73	7.6
TA replica	16.2	871	70	9.9

Reprinted with permission from [40]. Copyright (2011) Nature Publishing Group

Carbon-based materials (carbon nanotube and graphene) are regarded as alternatives to ITO. Carbon nanotube thin films used as transparent electrodes have attracted much attention because of their excellent properties, such as high elastic modulus (∼1–2 Tpa), high tensile strength (∼13–5 Gpa), and high conductivity (the theoretical value is 100 times higher than that of copper) [41].

Two kinds of commercial carbon nanotubes (HiPCO and arc-discharge nanotubes) which acted as electrodes for organic light-emitting diodes were investigated. It has been shown that films based on arc-discharge nanotubes are overwhelmingly better than HiPCO-nanotube-based films in all critical aspects, including surface roughness, sheet resistance, and transparency. Moreover, arc-discharge nanotube films can be further improved by PEDOT passivation for better surface smoothness and $SOCl_2$ doping for lower sheet resistance [42].

Transparent conductive films consisting of graphene/GO have been considered as replacements of ITO. Zheng et al. used Langmuir-Blodgett (LB) technique to deposit ultralarge graphene oxide (UL-GO). The as-fabricated UL-GO sheets with a close-packed flat structures exhibited exceptionally high electrical conductivity and transparency after thermal reduction and chemical doping treatments. It had been demonstrated that sheet resistance was ~500 Ω/sq at 90% transparency [43].

Graphene electrodes were deposited on quartz substrate, and the power conversion efficiency (PCE) was 0.4%, which was half the PCE of the cells based on ITO (0.84%). The transmittance and sheet resistance values were between 85 and 95% and 100 and 500 Ω/sq for graphene films of thickness between 4 and 7 nm [44].

Kim et al. reported graphene composing of a few layers as a transparent conductive layer to the p-contact layer in GaN UV-LED. It had been showed that the electrode exhibited great promise for use in UV-LED for which conventional ITO was less transparent than graphene [45].

Thirty inch graphene films for transparent electrodes were produced by roll-to-roll process. It had been shown that the films had sheet resistances as low as ~125 Ω/sq with 97.4% optical transmittance. The sheet resistance of a doped four-layer film was as low as ~30 Ω/sq at ~90% transparency, which was superior to commercial transparent electrodes such as indium tin oxides [46].

10.8.2 Transparent Metal Electrodes

Silver nanowire film as transparent conductive electrodes showed promising results. A simple spray-coating technique was used for to fabricate large-scale, highly uniform silver nanowires on arbitrary substrate. Combined with PDMS-assisted contact transfer technique, large-scale high-quality patterned films of silver nanowires were also obtained. A transparency of 85% and sheet resistance of 33 Ω/sq were achieved, which were comparable to that of tin-doped indium oxide (ITO) on flexible substrates [47].

A nanoimprinted transparent metal electrode process was developed by Guo's group, which can replace conventional ITO electrodes in solar cell fabrication. It had been demonstrated that the nanopatterned metal electrodes exhibited high optical transmittance in the visible range and high electrical conductivity. The nanopatterned electrodes can be processed by roll-to-roll process with a flexible stamp and a metal transfer process [48]. Figure 10.26 shows a SEM image of the Cu electrode transferred to the substrate.

Some different metal electrodes were investigated for organic solar cell. The current density as a function of voltage characteristics of organic solar cell is shown in Fig. 10.27. The performance of the solar cells is very similar to each other, and a power conversion efficiency of ~2% is obtained.

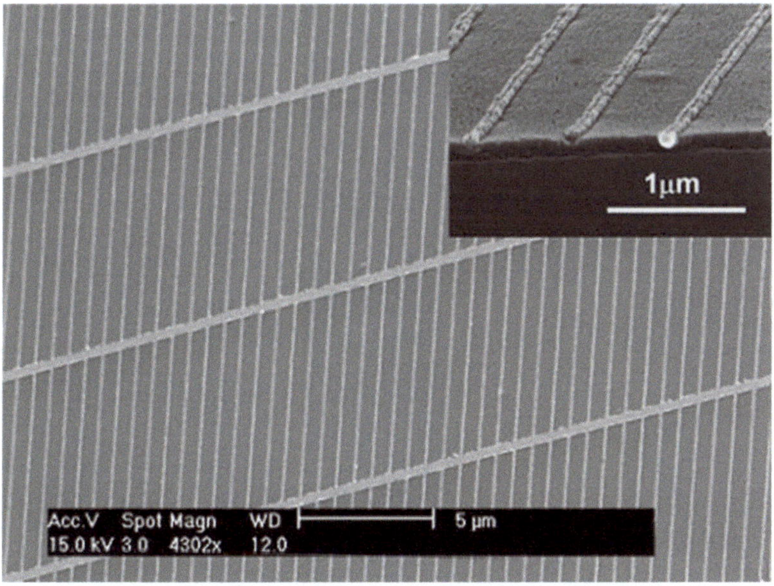

Fig. 10.26 SEM image of the Cu electrode after transfer to the PEDOT:PSS-coated substrate. The *inset* shows the cross section of the transferred metal electrode (Reprinted with permission from [48]. Copyright (2008) Wiley-VCH)

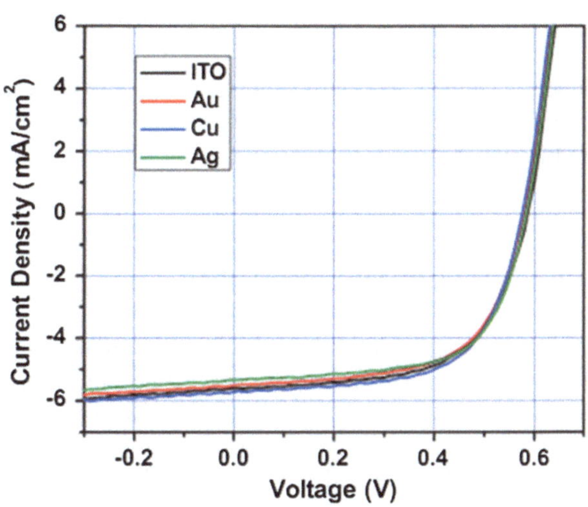

Fig. 10.27 Current density as a function of voltage characteristics of organic photovoltaic cells fabricated using different electrodes. Nanostructured Au, Ag, and Cu and conventional ITO electrodes (Reprinted from [49]. Copyright (2010), with permission from Elsevier)

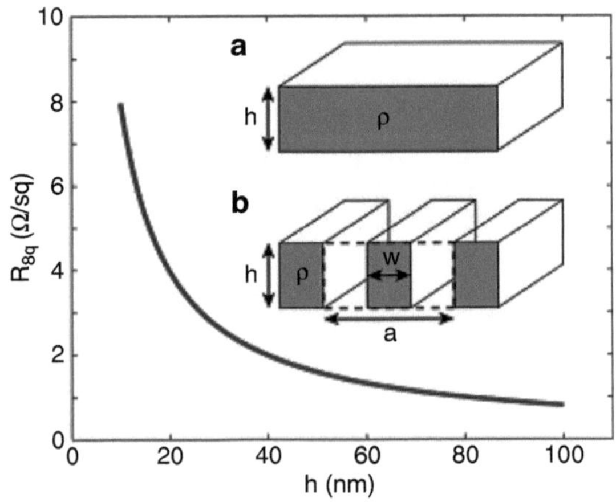

Fig. 10.28 Electrical sheet resistance of Ag metallic film electrodes as a function of film thickness. $\rho_{Ag} = 1.58 \times 10^{-8} \Omega m$, $a = 400$ nm and $w = 80$ nm. The *inset a* is schematic of homogeneous metallic film. The *inset b* is 1D nanostructure metallic film electrodes (Reprinted with permission from [50]. Copyright (2010) American Chemical Society)

The electronic properties of metal electrodes and metal films are different. The metallic film sheet resistance is given by

$$R_{sq} = \rho / h, \tag{10.10}$$

where h is thickness of metallic film and ρ is metal resistivity.

A sheet resistance of 1D nanopatterned metallic film is given by

$$R_{sq} = (\rho / h)(a/w), \tag{10.11}$$

where a is pitch, w is wire width, and h is wire thickness.

Considering Ag resistivities with different wire thickness, the results are shown in Fig. 10.28.

10.9 Microprism Substrates

Another possible enhancement of efficiency of solar cell is to take advantage of increased optical path lengths for inclined incident angles. The folded solar cell structure was achieved [51]. The organic materials (PEDOT:PSS, PHT:PCBM) were deposited on the surface of microprism structure, as shown in Fig. 10.29. The prism substrate with metal grid exhibited twofold reflection of the incident light because of the microprism structure. The short-circuit current was increased by ~60% because of the enhanced absorption in the photoactive layer. The obtained performance of the microprism solar cell was $V_{oc} = 0.455$ V, $J_{sc} = 7$ mA/cm^2, FF = 0.35, $\eta = 1.1\%$.

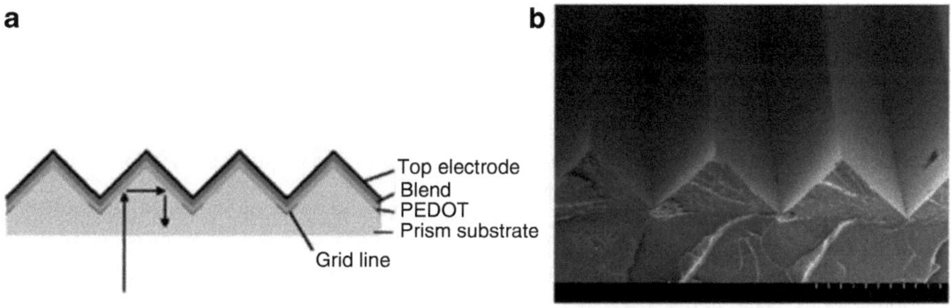

Fig. 10.29 (**a**) Cross section of an organic microprism solar cell. (**b**) Cross-sectional SEM image of a microprism structure (Reprinted from [51]. Copyright (2006), with permission from Elsevier)

References

1. Solar energy perspectives. http://www.iea.org/. Accessed 2 Mar 2012
2. Kurokawa K, Ikki O (2001) Residential PV system dissemination program. Sol Energy 70: 457–466
3. Green MA (2004) Third generation photovoltaics: advanced solar energy conversion. Springer, Heidelberg
4. Brown GF, Wu JQ (2008) Third generation photovoltaics. Laser Photonics Rev 3:394–405
5. Soga T (2006) Nanostructured materials for solar energy conversion. Elsevier, Amsterdam
6. Nelson J (2003) The physics of solar cell. Imperial College, London
7. Li JH, Yu HY, Wong SM, Li XC, Zhang G (2009) Design guidelines of periodic Si nanowire arrays for solar cell. Appl Phys Lett 95:243113–243115
8. Marin A, Stephan S, Dietrich H (2008) Energy harvesting using nanowires? Adv Mater 20:4021–4026
9. Maiolo JR, Kayes BM, Filler MA, Putnam MC, Kelzenberg MD, Atwater HA, Lewis NS (2007) High aspect ratio silicon wire array photoelectrochemical cells. J Am Chem Soc 129:12346–12347
10. Jackson JD (1998) Classical electrodynamics. Wiley, New York
11. Taflove A, Hagness SC (2000) Computational electrodynamics: the finite-difference time-domain method. Artech House, Boston
12. Hauser H, Berger P, Michl B, Müller C, Schwarzkopf S, Hermle M, Bläsi B (2010) Nanoimprint lithography for solar cell texturisation. Proc SPIE 7716:77160X-1-9
13. Zhu J, Hsu CM, Yu ZF, Fan SH, Cui Y (2010) Nanodome solar cells with efficient light management and self-cleaning. Nano Lett 10:1979–1984
14. Zhu J, Yu Z, Burkhard GF, Hsu CM, Connor ST, Xu Y, Wang Q, McGehee M, Fan S, Cui Y (2009) Optical absorption enhancement in amorphous silicon nanowire and nanocone arrays. Nano Lett 9:279–282
15. Sun CH, Jiang P, Jiang B (2008) Broadband moth-eye antireflection coatings on silicon. Appl Phys Lett 92:061112–061114
16. Chen Q, Hubbard G, Shields PA, Liu C, Allsopp DWE, Wang NW, Abbott S (2009) Broadband moth-eye antireflection coatings fabricated by low-cost nanoimprinting. Appl Phys Lett 94:263118–263120
17. Deniz H, Khudiyev T, Buyukserin F, Bayindir M (2011) Room temperature large-area nanoimprinting for broadband biomimetic antireflection surfaces. Appl Phys Lett 99: 183107-1-3
18. Gregg BA, Hanna MC (2003) Comparing organic to inorganic photovoltaic cells: Theory, experiment, and simulation. J Appl Phys 93:3605–3614

19. Liao KS, Yambem SD, Haldar A, Alley NJ, Curran SA (2010) Designs and architectures for the next generation of organic solar cells. Energies 3:1212–1250
20. Tvingstedt K (2008) Light trapping and alternative electrodes for organic photovoltaic devices. Dissertation, Linköping University
21. Fan ZY, Razavi H, Do JW, Moriwaki A, Ergen O, Chueh YL, Leu PW, Ho JC, Takahashi T, Reichertz LA, Neale S, Yu K (2009) Three-dimensional nanopillar-array photovoltaics on low-cost and flexible substrates. Nat Mater 8:648–653
22. Peng KQ, Wang X, Li L, Wu XL, Lee ST (2010) High-performance silicon nanohole solar cells. J Am Chem Soc 132:6872–6873
23. Hu L, Chen G (2007) Analysis of optical absorption in silicon nanowire arrays for photovoltaic applications. Nano Lett 7:3249–3252
24. Chen CK, de Castro ARB, Shen YB (1981) Surface-enhanced second-harmonic generation. Phys Rev Lett 46:145–148
25. García-Vidal FJ, Pendry JB (1996) Collective theory for surface enhanced Raman scattering. Phys Rev Lett 77:1163–1166
26. Ajmal KM, Huang H, Shanker B, Timothy H (2008) Surface plasmon structures for surface-enhanced raman scattering. Mater Res Soc Symp Proc 1055:45–52
27. Kruszewski S (2006) Surface enhanced Raman scattering phenomenon. Cryst Res Technol 41:562–569
28. Yang Y, Shi JL, Tanaka T, Nogami M (2007) Self-assembled silver nanochains for surface-enhanced Raman scattering. Langmuir 23:12042–12047
29. Atwater HA, Polman A (2010) Plasmonics for improved photovoltaic devices. Nat Mater 9:205–213
30. Ko DH, Tumbleston JR, Gadisa A, Aryal M, Liu YC, Lope RN, Samulski ET (2011) Light-trapping nano-structures in organic photovoltaic cells. J Mater Chem 21:16293–16303
31. Pillai S, Green MA (2010) Plasmonics for photovoltaic applications. Sol Energ Mat Sol C 94:1481–1486
32. Creighton JA, Eadon DG (1991) Ultra-visible absorption spectra of the colloidal elements. J Chem Soc 87:3881–3891
33. Catchpole KR, Polman A (2008) Design principles for particle plasmon enhanced solar cells. Appl Phys Lett 93:191113–191115
34. Nakayama K, Tanabe K, Atwater HA (2008) Plasmonic nanoparticle enhanced light absorption in GaAs solar cells. Appl Phys Lett 93:121904–121906
35. Tvingstedt K, Persson NK, Inganäs O, Rahachou A, Zozoulenko IV (2007) Surface plasmon increase absorption in polymer photovoltaic cells. Appl Phys Lett 91:113514–113516
36. Na SI, Kim SS, Jo J, Oh SH, Kim J, Kim DY (2008) Efficient polymer solar cells with surface relief gratings fabricated by simple soft lithography. Adv Funct Mater 18:3956–3963
37. Kang MG, Xu T, Park HJ, Luo XG, Guo LG (2010) Efficiency enhancement of organic solar cells using transparent plasmonic Ag nanowire electrodes. Adv Mater 22:4378–4383
38. Murdoch GB (2010) Transparent conductive oxides for organic electronics. Dissertation, University of Toronto
39. Minami T (2005) Transparent conducting oxide semiconductors for transparent electrodes. Semicond Sci Technol 20:S35
40. Battaglia C, Escarre J, So K et al (2011) Nanomoulding of transparent zinc oxide electrodes for efficient light trapping in solar cells. Nat Photonics 21:1–4
41. Hong S, Myung S (2007) Nanotube electronics: a flexible approach to mobility. Nat Nanotechnol 2:207–208
42. Zhang DH, Ryu K, Liu XL, Polikarpov E, Ly J, Tompson ME, Zhou CW (2006) Transparent, conductive, and flexible carbon nanotube films and their application in organic light-emitting diodes. Nano Lett 6:1880–1886
43. Zheng QB, Ip WH, Lin XY, Yousefi NM, Young KK, Li ZG, Kim JK (2011) Transparent conductive films consisting of ultralarge graphene sheets produced by Langmuir Blodgett Assembly. ACS Nano 5:6039–6051

44. Wu JB, Becerril HA, Bao ZN, Liu ZF, Chen YS, Peumans P (2008) Organic solar cells with solution-processed graphene transparent electrodes. Appl Phys Lett 92:263302–263304
45. Kim BJ, Mastro MA, Hite J, Eddy CR (2010) Transparent conductive graphene electrode in GaN-based ultra-violet light emitting diodes. Opt Express 18:23030–23034
46. Bae S, Kim H, Lee YB, Xu XF et al (2010) Roll-to-roll production of 30-inch graphene films for transparent electrodes. Nat Nanotechnol 5:574–578
47. Madaria AR, Kumar A, Zhou CW (2011) Large scale, highly conductive and patterned transparent films of silver nanowires on arbitrary substrates and their application in touch screens. Nanotechnology 22:245201–245206
48. Kang MG, Kim MSJ, Kim JS, Guo LJ (2008) Organic solar cells using nanoimprinted transparent metal electrodes. Adv Mater 20:4408–4413
49. Kanga MG, Parkb HJ, Ahnc SH, Guo LJ (2010) Transparent Cu nanowire mesh electrode on flexible substrates fabricated by transfer printing and its application in organic solar cells. Sol Energ Mat Sol C 94:1179–1184
50. Catrysse PB, Fan SH (2010) Nanopatterned metallic films for use as transparent conductive electrodes in optoelectronic devices. Nano Lett 10:2944–2949
51. Niggemann M, Glatthaar M, Lewer P, Müller C, Wagner J, Gombert A (2006) Functional microprism substrate for organic solar cells. Thin Solid Films 511–512:628–633

Index

DOI 10.1007/978-3-642-34428-2,

MIX
Papier aus verantwortungsvollen Quellen
Paper from responsible sources
FSC® C105338

If you have any concerns about our products,
you can contact us on
ProductSafety@springernature.com

In case Publisher is established outside the EU,
the EU authorized representative is:
Springer Nature Customer Service Center GmbH
Europaplatz 3, 69115 Heidelberg, Germany

Printed by Libri Plureos GmbH
in Hamburg, Germany